Imaging Physics

CASE REVIEW SERIES

Volumes in the CASE REVIEW Series

Brain Imaging
Breast Imaging
Cardiac Imaging
Duke Review of MRI Physics
Emergency Radiology
Gastrointestinal Imaging
General and Vascular Ultrasound
Genitourinary Imaging
Head and Neck Imaging
Imaging Physics
Musculoskeletal Imaging
Neuroradiology
Non-Interpretive Skills for Radiology
Nuclear Medicine and Molecular Imaging
Obstetric and Gynecologic Ultrasound
Pediatric Imaging
Spine Radiology
Thoracic Imaging
Vascular and Interventional Imaging

Imaging Physics

CASE REVIEW SERIES

R. Brad Abrahams, DO, DABR
Interventional and Diagnostic Radiologist
Radiology Consultants of the Midwest
Assistant Clinical Professor of Radiology
Creighton University
Omaha, Nebraska

Walter Huda, PhD
Professor of Radiology
Geisel School of Medicine at Dartmouth College
Director of Physics Education
Dartmouth-Hitchcock
Lebanon, New Hampshire

William F. Sensakovic, PhD
Diagnostic Medical Physicist
Florida Hospital
Associate Professor of Medical Education
University of Central Florida
Orlando, Florida

ELSEVIER

IMAGING PHYSICS: CASE REVIEW SERIES

ISBN: 978-0-323-42883-5

Library of Congress Cataloging-in-Publication Control Number: 2018952425

Content Strategist: Kayla Wolfe
Senior Content Development Specialist: Rae Robertson
Publishing Services Manager: Catherine Jackson
Book Production Specialist: Kristine Feeherty
Design Direction: Amy Buxton

Printed in China

Last digit is the print number: 9   8   7   6   5   4   3   2   1

ELSEVIER

1600 John F. Kennedy Blvd.
Ste 1600
Philadelphia, PA 19103-2899

Working together
to grow libraries in
developing countries

www.elsevier.com • www.bookaid.org

I am so excited about the launch of this new edition of the Case Review Series. We had not, until now, been able to bring a comprehensive imaging physics volume to the series. We have, however, had the excellent Duke Review of MRI physics edition that was published in 2012 and is due for another release shortly. There was a gap.

Who better to write the content for this new volume than the person whose name is associated with great teaching on the subject—Professor Walter Huda. Huda and physics go together like salt and pepper. He was the best "catch" of the Case Review Series, and we have benefited from it. He asked for the support of Drs. R. Brad Abrahams and William Sensakovic to create the dynamic trio. We feel certain that these stellar authors have created an edition that will assist residents and trainees at all levels

in preparing to tackle radiological physics…and the American Board of Radiology Core and Certifying exams.

The approach here is unique in that it is case-based and relies on imaging studies to make the teaching points, as opposed to lots of equations and diagrams and dry text. Perfect for the Case Review Series.

Welcome to the series, gentlemen!

Please enjoy!

David M. Yousem, MD, MBA
Professor of Radiology
Director of Neuroradiology
Russell H. Morgan Department of Radiology Science
The Johns Hopkins Medical Institutions
Baltimore, Maryland

The incorporation of radiological physics into the American Board of Radiology Core and Certifying examinations has made a dramatic impact on radiology physics education. This has been a welcome change to the many clinically minded trainees and educators in the field of radiology. Despite the abundance of high-quality physics resources, few authors have framed the discussion of physics using a case-based and image-rich approach. We hope that the addition of this book to the Case Review Series will fill this gap.

Although physics is the foundation of radiology, many of us often overlook its importance in our daily clinical work. There can sometimes be a disconnect between the granularity of physics and the real-world needs of patient X on our exam table. Even though there is not always a direct connection between every clinical image and physics principle, understanding the underlying concepts of radiological physics will pay dividends to the reader for years to come. Throughout your career you will be tasked with optimizing image quality, troubleshooting artifacts, purchasing equipment, and building on your knowledge as new technologies emerge.

This book follows the familiar structure of other books in the Case Review Series. The chapters are generally broken down into modalities, with each case highlighting a specific topic. The case begins with a clinical image and several multiple-choice questions pertaining to the physics principle of interest. The answers and explanations are shown on the reverse page, followed by a short discussion. There is an illustrated figure or artifact simulation paired with each discussion to complement the topic. Despite the intimate relationship of physics and mathematics, we have attempted to limit the appearance of numbers and equations in this book. Although the memorization of certain numbers is unavoidable, most of the numbers provided in the following cases were added to give a "ballpark" idea of values or to illustrate specific examples.

There is an ever-increasing focus on patient safety in medicine, and it is our responsibility to balance the clinical benefit and the negative consequences of imaging studies. While the first 12 chapters of this book focus on the physics of radiological modalities, the last two chapters are dedicated to radiation doses and safety topics. These chapters will be a great resource for both exam preparation and the clinical practice of radiology.

Unlike other topics in radiology, radiological physics is not something that can be predominantly learned at the workstation. Combining didactics, reading, self-study, practice questions, and clinical learning is the best approach to mastering the basic principles of radiological physics. We hope that you enjoy this case-based approach to learning and wish you the best in your exciting career.

R. Brad Abrahams
Walter Huda
William F. Sensakovic

CONTENTS

CASE 1.1

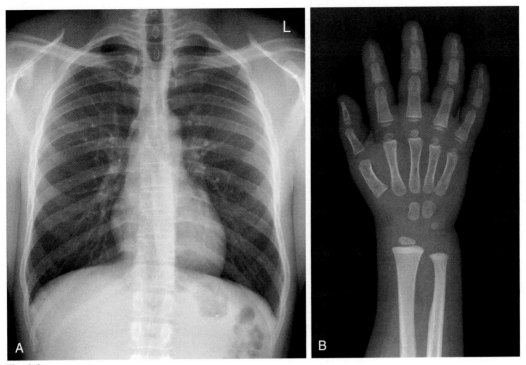

Fig. 1.1

1. What is most likely to be affected by the choice of a radiographic focal spot size?
   A. Radiographic mottle
   B. Lesion contrast
   C. Spatial resolution
   D. Image artifacts

2. Increasing what parameter will most likely result in the largest increase in focal spot blur?
   A. Geometric magnification
   B. X-ray tube voltage
   C. X-ray tube current
   D. Exposure time

3. What is most likely reduced when focal spot size is increased?
   A. Grid artifacts
   B. Image mottle
   C. Lesion contrast
   D. Motion blur

4. What radiographic examination most likely uses a small focal spot?
   A. Extremity
   B. Skull
   C. Chest
   D. Abdomen

CASE 1.1

Focal Spot

Fig. 1.1 Chest radiograph in an adult patient (A) using a large focal spot and hand radiograph of a pediatric patient (B) using a small focal spot.

1. **C.** Spatial resolution is the image quality metric that is affected by the focal spot because a larger focus increases focal spot blur, especially when there is geometric magnification. The focal spot has no effect at all on mottle, contrast, or artifacts.

2. **A.** Focal spot blur always increases with increasing geometric magnification. When magnification mammography is performed, it is essential to use a very small focal spot (0.1 mm) to reduce focal spot blur. Increasing the tube voltage, tube current, and exposure time will have no significant effect on focal spot blur.

3. **D.** When the focal spot is increased, this usually means that the power incident on the target can be increased. Increasing power will mean that exposure time can be reduced, which helps to minimize motion blur. Focal spot size will not affect grid artifacts, mottle, or contrast in radiographic images.

4. **A.** Use of a small focal spot in an extremity exam is important to reduce focal blur and improve the chance of detecting (small) hairline fractures. Large focal spots are used in the skull, chest, and abdomen to reduce exposure times, and where detection of small features is unlikely to be a major diagnostic task.

Comment

The focal spot in radiography is the size of the x-ray tube region that produces x-rays. When electrons hit the target located in the x-ray tube anode, x-rays are produced. The power used in x-ray tubes is much greater than in an average domestic home (e.g., 100 kW vs. 4 kW), and overheating issues can be important in radiologic imaging. For this reason, the target is arranged to be at a small angle (e.g., 15 degrees) so that a large area is irradiated but appears relatively small when viewed from the patient perspective (Fig. 1.2). The size of the focal spot influences image sharpness (blur), with larger focal spots typically resulting in blurrier images than small ones (Fig. 1.3). The focal spot poses a fundamental limit on the sharpness of any image. Once such a limit is reached, use of smaller pixels will not translate into improved imaging resolution performance.

The amount of focal blur is critically dependent on the amount of geometric magnification. The amount of geometric magnification is determined by the distance from an object to the image receptor, relative to the distance from the focal spot to the image receptor. As geometric magnification increases, focal spot blur also increases, requiring the use of smaller focal spots (see Fig. 1.3). Geometric magnification is sometimes used in interventional neuroradiology and requires special x-ray tubes with exceptionally small focal spots.

Most conventional x-ray tubes use two focal spot sizes, a small one (0.6 mm) and a large one (1.2 mm). The power limit on the large focal spot (100 kW) is four times greater than the

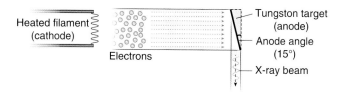

Fig. 1.2 Electrons flow from the cathode to the anode, colliding with the tungsten target and producing x-rays.

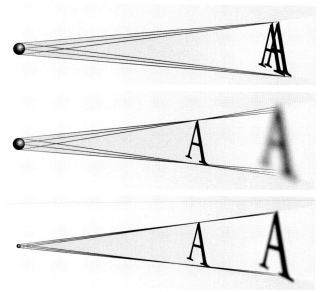

Fig. 1.3 With no geometric magnification, there is no focal spot blur *(upper image)*. When the object-to-detector distance is increased, the image is magnified and blurred *(middle image)*. Switching from a large focal spot to a small focal spot is essential to reduce image unsharpness with geometric magnification *(lower image)*.

small focal spot power limit. In most of radiography, the key issue is a reduction of the exposure time and thereby the corresponding amount of motion blur. Accordingly, a large focal spot is essential where exposure times must be minimized (chest).

When the amount of image detail becomes important, such as detection of hairline fractures in extremity imaging, a small focal spot is used. When the imaged body part is not very attenuating, the amount of radiation required to penetrate the patient will be less. Although using a small focal spot in extremity imaging increases exposure time, this is counteracted by improved radiation penetration so that motion blur does not become problematic.

References

Bushberg JT, Seibert JA, Leidholdt EM Jr, Boone JM. X-ray production, x-ray tubes, and x-ray generators. In: *The Essential Physics of Medical Imaging*. 3rd ed. Philadelphia: Wolters Kluwer; 2012:181–184.

Huda W. Image quality. In: *Review of Radiological Physics*. 4th ed. Philadelphia: Wolters Kluwer; 2016:36–37.

Huda W. Radiography. In: *Review of Radiological Physics*. 4th ed. Philadelphia: Wolters Kluwer; 2016:91–99.

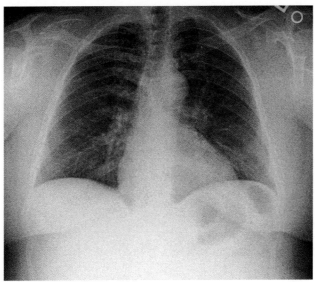

Fig. 1.4

1. What is impacted the most when milliampere-seconds (mAs) is adjusted?
   A. Mottle
   B. Contrast
   C. Sharpness
   D. Artifacts

2. Which radiographic examination most likely uses the lowest mAs?
   A. Skull
   B. Extremity
   C. L-spine
   D. Abdomen

3. Which of the following detectors is expected to be quantum mottle limited for chest radiography?
   A. Scintillator (cesium iodide)
   B. Photoconductor (selenium)
   C. Photostimulable phosphor (PSP) (BaFBr)
   D. A, B, and C

4. How will quadrupling the mAs used to perform a bedside radiograph affect the amount of noise (mottle) in the image?
   A. Quadrupled
   B. Doubled
   C. Halved
   D. Quartered

CASE 1.2

Tube Output and Mottle

Fig. 1.4 Chest radiograph with extremely low mAs. Note the visible noise in the image.

1. **A.** The number of photons used to create any radiographic image will affect only the amount of mottle. Quadrupling the number of photons will halve the amount of mottle. The mAs generally has no effect on contrast, sharpness, or the artifacts in the resultant image.

2. **B.** An extremity would use an mAs that is very low and much less than a skull, L-spine, or abdomen. Two useful benchmarks in radiography are 1 mAs for a chest radiograph (posteroanterior [PA]) and 20 mAs for an abdomen (anteroposterior [AP]). The former is one of the low values implemented in radiography, and the latter is one of the high values.

3. **D.** All medical imaging systems are normally operated at exposure levels that guarantee that they are quantum mottle limited. What this means is that the only technical way to reduce mottle is to use more photons to create the image, which will also increase the patient dose.

4. **C.** The noise is halved when the number of photons is quadrupled. When four images are added together, the signal is quadrupled. Because noise is random, adding four images together will only double the noise. The overall improvement by adding N images together in any imaging modality is usually given by $N^{0.5}$.

Comment

Consider a conventional chest x-ray where the selected x-ray tube voltage is set at 120 kV. The x-ray tube current will be several hundred mA (e.g., 400 mA), and the exposure time will be very short (e.g., 2.5 ms). The total x-ray intensity (x-ray tube output) is the product of the tube current and exposure time and, in this instance, equal to 1 mAs. For an abdominal radiograph, the tube current would likely be several hundred mA (say, 400 mA), but the exposure time would be much longer to penetrate the much thicker body part (say, 50 ms). The total x-ray intensity in this abdominal radiograph would be 20 mAs, or 20 times greater than the chest x-ray. When more radiation is required, this is achieved by increasing the tube current, the exposure time, or both. The ideal scenario is to increase the tube current, but because focal spot power loading (i.e., heating) will also increase, this may not always be possible.

The mAs is a relative indicator of x-ray tube output and is increased whenever "more radiation" is required to create a radiologic image. It is important to note that knowledge of the mAs does not permit a definitive determination of the amount of radiation that is being emitted. X-ray tube output depends on a number of additional factors, including x-ray tube design, tube voltage, and x-ray beam filtration. The absolute amount of radiation incident on the patient is given by the entrance air kerma (K_{air}), which is approximately 1 mGy for a lateral skull x-ray. The absolute amount of radiation incident on the image receptor is much lower than entrance K_{air} and typically 3 μGy. The radiation intensity at the image receptor is much lower because of patient attenuation, increased distance from the focus, and losses in the antiscatter grid. It is helpful to think of mAs as analogous to the tachometer in a car, where increasing

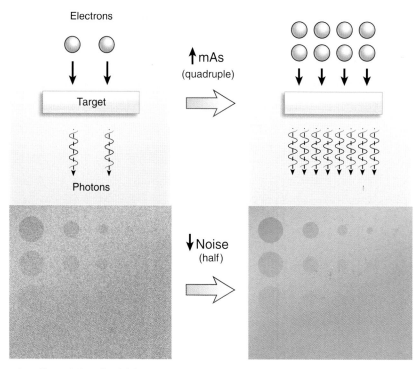

Fig. 1.5 Quadrupling the mAs will result in a fourfold increase in the number of x-rays produced but will reduce image noise only by half. Increasing mAs will not increase the photon energy or beam quality, and so the contrast of all lesions will remain exactly the same. On the left, lesion contrast is the same as on the right, but lesion visibility is inferior because of higher noise.

revolutions per minute (rpm) will increase car speed. However, it is the speedometer that tells you the absolute speed, which depends on engine design and the gear you are in ($\equiv K_{air}$).

Noise is any unwanted signal in a medical image and can be considered to be structured or random. A rib cage in a chest radiograph may mask a lung lesion, as an example of structured (or anatomic) noise. When a detector is uniformly irradiated, not every pixel has exactly the same value, and there will be a salt-and-pepper appearance that is called mottle (noise). Quantum mottle results from the discrete nature of x-ray photons and is the only important source of random noise in x-ray–based imaging. The only technical way to reduce the amount of mottle in an x-ray–based image is to increase the number of photons interacting with the detector, either by improving detection efficiency or by using more photons (Fig. 1.5). The latter method of using more photons by increasing the mAs used will also increase patient dose. It is possible to reduce the apparent mottle using image-processing techniques, but these come at a cost in terms of imaging performance. For example, averaging four pixels together will result in less random inter-pixel fluctuations, but spatial resolution performance will also be reduced.

Random noise is important because it limits the visibility of low-contrast lesions. Noise is irrelevant when the lesion has high contrast, irrespective of whether the lesion is large or small. However, to assess the visibility of any lesion, it is important to take into account both the amount of contrast and the corresponding level of noise. The contrast-to-noise ratio (CNR) is a number that is used to assess how lesion visibility is affected when radiographic techniques (kV and mAs) are modified. When mAs is increased, contrast is unchanged and noise is reduced, so the CNR increases and the lesion becomes more visible. When automatic exposure control (AEC) systems are used, noise is "fixed" so that lesion CNR is solely determined by the choice of voltage. For AEC exposures, high voltages reduce contrast and vice versa.

References

Huda W. Image quality. In: *Review of Radiological Physics*. 4th ed. Philadelphia: Wolters Kluwer; 2016:33–40.

Huda W, Abrahams RB. Radiographic techniques, contrast, and noise in X-ray imaging. *AJR Am J Roentgenol*. 2015;204(2):W126–W131.

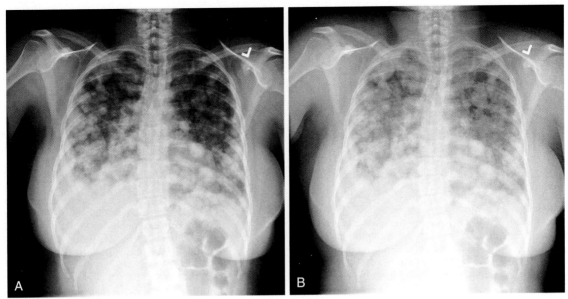

Fig. 1.6

1. X-ray beam quality is *independent* of which x-ray beam factor?
   A. Tube voltage (kV)
   B. Tube output (mAs)
   C. Filter thickness (mm)
   D. Filter atomic number (Z)

2. What will happen to patient dose and lesion contrast when 0.1 mm Cu is added as a filter in a pediatric x-ray examination performed using an AEC system?
   A. Increase; Increase
   B. Increase; Decrease
   C. Decrease; Increase
   D. Decrease; Decrease

3. Which parameter is generally independent of lesion contrast?
   A. Tube output (mAs)
   B. Tube voltage (kV)
   C. Beam filtration (mm)
   D. Grid ratio (10:1)

4. How will reducing the display window width impact lesion contrast in the image?
   A. Increase
   B. No effect
   C. Reduce

CASE 1.3

Beam Quality and Contrast

Fig. 1.6 (A) and (B) Frontal chest radiograph in a patient with multiple metastatic pulmonary masses. The image taken at higher kV (B) has lower soft tissue contrast.

1. **B.** Tube output (mAs) has no effect on x-ray beam quality. Voltage and beam filtration are important variables that radiologists have to adjust the beam quality in any radiographic examination. Average x-ray beam energy can also be adjusted by changing the filter thickness or x-ray beam filter atomic number.

2. **D.** The addition of 0.1 mm Cu to an x-ray beam will preferentially eliminate low-energy photons, resulting in a more penetrating beam. To maintain the same radiation intensity at the image receptor, less radiation is required to be incident on the patient, so the dose is reduced. The difference in transmitted radiation between a lesion and the surrounding background is always reduced when energy increases, so contrast goes down.

3. **A.** The mAs used to generate a radiographic image never affects contrast. If a lesion transmits 10% less radiation than do the surrounding tissues, this will be true at 1 mAs and at 1000 mAs. Contrast is reduced when voltage and filtration increase and is reduced when the amount of scatter in an image is increased, which occurs at lower grid ratios.

4. **A.** When the window width is reduced, contrast in the displayed image will increase. Conversely, wide windows always reduce displayed image contrast.

Comment

When x-rays interact with patients, the two most important processes are photoelectric absorption and Compton scatter. As photon energy increases in the human body, the likelihood of an interaction decreases. As a result, increasing photon energy will increase the amount of radiation that is transmitted through the patient. The penetrating power of an x-ray beam, which reflects its average energy, is quantified by the thickness of aluminum (mm) that attenuates half of the x-ray beam intensity. For example, the average voltage in abdominal radiography is 80 kV, where the corresponding beam quality is approximately 3 mm aluminum. If the average energy is reduced, less aluminum is required to halve the x-ray beam intensity, and when the average energy increases, more aluminum would be required to halve the x-ray beam intensity. X-ray beam quality affects patient dose for examinations performed using AEC with a fixed K_{air} at the image receptor. As the x-ray beam becomes more penetrating, fewer photons need to be used to achieve the required K_{air} at the image receptor, resulting in a decrease in patient dose.

A lesion must transmit more or less radiation than the surrounding tissues to be observable. The lesion will appear darker when more radiation is transmitted. Conversely, the lesion will appear lighter when less radiation is transmitted. Photon energy is the most important determinant of subject contrast in radiography, with lower photon energies increasing contrast and vice versa. Photon energy increases with increasing x-ray tube voltage, as well as increasing x-ray beam filtration. These increases would generally be expected to reduce the amount of contrast in all radiographic images (Fig. 1.7).

Changes in contrast are highly dependent on the atomic number of the lesion relative to that of soft tissues ($Z = 7.5$).

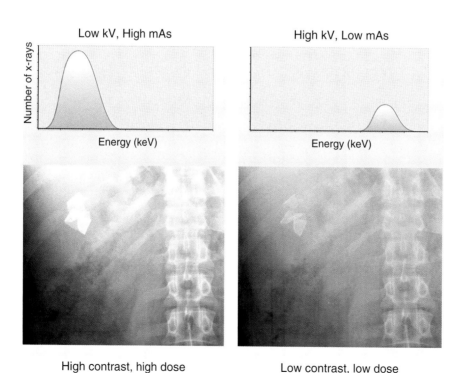

Low kV, High mAs

High kV, Low mAs

Number of x-rays

Energy (keV)

Energy (keV)

High contrast, high dose

Low contrast, low dose

Fig. 1.7 This retained surgical lap pad can be easily seen in the left image, where kV is low and mAs is high. Patient dose is high because of the need to use a lot of radiation to achieve a given amount at the image receptor. Raising the kV markedly reduces the contrast between the sponge and surrounding tissue, but very little radiation is now needed because of the increased penetration to achieve the same radiation intensity at the image receptor.

When the lesion atomic number is similar to that of tissue, changes in contrast with x-ray photon energy will be modest. However, for high (or low) atomic number lesions, increasing photon energy will markedly reduce lesion contrast. For example, angiographic studies performed at high voltages (120 kV) will have very poor contrast in comparison with those performed at an optimal voltage of 70 kV that matches the average photon energy to that of the iodine K-edge (i.e., 33 keV).

Scatter reduces contrast but does not influence resolution or mottle. A lesion that transmits 50 photons compared with 100 for surrounding tissues has a 50% contrast. Adding 100 scatter photons to all locations of this image would reduce the contrast to 25% because there are now 150 in the lesion compared with 200 in the surrounding tissues. The appearance of a lesion in any radiographic image can always be adjusted by modifying the display characteristics (window/level settings). In chest computed tomography (CT), a narrow window is used to visualize soft tissues. When a wide chest CT window is used to see all the tissues within the lung, perceived differences between soft tissues in the displayed image will "disappear."

References

Huda W. Image quality. In: *Review of Radiological Physics*. 4th ed. Philadelphia: Wolters Kluwer; 2016:31-33.

Huda W, Abrahams RB. Radiographic techniques, contrast, and noise in X-ray imaging. *AJR Am J Roentgenol*. 2015;204(2):W126-W131.

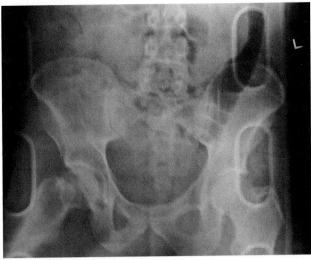

Fig. 1.8

1. What is most likely to be adversely affected by an increase in exposure time in radiography performed using AEC?
   A. Noise (mottle)
   B. Patient dose
   C. Geometric magnification
   D. Image sharpness

2. What improves when exposure time is increased on a table-top examination (i.e., no AEC)?
   A. Lesion contrast
   B. Image mottle
   C. Gridline artifacts
   D. Motion blur

3. What is the typical exposure index (EI) for an adult body radiograph?
   A. 3
   B. 30
   C. 300
   D. 3000

4. Which radiographic examination is most likely to make use of AEC?
   A. Skull
   B. Extremity
   C. Infant body
   D. Bedside chest

CASE 1.4

Exposure Time and Automatic Exposure Control

Fig. 1.8 Image blur from patient motion in a trauma patient. The stationary backboard is sharp, whereas the projection of the patient is blurred due to motion.

1. **D.** Image sharpness may be lost when exposure times increase, because of voluntary and involuntary motion by the patient. Noise and patient dose stay the same when an AEC system is used because this fixes the amount of radiation at the receptor. With an AEC, a lower tube current would result in a longer exposure and vice versa (i.e., fixed mAs). Geometric magnification is not dependent on exposure time.

2. **B.** A tabletop examination does not use an AEC system, so increasing the exposure time means that more radiation is incident on the patient and more is used to create the image. As the mAs increases, mottle is reduced and patient dose will also increase. Contrast will not be affected, whereas resolution due to motion blur (artifacts) will get worse, not better.

3. **C.** The radiation at the image receptor for most radiographic examinations such as skull, chest, and abdomen is approximately 3 μGy, which will be recorded in the DICOM header as an EI of 300 and can be displayed on any PACS workstation.

4. **A.** Skull x-rays are generally performed using AEC systems. Extremity and infants can be exposed on the tabletop with no grid or AEC. AEC systems are impractical at the patient bedside in an intensive care unit (ICU) setting.

Comment

The total time electrons are hitting the target to create x-rays is called the exposure time. This is an important protocol parameter that impacts image quality, as well as the radiation dose received by the exposed patient. Increasing the exposure time increases the number of photons that are incident on the patient. Quadrupling the exposure time therefore will quadruple the patient dose but will also halve the amount of mottle in the resultant image. However, increasing exposure time also increases the likelihood of motion by the patient and is therefore more likely to result in a more blurred image. Chest x-rays have exposure times of a few milliseconds to minimize unavoidable heart motion, which goes through a complete cardiac cycle in approximately 0.5 to 1 second. For abdominal radiographs, typical exposure times are tens of milliseconds and generally result in images with negligible motion blur.

Most radiographic examinations are performed using an AEC system (Fig. 1.9). The x-ray beam quality is selected by the operator by adjusting the tube voltage (kV) and adding or removing filtration at the collimator. The examination is terminated when a radiation detector located at the image receptor registers a predetermined amount of radiation based on anatomy of interest (e.g., 3 μGy). This mechanism ensures that the right amount of radiation is used to generate each radiographic image and avoids the problems of increased mottle (underexposed), patient dose, and risk (overexposed).

In 2010 an agreement was reached between vendors and the imaging science community to standardize the way that EI is defined and transmitted to clinical practitioners. A radiation intensity of 1 μGy would be expressed as an EI of 100, which is directly proportional to the image receptor K_{air}. For each radiographic examination the responsible radiologist should identify the target K_{air} with appropriate input from technologists, vendors, and imaging scientists. All state-of-the-art commercial systems will generate a number (deviation index [DI]) to inform operators how closely a given radiograph actually met the target EI value (e.g., 300). A DI of −3 means that half the required radiation was used, 0 means the target value was met, and +3 means that twice the required amount was used. Values of EI and DI can be selected for display on workstations and are also located in the DICOM header associated with each radiographic image.

A radiation intensity (K_{air}) of 3 μGy would generate a good radiographic image and is commonly used in all digital radiography of the head and body. Accordingly, the target EI value for most radiographic examinations is 300. If 1 μGy (EI = 100) were used, the mottle would be too high. If 6 μGy (EI = 600) or more were to be used, the patient would have been exposed to twice the radiation than is actually needed for diagnostic purposes. In extremity radiography, the EI value is usually set higher, at approximately 1000. Extremities use very low-energy photons generated at 60 kV and deposit less energy into the x-ray detectors. The higher EI ensures that the image quality (mottle) in extremities is comparable to that in conventional (screen-film) radiographic imaging by ensuring that the same amount of x-ray energy is absorbed in the radiographic detector.

References

American College of Radiology. ACR–AAPM–SIIM practice parameter for digital radiography. Revised 2017. https://www.acr.org/-/media/ACR/Files/Practice-Parameters/rad-digital.pdf?la=en. Accessed March 1, 2018.

Huda W. X-rays. In: *Review of Radiological Physics*. 4th ed. Philadelphia: Wolters Kluwer; 2016:11–12, 96.

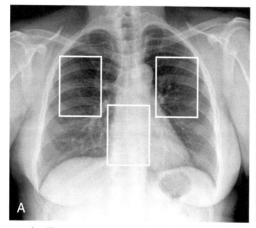

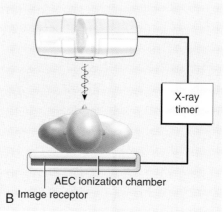

Fig. 1.9 (A) The rectangles illustrate the typical location of the automatic exposure control *(AEC)* detectors in chest radiography. (B) The AEC detects the radiation intensity at the image receptor (air kerma) in predefined regions, which enables the exposure to be terminated when the correct amount of radiation has been reached. In this way, the amount of mottle in the image is always fixed so that the right amount of radiation is always used to create an image.

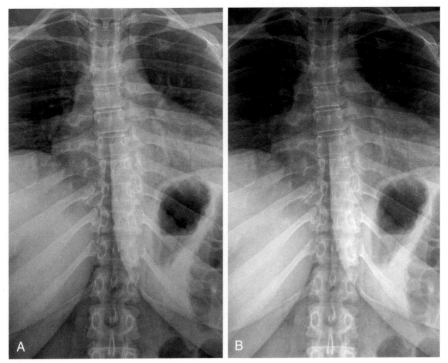

Fig. 1.10

1. What part of an x-ray tube produces differential x-ray attenuation that is called the heel effect?
   A. Cathode
   B. Housing
   C. Filter
   D. Target

2. Changing what parameter is unlikely to influence the magnitude of the heel effect in radiographic imaging?
   A. Tube output (mAs)
   B. Source-to-image receptor distance (SID)
   C. X-ray tube anode angle
   D. Radiographic receptor size

3. What should be reduced to minimize the heel effect in radiography?
   A. SID
   B. Anode angle
   C. Cassette size
   D. A, B, and C

4. In what type of radiograph would x-ray tube orientation be most important?
   A. Chest
   B. Skull
   C. Abdomen
   D. Extremities

CASE 1.5

Heel Effect

Fig. 1.10 (A) and (B) Frontal views of the abdomen demonstrating how the appearance of an image changes when an x-ray tube is rotated through 180 degrees. The top part of image B is more exposed *(darker)* due to increased exposure at the cathode side of the x-ray tube (i.e., heel effect).

1. **D.** It is the differential attenuation in the tungsten target that causes the heel effect. X-rays traveling in the anode direction traverse much more of the strongly attenuating tungsten ($Z = 74$), which results in a weaker x-ray intensity at the anode side of the x-ray tube (Fig. 1.10). The filter will attenuate all x-rays equally, and the cathode and housing play no role in x-rays that are directed toward the patient.

2. **A.** The heel effect is not affected by the mAs used in radiography. When the anode side is 25% weaker than the central beam, this will be true at 1 mAs and 1000 mAs. The heel effect increases with reduced SID, reduced anode angle, and for larger image receptors based on irradiation geometry.

3. **C.** When the cassette size is reduced, it is only the central region of the x-ray beam that is being used, which is more uniform than the larger beam. Reducing the SID and anode angle will increase the heel effect.

4. **A.** In a chest x-ray, the anode should point toward the upper chest region, which is much less attenuating. In all the other examinations, the patient is "more uniform" than the chest, and the orientation of the x-ray tube (heel effect) is thus of less importance.

Comment

X-rays are produced when energetic electrons interact with tungsten atoms in the target that are embedded within an anode. These energetic photons are produced at some depth (<1 mm) in the tungsten and are attenuated as they emerge from the production site and travel toward the patient. Because the target is angled at approximately 15 degrees (Fig. 1.11), x-rays that travel toward the cathode pass through less of the target than those that travel toward the anode. The tungsten target has a very high atomic number ($Z = 74$), which attenuates almost as much as lead (Pb) ($Z = 82$). As a result, the x-ray beam intensity at the anode side of an x-ray tube has a reduced intensity when compared with the corresponding radiation intensity at the cathode side of an x-ray tube. All commercial x-ray tubes label the anode side on the tube housing to enable operators to "know" the direction with the highest (cathode) and lowest (anode) intensities.

As the anode angle is reduced, there is a greater differential distance in path lengths that travel in the anode and cathode directions, which increases the heel effect. Conversely, increasing the anode angle will reduce the magnitude of the heel effect, but this is never eliminated (see Fig. 1.11). The heel effect increases with increasing distance from the central axis, so it is much more pronounced with large cassette sizes. A small cassette size will capture the central region of the x-ray beam where differences in path length between opposite edges will be reduced (smaller heel effect). As the SID increases, an image receptor will increasingly capture only the central part of the x-ray beam, where the magnitude of the heel effect is smaller. It is only at shorter SIDs that the heel effect becomes of increased importance.

The heel effect is always present in all x-ray–based imaging systems and is used to improve the resultant imaging performance. In chest radiography, the more intense cathode side is directed toward the abdomen, which attenuates much more than the chest and benefits from increased x-ray intensities. The less intense anode side is directed toward the upper thoracic region, where x-ray attenuation is much lower than in the abdomen. Similarly, in mammographic imaging, the cathode side points toward the chest wall, whereas the anode side is directed toward the nipple. In CT, the anode-cathode axis is oriented perpendicular to the axial image plane to ensure that the detected projections are not influenced by the heel effect.

References

Bushberg JT, Seibert JA, Leidholdt EM Jr, Boone JM. X-ray production, x-ray tubes, and x-ray generators. In: *The Essential Physics of Medical Imaging.* 3rd ed. Philadelphia: Wolters Kluwer; 2012:184–186.

Huda W. Radiography. In: *Review of Radiological Physics.* 4th ed. Philadelphia: Wolters Kluwer; 2016:92–94.

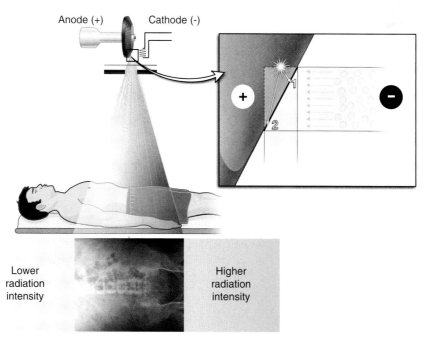

Fig. 1.11 In the upper right image, the heel effect occurs because photons produced in the tungsten target are attenuated as they exit and move toward the patient. If the photon exits toward the cathode *(1),* there is less tungsten and therefore less attenuation along its path. If the photon exits toward the anode *(2),* it is attenuated by a longer path within the target. In the radiograph, the lower intensity on the anode side is directed to the less attenuating part of the patient (chest), and the higher intensity cathode side is directed to the more attenuating part of the patient (pelvis).

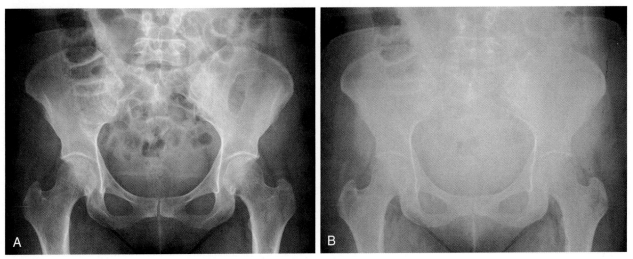

Fig. 1.12

1. Which radiographic examination is most likely to use a scatter removal grid?
   A. Skull
   B. Extremity
   C. Infant body
   D. Bedside chest

2. Which grid ratio is typically used to perform a chest x-ray on a dedicated departmental imaging system?
   A. 2:1
   B. 5:1
   C. 10:1
   D. 20:1

3. What percentage (%) of primary x-ray photons is typically lost in grids used when performing adult abdominal radiography?
   A. 10
   B. 30
   C. 70
   D. 90

4. What is the most likely dose reduction (%) in infant radiation dose when a grid is removed for a follow-up AEC radiograph?
   A. 5
   B. 15
   C. 50
   D. 90

CASE 1.6

Grids

Fig. 1.12 Frontal views of the pelvis with a grid (A) and simulated without a grid (B), where the latter showed a dramatic reduction in contrast because of high scatter. Without the grid, for every primary photon transmitted there are five scattered photons that dramatically reduce image contrast.

1. **A.** Skull x-rays are performed with a grid, likely 10:1. Grids are optional for thin anatomy such as an infant or an extremity, where the amount of scatter radiation is relatively low. Aligning a grid at the bedside is very difficult, and scatter is usually reduced by the use of a lower x-ray tube voltage (80 vs. 120 kV).

2. **C.** Most departmental radiographs, including chest x-rays on a dedicated unit, would use a 10:1 grid. A 5:1 grid is used in mammography, and there are no 2:1 or 20:1 grids in current clinical practice.

3. **B.** Approximately 70% of the primary x-rays that are incident on a grid will be transmitted through the gaps, and the remaining 30% are lost in the Pb or W strips (Fig. 1.13). Losses could never be as low as 10% or as high as 70% in any clinical scatter removal grid.

4. **C.** For an infant, the Bucky factor (BF), which is defined as the incident/transmitted radiation through a grid, is approximately 2. When the grid is removed, there is twice as much radiation incident on the AEC, so the radiation has to be reduced to 50% of the initial value.

Comment

The amount of scatter radiation in abdominal radiography is very high, amounting to up to five scattered photons for every primary photon transmitted through the patient. This amount of scatter markedly reduces image contrast and would render most radiographs "nondiagnostic." Antiscatter grids are used to reduce the amount of scatter radiation reaching an image receptor, which dramatically improves image contrast. Grids consist of strips (septa) of Pb or other high attenuating material (e.g., tungsten) that are separated from each other with an intergrid material such as aluminum that will transmit most of the incident primary photons. The strips are angled to point toward the source of x-ray photons (focus). Artifacts may be caused by using the grids at distances other than the focal distance. Grids are not observed on radiographic images because these move (oscillate) during the exposure in a device known as a Bucky system. Grid suppression image processing also helps to remove gridlines from images.

The septa have a length of approximately 1 mm or so, a septal thickness of approximately 50 μm, and an interspace gap of approximately 0.1 mm. The grid ratio is the most important grid characteristic, is defined as the length divided by the interspace gap, and is generally approximately 10 or so in radiographic imaging. The grid frequency is the number of grid septa per centimeter, which is generally approximately 60 lines/cm. Typical grids have 10:1 ratios, which will remove 90% of the

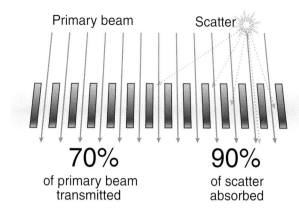

Fig. 1.13 Grid septa transmit 70% of the primary beam while absorbing 90% of scatter radiation. If the grid ratio is increased, more of the primary and scatter radiation will be absorbed by the grid and vice versa.

scatter photons incident on the image receptor. When primary photons hit the septa, these are invariably lost, and in a 10:1 grid approximately 30% of the primary photons are attenuated (lost) so that only approximately 70% are transmitted through to the image receptor (see Fig. 1.13). Increasing the grid ratio improves scatter removal but also increases primary photon losses and vice versa. The grid ratio is a useful guide to imaging performance but does not account for the overall size of the septa. An alternative approach to assessing grid performance is to consider the amount of Pb used (areal thickness in gm/cm^2), where increased Pb content would likely improve scatter removal.

In abdominal radiography, where the scatter-to-primary ratio is approximately 5:1, the use of a grid markedly reduces how much of the radiation intensity exiting the patient gets transmitted through to the image receptor. The BF is the ratio of the incident radiation to that transmitted through the grid. A typical BF in adult abdominal radiography is approximately 5, and this value has a special importance for radiographic imaging. Consider an AEC radiograph obtained without a grid and then repeated with a grid. Because only one-fifth of the radiation is transmitted when the BF is 5, the mAs must now be increased fivefold to achieve the required K_{air} at the image receptor. The BF is therefore correlated to the increase in patient dose when a grid is introduced in AEC radiography. Grids are optional when the patient thickness is less than 12 cm, and in infants the BF is typically approximately 2. Accordingly, adding a grid in an infant AEC radiograph to improve image quality (contrast) will double the infant's radiation exposure.

References

Huda W. Radiography. In: *Review of Radiological Physics*. 4th ed. Philadelphia: Wolters Kluwer; 2016:97–98.

Huda W. X-ray imaging. In: *Review of Radiological Physics*. 4th ed. Philadelphia: Wolters Kluwer; 2016:18–19.

Radiological Society of North America. RSNA/AAPM physics modules. Projection x-ray imaging: basic concepts in radiography. https://www.rsna.org/RSNA/AAPM_Online_Physics_Modules_.aspx. Accessed March 1, 2018.

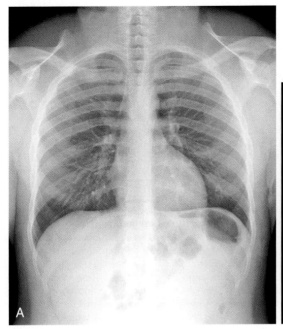

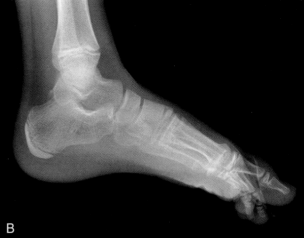

Fig. 1.14

1. Which type of detector is most likely to result in the lowest patient dose in an abdominal radiograph (80 kV)?
   A. Gas chamber (xenon)
   B. Photoconductor (selenium)
   C. Scintillator (cesium iodide)
   D. PSP (BaFBr)

2. Which type of detector is most likely to result in the sharpest images?
   A. Scintillator
   B. PSP
   C. Photoconductor
   D. A ≈ B ≈ C

3. Image sharpness in digital radiographic imaging is independent of which parameter?
   A. Focal spot size
   B. Pixel size
   C. Scintillator thickness
   D. Technique (kV and mAs)

4. Increasing what parameter is most likely to improve spatial resolution?
   A. Matrix size
   B. Focal spot size
   C. Scintillator thickness
   D. Imaging time

CASE 1.7

Imaging Plates and Resolution

Fig. 1.14 The chest radiograph (A) uses a larger physical cassette size compared with the foot radiograph (B). With a similar matrix size of 2000 × 2500, the smaller cassette offers improved spatial resolution (5 line pairs [lp]/mm vs. 3 lp/mm).

1. **C.** A cesium iodide scintillator is an excellent absorber of incident x-rays because the K-edges of Cs and I (i.e., 36 and 33 keV, respectively) generally match a typical radiographic beam generated at 80 kV. A typical cesium iodide flat panel detector would likely absorb most (>80%) of the incident radiation, far higher than current xenon, selenium, and BaF-Br detectors.

2. **C.** When a photoconductor absorbs x-rays, the charge that is produced is collected by an electric field. This charge does not spread out in the way light spreads out in a scintillator or a PSP.

3. **D.** kV, mAs, and beam quality have no practical direct effect on the maximum achievable spatial resolution. In all radiographic imaging, the imaging system resolution is affected by the focal spot and the important detector characteristics of detector thickness, as well as the pixel size.

4. **A.** Only a larger matrix in the same field of view (i.e., smaller pixels) may improve resolution. However, at some matrix size, one must reach the intrinsic limits to resolution based on focal blur and the detector characteristics (i.e., thickness). As focal spot size, scintillator thickness, and imaging time increase, the resultant images will likely be less sharp.

Comment

Scintillators (cesium iodide) absorb x-rays and convert the absorbed energy into light photons. The magnitude of the detected light signal, which is then converted into charge for quantitative detection, is directly proportional to the absorbed x-ray energy. Photoconductors (selenium) absorb x-rays and produce a charge that is collected in a "charge detector" by the application of a voltage across the photoconductor. Charge produced in photoconductors is *directly* measured, whereas in scintillators the light produced is subsequently converted into charge (i.e., indirect detection). PSPs (BaFBr) absorb and store a fraction of the absorbed x-ray energy incident during a radiographic examination. The stored energy is released by the application of a (red) laser and emits light that is blue. The detected blue light is a measure of the energy absorbed in each pixel and is used to generate the radiographic image.

The type of detector used will influence the amount of detail in the resultant image. Photoconductors are most likely to result in the sharpest images because the charge that is produced at the x-ray interaction site does not diffuse very far before it is collected by the "charge detectors." Scintillators generally have average resolution because the light produced at the interaction site diffuses before being detected, which results in blurred images. The thicker the scintillator, the greater the light spreading before being intercepted by the "light detectors" and vice versa (Fig. 1.15). PSPs have the poorest resolution because the incident light used to "read out" the stored data is also subject to scattering and can therefore release light in adjacent pixels, reducing the resultant image sharpness. Because of the light scatter problems, there is a limit to the thickness of any PSP detector used in radiographic imaging.

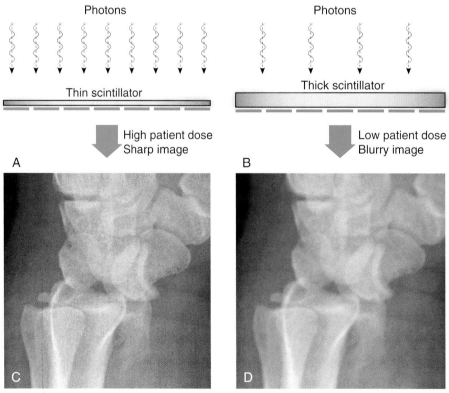

Fig. 1.15 When the number of absorbed photons in the detector is kept constant as in the above examples, the patient dose will be lower with the thicker scintillator (B) than with the thinner scintillator (A). The thin scintillator image (C) is sharper than the thick scintillator image (D) because of lower light spreading.

Different types of detectors have different efficiencies in absorbing photons, which will influence the radiation dose because a specified number of absorbed photons are needed to create an image. Cesium iodide is an excellent x-ray absorber because these atoms have K-edge energies of 33 and 36 keV, which are close to the average photon energies generated by 80-kV x-ray voltages. On the other hand, selenium is a very poor absorber because the K-edge is only 13 keV, and this material has poor x-ray absorption, especially at higher photon energies. BaFBr in PSP plates is intermediate between the good absorption of cesium iodide and the poor absorption of selenium at energies used in radiography. Although the atomic number of BaFBr is much higher than that of selenium, the read-out mechanism (light lasers) limits the thickness of material that may be used to ensure adequate spatial resolution performance. In radiographic imaging normally performed using x-ray tube voltages ranging between 60 and 120 kV, indirect cesium iodide detectors will likely result in the lowest patient doses and direct selenium detectors will likely result in the highest doses when image quality (mottle) is kept fixed.

Most digital detectors used in chest radiography (35 × 43 cm) have 175-μm pixels, which results in a limiting resolution of 3 lp/mm. When the cassette physical size is reduced, the matrix size is normally unchanged (2000 × 2500) so that a 20 × 25 cm cassette has smaller pixels (100 μm) and improved spatial resolution (5 lp/mm). To appreciate these spatial resolution values, the human eye has a limiting resolution of 5 lp/mm at a normal viewing distance of 25 cm (approximately arm's length). Radiographic technique factors of kV and mAs, which are of paramount importance in influencing a lesion CNR, have no direct effect on spatial resolution in radiographic imaging.

References

Bushberg JT, Seibert JA, Leidholdt EM Jr, Boone JM. Radiography. In: *The Essential Physics of Medical Imaging*. 3rd ed. Philadelphia: Wolters Kluwer; 2012:231–235.

Huda W. X-ray imaging. In: *Review of Radiological Physics*. 4th ed. Philadelphia: Wolters Kluwer; 2016:22–25.

Huda W, Abrahams RB. X-ray-based medical imaging and resolution. *AJR Am J Roentgenol*. 2015;204:W393–W397.

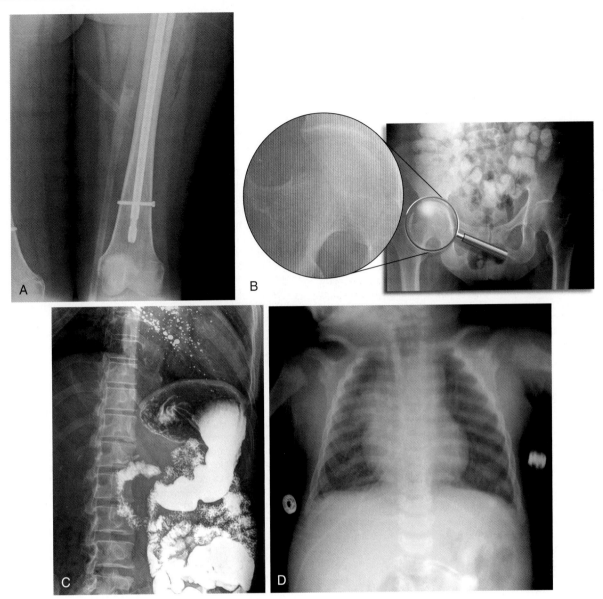

Fig. 1.16

1. What is the most likely cause of the artifact depicted in Fig. 1.16A?
   A. Ghosting
   B. Grid
   C. Barium (Ba) spill
   D. Motion

2. What is the most likely cause of the artifact depicted in Fig. 1.16B?
   A. Ghosting
   B. Grid
   C. Ba spill
   D. Motion

3. What is the most likely cause of the artifact depicted in Fig. 1.16C?
   A. Ghosting
   B. Grid
   C. Ba spill
   D. Motion

4. What is the most likely cause of the artifact depicted in Fig. 1.16D?
   A. Ghosting
   B. Grid
   C. Ba spill
   D. Motion

CASE 1.8

Artifacts

Fig. 1.16 A) Radiograph of the femur demonstrating image ghosting. (B) Radiograph of the pelvis demonstrating grid artifact. (C) Radiograph of abdomen demonstrating a barium spill. (D) Radiograph of a pediatric chest demonstrating motion artifact.

1. **A.** Radiograph of the femur demonstrating image ghosting. A highly attenuating implant from a previous exam is superimposed on top of the current image. When a region of an image has received an unexpectedly high (or low) exposure, this area of the receptor can have a different sensitivity for a short period after the exposure. When the next radiograph is taken, the different sensitivities cause a "ghost" image of the previous exposure to appear in the current image.

2. **B.** Radiograph of the pelvis demonstrates grid artifact. Grid artifacts can manifest as repeating coarse or fine lines (see vertical lines in the inset image) and nonuniform intensity across the image (note overall intensity from right to left side of the image), caused by grid cutoff when the SID does not match that of the grid or when the grid is not aligned correctly. Grids need to be correctly aligned to transmit primary x-rays over the entire image.

3. **C.** This radiograph of the abdomen demonstrates a Ba spill. The highly attenuating Ba appears bright white on the image and may render an image nondiagnostic if it obscures or fails to opacify the desired anatomy. Objects worn by patients (false teeth, necklaces, hairpins) can also mask important anatomy or even simulate pathology.

4. **D.** This radiograph of a pediatric chest demonstrates motion artifact. The image blurs as the patient moves during acquisition. Although it may appear that a radiograph is created instantaneously, in reality it takes several to tens of milliseconds. If the patient moves during this time, then the anatomy will smear along the direction of motion like a blurred photograph. Minimizing exposure time is crucial to minimizing motion artifacts.

Comment

An artifact is something observed that is not actually present. The creation of a diagnostic image requires that all components of the imaging system are functioning correctly and that all aspects of the imaging chain are properly implemented. Failure of one or more aspects can result in artifacts that degrade image quality and may even result in missed or false diagnoses.

When the technologist or radiologist uses an x-ray system, they assume that it is functioning properly. A malfunctioning detector can result in ghost images (see Fig. 1.16A), patches of uneven uniformity, or even areas of complete signal loss (dead pixels). These problems are often corrected by recalibrating the detector and image processing (usually performed by the physicist or vendor service). Damage from physical collisions can result in artifactual marks on the image due to bent grids, gouged detector covers, or debris in the x-ray tube. If these artifacts are severe and impede diagnosis (or mimic pathology), then the damaged components must be replaced. Proper quality control carried out by technologists and medical physicists is vital to identifying and correcting system malfunctions before they create artifacts in clinical images.

A properly functioning system does not guarantee an artifact-free image. The radiologist or technologist must position the patient correctly and ensure that proper technique is used throughout the imaging procedure. Poor positioning can result in grid cutoff (see Fig. 1.16B), too much or too little anatomy in the field of view, or even patient motion due to discomfort. When such artifacts occur in clinical practice, it is the decision of the radiologist whether to retake the images to ensure diagnostic quality at the cost of extra patient dose. When contrast media is used, suboptimal images can occur due to spills, extravasation, and incorrect volume or injection rates. Finally, the human visual system can create the perception of artifacts that are not actually present in the images. Mach bands are perceived bands of light and dark near high-contrast edges that are created in our eye but are not actually present in the pixel values of the images.

Technique can impact the appearance of artifacts. Longer exposure times can lead to increased motion artifact (see Fig. 1.16D); thus short acquisitions are especially important in noncompliant patients such as infants and those with diminished intellectual capabilities. Proper kV must be selected to ensure that the K-edge of contrast media is used. Image processing can impact contrast, resolution, and noise, causing any one of them to be suboptimal, depending on the settings. Grid artifacts may be present if the grid is not aligned and set up properly and if grid suppression image processing is not active (see Fig. 1.16B). Grid cutoff may also occur if the grid is installed upside down (Fig. 1.17). Image processing can also create artifactual "halos" around metal implants if edge enhancement is improperly set.

References

Huda W. Radiography. In: *Review of Radiological Physics.* 4th ed. Philadelphia: Wolters Kluwer; 2016:100–101.

Walz-Flannigan A, Magnuson D, Erickson D, Schueler B. Artifacts in digital radiography. *AJR Am J Roentgenol.* 2012;198:156–161.

Zylak CM, Standen JR, Barnes GR, Zylak CJ. Pneumomediastinum revisited. *RadioGraphics.* 2000;20(4):1043–1057.

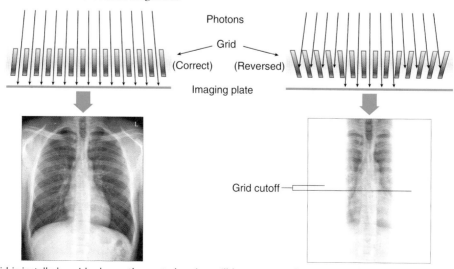

Fig. 1.17 When a grid is installed upside down, the central region will have a normal appearance but appear white (i.e., grid cutoff) toward the edges.

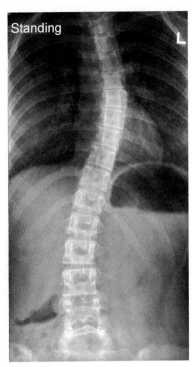

Fig. 1.18

1. Radiographic examination of what body part likely has the lowest x-ray tube voltage?
   A. Skull
   B. Chest
   C. Abdomen
   D. Extremity

2. Radiographic examination of what body part likely has the highest x-ray tube voltage?
   A. Skull
   B. Chest
   C. Abdomen
   D. Extremity

3. What type of follow-up examination is most likely to use decreased x-ray tube output (mAs)?
   A. Scoliosis
   B. ICU chest
   C. Extremity
   D. Abdomen

4. Decreasing what parameter is most likely to increase the lesion CNR in radiographic imaging performed using AEC?
   A. Current (mA)
   B. Voltage (kV)
   C. Exposure time (s)
   D. Focus (mm)

CASE 1.9

Radiographic Techniques and Diagnostic Task

Fig. 1.18 Frontal view of the spine in a patient with idiopathic scoliosis. Because the clinical question pertains only to high-contrast bony anatomy, radiation exposure can be reduced because increased noise will not impact diagnosis.

1. **D.** Extremity exams are generally performed at 55 to 65 kV, which is lower than for a skull (80 kV), dedicated chest (120 kV), or abdomen (80 kV). The thin extremities generally are easy to penetrate, and thus low kV can be used. Thicker structures such as the abdomen require higher kV.

2. **B.** Dedicated chest exams are generally performed at 120 kV with heavy filtration, which is higher than an extremity (60 kV) or skull/abdomen (80 kV). Although the attenuation of the lungs is relatively low, a high kV is used in digital radiography to reduce attenuation of ribs and thus improve visibility of underlying tissue.

3. **A.** A follow-up scoliosis examination is primarily concerned with any changes in the curvature of the spine, which are very easy to see. If the radiation intensity (mAs) was reduced by 90%, this would be unlikely to affect diagnostic performance. Because this examination is most often performed in children, this offers substantial and worthwhile patient dose savings.

4. **B.** When an examination is performed using AEC, the mottle in the image will always be exactly the same, no matter how kV, mA, or s is changed. The only change can be to the amount of contrast. If the kV goes down and the AEC is working, noise will stay the same but contrast will increase. As a result, CNR will also increase.

Comment

One of the most important lessons provided by imaging scientists is that, in the absence of a defined imaging task, there is no such thing as "image quality." This can be illustrated by considering a very mottled image where it would be virtually impossible to detect a subtle low-contrast lesion but simple to measure the dimension of some anatomic structure (Fig. 1.19). The image is "terrible" for detecting a subtle lesion but perfectly adequate if the diagnostic imaging task happens to be obtaining the anatomic dimension for a surgeon.

Imaging scientists have investigated the issue of whether a given lesion would likely be detected by a trained observer (radiologist). This question can be answered in an objective manner provided that the key aspects of imaging performance (lesion contrast, noise, and resolution performance), as well as the characteristics of the human visual system, are taken into account. For a specified lesion, a radiographic signal-to-noise ratio (SNR) can be computed, where this number provides an objective measure of the likelihood of an observer detecting the lesion. Imaging scientists would consider a lesion with a radiographic SNR of 1 to be "invisible" but a radiographic SNR of 5 to be detected by any human observer.

SNR is a highly technical absolute measure of the chance of a specified lesion being detected. Conversely, the CNR is a relative measure of lesion visibility that is influenced by choice

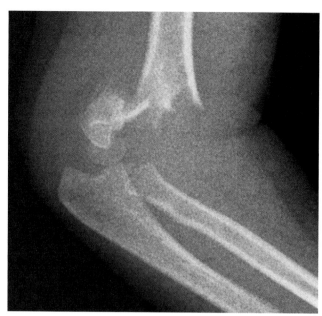

Fig. 1.19 Although high image noise may prevent detection of soft tissue abnormalities, this displaced supracondylar fracture is easily detected. Image quality would thus be very poor for detection of soft tissue abnormalities but adequate for detection of a major fracture.

of radiographic techniques and whose absolute value has no significance whatsoever. X-ray beam quality (kV) influences the amount of radiation that penetrates the patient, lesion contrast, and the amount of scatter that is produced. Low beam qualities will reduce penetration and increase lesion contrast. On the other hand, quantity (mAs) influences only the amount of noise in the resultant image. Changing the quantity of radiation used will have absolutely no effect on lesion contrast. When the effect of any technique changes on lesion CNR is fully taken into account, an increase in this metric means the lesion is more visible and vice versa.

A diagnostic departmental chest x-ray needs to be high quality to diagnose a range of pathologies and accurately assess the size of the heart. High voltages (120 kV) and high filtration reduce the patient dose and help visualize a wide range of tissues types, and the SID in PA projections is increased to 180 cm to minimize geometric magnification of the heart. A 10:1 grid is used to minimize scatter, and phototiming is used to guarantee the right intensity at the image receptor (K_{air}). Bedside radiographs are obtained primarily to monitor the placement of tubes, lines, and catheters. An AP projection is used with a reduced SID (100 cm) and manual techniques. Because of alignment difficulties, grids are generally not used and the tube voltage is reduced to 80 kV to help minimize scatter. Optimal performance would also process the resultant images using unsharp mask enhancement to improve the visibility of sharp edges associated with all inserted devices.

References

Huda W. Radiography. In: *Review of Radiological Physics*. 4th ed. Philadelphia: Wolters Kluwer; 2016:94–96.

Huda W. Understanding (and explaining) imaging performance. *AJR Am J Roentgenol.* 2014;203(1):W1–W2.

Huda W. X-rays. In: *Review of Radiological Physics*. 4th ed. Philadelphia: Wolters Kluwer; 2016:11–13.

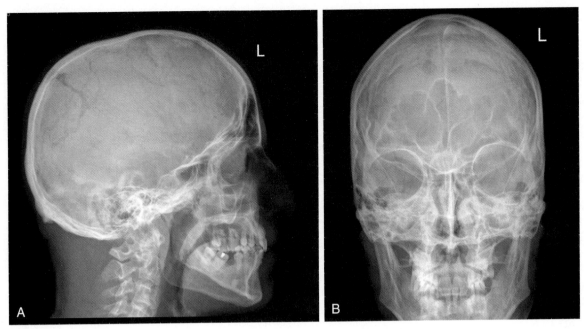

Fig. 1.20

1. What is the patient entrance K_{air} (mGy) for a lateral skull radiograph?
   A. 0.1
   B. 1
   C. 10
   D. 100

2. What is the kerma area product (KAP) (Gy-cm²) for an AP skull x-ray examination?
   A. 0.1
   B. 1
   C. 10
   D. 100

3. How many adult body radiographic examinations would be needed to give the same amount of radiation as a typical fluoroscopy-guided gastrointestinal (GI) study?
   A. 1
   B. 10
   C. 100
   D. 1000

4. How many adult body radiographic examinations would be needed to give the same amount of radiation as a typical interventional radiologic procedure?
   A. 1
   B. 10
   C. 100
   D. 1000

CASE 1.10

Incident Air Kerma and Kerma Area Product

Fig. 1.20 Lateral (A) and frontal (B) radiographs of the skull in a patient with Gorlin syndrome.

1. **B.** For an adult lateral skull x-ray, 1 mGy is most likely incident on the patient. The amount that reaches the detector will be much lower (0.003 mGy) because of attenuation in the patient and inverse square law drop in intensity, as well as loss of primary photons in the grid.

2. **B.** An AP skull x-ray examination in an adult will likely use 1 Gy-cm² of radiation. Chest x-rays would likely be two or three times lower and abdominal radiographs two or three times higher.

3. **B.** A KAP in a GI/genitourinary examination is at least 10 Gy-cm², which is an order of magnitude higher than in radiography.

4. **C.** A KAP in an IR procedure is at least 100 Gy-cm², which is two orders of magnitude higher than in radiography.

Comment

Patient entrance K_{air} values generally depend on x-ray tube characteristics, x-ray technique (mAs and kV), and distance from the focus to the patient entrance. The radiation intensity from any source falls off according to the Inverse Square Law ($1/[\text{distance}]^2$), where doubling the distance reduces the intensity by a factor of 4. The entrance K_{air} for a lateral skull x-ray is approximately 1 mGy and about twice this value for the more attenuating AP projection when radiographs are obtained with systems that employ an AEC. Radiographs of much less attenuating body parts (PA chest) require an entrance K_{air} of approximately 0.1 mGy, whereas those for much more attenuating regions (lateral lumbar spine) would require a K_{air} of about 10 mGy. K_{air} is independent of the x-ray beam cross-sectional area and therefore cannot account for the total amount of radiation used in any radiographic examination.

When the average K_{air} (mGy) value of the x-ray beam incident on the patient is multiplied by the corresponding x-ray beam cross-sectional area (cm²), one obtains the KAP. The KAP, also known as the dose area product (DAP), measures the total amount of radiation incident on the patient because it accounts for the beam area (Fig. 1.21). When the value of K_{air} is 1 mGy, and the x-ray beam has an area of 1000 cm² (approximately 30 × 30 cm), the KAP is 1 Gy-cm². The average KAP for a complete radiographic examination is 1 Gy-cm². Chest x-ray examinations are approximately a factor of 3 lower than this average value, and abdominal x-ray examinations are approximately a factor

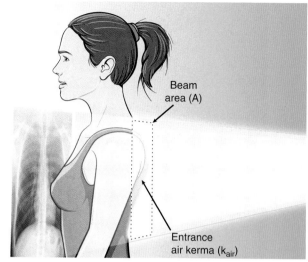

Kerma Area Product (KAP) = k_{air} x Area

Fig. 1.21 The intensity of the x-ray beam (i.e., photons per mm²) is given by air kerma (k_{air}) and is measured in mGy. The corresponding x-ray beam area *(A)* is measured in cm². The kerma area product *(KAP)* is expressed in Gy-cm², where 1 Gy-cm² is 1000 mGy-cm². The KAP is often referred to as the dose area product (DAP), and these two terms are synonymous.

of 3 higher. In pediatric radiography, entrance K_{air} is lower because patients are thinner and KAP is much lower because the corresponding x-ray beam areas are also lower.

Radiologists and technologists are responsible for the x-ray beam quantity and quality that is incident on a patient. These factors will depend on patient characteristics, and the specified diagnostic imaging task for which the examination is being performed. Medical physicists can convert values of incident radiation into corresponding patient doses and risks. To do this, it is essential that account is taken of both technical factors such as the x-ray beam quality and quantity, as well as exam type that includes the body region (e.g., chest) and the specific projection used (e.g., AP). A complete assessment of patient doses and risks should also include the important patient characteristics including size and patient demographics.

References

Huda W. Kerma-area product in diagnostic radiology. *AJR Am J Roentgenol.* 2014;203(6):W565–W569.

Huda W. Patient dosimetry. In: *Review of Radiological Physics.* 4th ed. Philadelphia: Wolters Kluwer; 2016:47–48.

Huda W. Radiography. In: *Review of Radiological Physics.* 4th ed. Philadelphia: Wolters Kluwer; 2016:100.

CASE 2.1

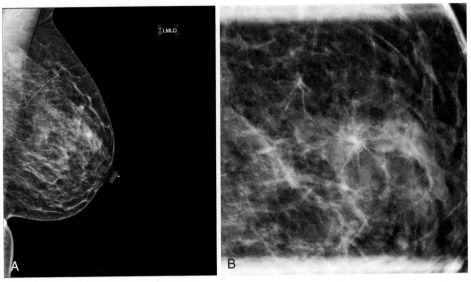

Fig. 2.1

1. What is the number of breast cancers most likely detected when 1000 (asymptomatic) women undergo screening mammography?
   A. 1
   B. 4
   C. 16
   D. 64

2. Reducing which factor is the most important when visualizing a soft tissue mass in mammography?
   A. Photon energy
   B. Receptor exposure
   C. Pixel sizes
   D. Focal spot

3. What needs to be done to permit the detection and characterization of microcalcification clusters in mammography?
   A. Reduce focal spot size
   B. Immobilize the breast
   C. Reduce pixel size
   D. A, B, and C

4. What should be reduced as much as possible for a diagnostic screening test in an asymptomatic population?
   A. Exam cost
   B. Imaging time
   C. Report time
   D. Dose (risk)

CASE 2.1

Diagnostic Task

Fig. 2.1 (A and B) Mediolateral oblique mammogram and spot magnification view showing a spiculated mass with associated fine, pleomorphic calcifications.

1. **B.** One in eight women in the United States is expected to get breast cancer during their lifetime. Breast cancer mortality is estimated to be approximately 25% and has been declining over the past decade. This decline in breast cancer mortality has been attributed to increased screening, better treatment, or both. The number of cancers detected per 1000 women is currently estimated to be approximately four.

2. **A.** The attenuation properties of fibroglandular tissues and malignant tissues are very similar, so it is essential to reduce the photon energy to increase lesion contrast. Reducing detector exposure will increase the amount of mottle and thereby reduce lesion visibility. Reducing pixel size and focal spot will improve resolution, which is important for detecting/characterizing microcalcifications but not (large) masses.

3. **D.** Visualizing and characterizing microcalcifications require exceptional spatial resolution performance. Resolution can be improved by reducing the focal spot size (which reduces focal blur) and immobilizing the breast to reduce motion blur. Smaller pixels and thinner detectors will also increase image sharpness. Therefore all three factors in this question are essential to improve detection and delineation of small features such as clusters of microcalcifications.

4. **D.** As described in Question 1, 996 out of 1000 screened women will not have cancer. Accordingly, it is essential that the radiation used to perform screening is as low as reasonably achievable (ALARA) to minimize radiation risks for women who do not have cancer (i.e., >99% of screened population).

Comment

When 1000 asymptomatic women undergo screening mammography, the number of malignant cancers that are likely to be detected is very low (typically four). Of those detected, only approximately 25% are fatal. This means that approximately 996 per 1000 (>99%) of the screened population are healthy but are exposed to ionizing radiation. The radiation dose associated with screening mammography therefore has to be very low, and the risk of inducing a breast cancer from the mammographic examination must be low enough to ensure that there is a net benefit of this examination (see Case 2.10). As a screening examination for cancer, the sensitivity of mammography is more important than specificity. This is because the outcome of a missed diagnosis may be death, whereas the outcome of a false diagnosis is an unnecessary biopsy. Many suspicious lesions are biopsied after being detected on a screening mammogram, but the majority of such lesions are benign. The cost of the unnecessary biopsies is balanced by the cost that would occur if the number of biopsies was reduced and more breast cancers were left undetected.

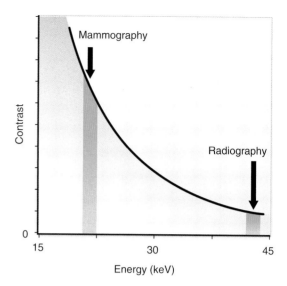

Fig. 2.2 The contrast at the low energies used in mammography is much higher than the contrast obtained at the higher energies in radiography. A typical mammography beam (W target + Rh filter) has a half-value layer of only 0.5 mm aluminum, whereas a radiography beam (80 kV + 3 mm aluminum) has a half-value layer of 3 mm aluminum.

The attenuation coefficient of breast carcinomas is only slightly higher than that of normal glandular tissue. What this means is that tumors embedded in glandular tissue transmit very similar amounts of radiation. At photon energies typically used in standard radiography, 80 kV (i.e., ~40 keV), the lesion contrast would be so low as to render the mass invisible. Subject contrast can be improved by reducing the photon energy, although this also reduces the penetration of the beam. Reducing the energy is an option in mammography but not other body parts because of the relatively low thickness of a compressed breast. To achieve low photon energies, special "k-edge" filters are used that transmit photons just below their k-edges. The average photon energy in mammography is very low, with the aluminum half-value layer (HVL) being approximately 0.5 mm Al (vs. 3 mm aluminum in radiography performed at 80 kV). Use of such low energies increases lesion contrast by up to an order of magnitude compared with standard radiography (Fig. 2.2).

Important diagnostic tasks in mammography include the detection and characterization of microcalcifications, masses, and architectural distortion. Microcalcifications are a few hundred micrometers. Their detection requires exceptional spatial resolution (7 line pairs per mm [lp/mm] or more), which is better than that of the human eye at a normal viewing distance of 25 cm (5 lp/mm). To achieve this, a small focal spot (0.3 mm) is used to reduce focal spot blur (Fig. 2.3). Compression paddles are used to minimize motion blur, and specialized detectors with very small pixels are used to reduce detector blur. The total number of pixels in a mammogram is greater than can be displayed on a specialized mammography workstation (5 million pixels), and to achieve the full resolution, the "zoom" function must be used.

	Mammography	Radiography
Resolution	‖‖‖ 7 lp/mm	‖ ‖ 3 lp/mm
Focal spot	• 0.3 mm	● 1.2 mm
Pixel size	▦ 70 μm	▦ 170 μm
Compression paddle	▭ Yes	No

Fig. 2.3 Key imaging characteristics in mammography and radiography.

References

American College of Radiology. ACR practice parameter for the performance of screening and diagnostic mammography (resolution 11), revised 2013. https://www.acr.org/-/media/ACR/Files/Practice-Parameters/screen-diag-mammo.pdf?la=en. Accessed March 1, 2018.

Huda W. Mammography. In: *Review of Radiological Physics,* 4th ed. Philadelphia: Wolters Kluwer; 2016:105–114.

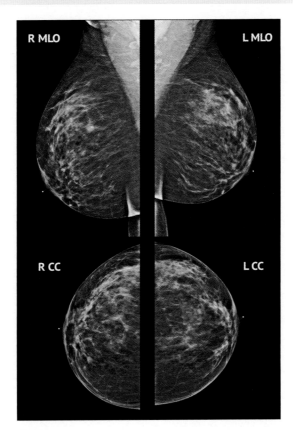

Fig. 2.4

1. What is the relative size of the focal spot used in contact mammography compared with the large focal spot used in radiography (%)?
   A. 5
   B. 10
   C. 25
   D. 50

2. What material is typically used as the x-ray output window to increase transmission of low-energy x-rays in mammography?
   A. Beryllium
   B. Aluminum
   C. Glass
   D. Molybdenum

3. What part of the compressed breast is the anode end of a mammography x-ray tube directed to?
   A. Chest wall
   B. Nipple
   C. Central axis of the breast
   D. Axilla

4. What filter is never used in mammographic imaging systems?
   A. Molybdenum
   B. Rhodium
   C. Silver
   D. Copper

CASE 2.2

X-Ray Tube

Fig. 2.4 Four-view screening mammogram.

1. **C.** The small and large focal spots in conventional radiography are 0.6 and 1.2 mm, respectively. In mammography, it is important to generate sharper images than in conventional radiography, so focal spots have to be reduced. In contact mammography, a 0.3-mm focal spot is used, which is 25% of that used in chest radiography.

2. **A.** Beryllium ($Z = 4$) is used to replace glass to improve the transmission of low-energy x-rays. Aluminum ($Z = 13$) and molybdenum ($Z = 42$) attenuate more than glass ($Z \sim 10$) because of their higher atomic number.

3. **B.** Less radiation emerges from the anode size of any x-ray tube because of the heel effect. The anode side is thus directed to the least attenuating part of the breast, namely the nipple. Because more radiation emerges from the cathode side of an x-ray tube, this would be directed toward the more attenuating chest wall part of the compressed breast.

4. **D.** Mammography uses special k-edge filters to control the average photon energy. These filters remove both low-energy photons that irradiate patients but do not contribute to the image as well as high-energy photons that reduce contrast. Three common k-edge filters used are molybdenum (20 keV), rhodium (23 keV), and silver (26 keV), and where the average transmitted photon energy is "below" the filter k-edge energy and is thus not primarily determined by the tube voltage. Copper is a filter whose k-edge energy (9 keV) is irrelevant for controlling the x-ray beam energy.

Comment

Historically, mammography x-ray tubes have used molybdenum targets, which offer intense characteristic x-rays with an average energy that corresponds to a beam HVL of only 0.3 mm aluminum. There are advantages to using slightly higher x-ray photon energies, which reduce patient doses because of the improved penetration through the breast. With screen-film systems, higher energies invariably reduce lesion contrast, but with digital systems this can be partly offset by the use of digital image processing. Modern mammography systems typically use tungsten (W) targets and vary the energy of the beam by the use of special k-edge filters (e.g., rhodium and silver). The switch to tungsten from molybdenum targets is the main reason for the continued reduction in patient doses in digital mammography, which are now substantially lower than doses from screen-film target/filter combinations such as molybdenum/molybdenum and molybdenum/rhodium.

Focal spots are much smaller in mammography than in standard radiography (0.3 vs. 1.2 mm) and are essential to ensure the exceptional resolution that is required to image the breast. The power loading that can be tolerated by a small 0.3-mm focal spot is relatively low (3 kW), which results in long exposure

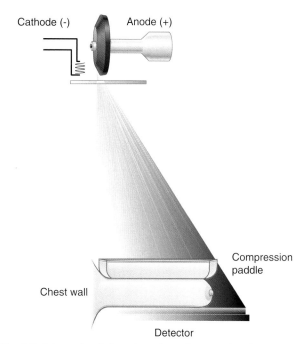

Fig. 2.5 Schematic of the orientation of x-ray tubes in mammography where the low-intensity output on the anode side of the x-ray tube (heel effect) is directed to the nipple, and the higher intensity at the cathode side is directed to the more attenuating chest wall region.

times of up to 1 second. The small focal spot is only 0.1 mm in size and must be used to reduce focal spot blur in magnification mammography. The power loading that the 0.1 mm focal spot can tolerate is only a quarter of that tolerated by the standard 0.3-mm focal spot, which results in very long exposures (>1 second).

The heel effect is caused by the angulation of the anode to approximately 15 degrees from the central axis and reduces the x-ray beam intensity at the anode side of the x-ray tube. Because the cathode side of the x-ray tube has a greater intensity, it is directed toward the more attenuating chest wall (Fig. 2.5). The anode side of the x-ray tube is directed toward the less attenuating nipple. A glass window would attenuate too many of the valuable low-energy x-rays produced by a mammography x-ray tube, which would further increase the long exposure times in mammography. As a result, a low atomic number ($Z = 4$) beryllium window is used.

References

Bushberg JT, Seibert JA, Leidholdt Jr EM, Boone JM. Mammography. In: *The Essential Physics of Medical Imaging*. 3rd ed. Philadelphia: Wolters Kluwer; 2012:238–281.

Huda W. Mammography. In: *Review of Radiological Physics*. 4th ed. Philadelphia: Wolters Kluwer; 2016:106–107.

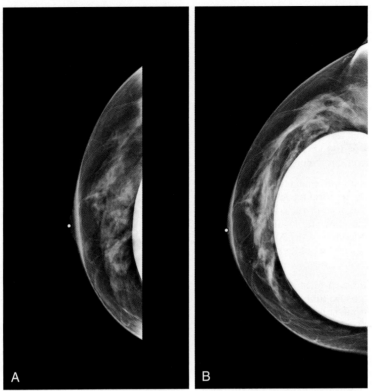

A

B

Fig. 2.6

1. How much power (kW) is most likely used in a clinical mammography system?
   A. 3
   B. 10
   C. 30
   D. 100

2. Which is the most likely x-ray tube voltage (kV) in a screening mammography examination?
   A. 20
   B. 30
   C. 40
   D. 60

3. What are the most likely tube currents (mA) and output values (mAs) values in screening mammography?
   A. 100 and 100
   B. 100 and 1000
   C. 1000 and 100
   D. 1000 and 1000

4. What is the most likely exposure time required to generate a satisfactory clinical contact mammogram (ms)?
   A. 1
   B. 10
   C. 100
   D. 1000

CASE 2.3

X-Ray Generator

Fig. 2.6 Implant-displaced (A, Eklund view) and traditional cranio-caudal (B) views in a patient with breast implants.

1. **A.** The typical x-ray tube voltage is 30 kV, and the corresponding tube current is 100 mA, which translates into a power of 3 kW. This is much less than in conventional radiography where voltages of 100 kV and tube currents of 1000 mA (i.e., 100 kW power) are common.

2. **B.** Most mammography uses approximately 30 kV voltages. The tube voltage is of minor importance in controlling the average photon energies because of the use of k-edge filters. A voltage of 20 kV is too low and would have spectra lower than the lowest k-edge filter (Mo 20 keV). Voltages of 40 and 60 kV are too high and would result in too many high-energy photons that reduce image contrast.

3. **A.** Tube currents for the normal focal spot (0.3 mm) are generally 100 mA, and exposure times are typically 1 second. Accordingly, the output (tube current × exposure time) is approximately 100 mAs.

4. **D.** Exposure times in mammography range between 500 and 1000 ms. Exposure times are much longer in magnification mammography, where tube current is reduced because of the need to use a small focal spot (0.1 vs. 0.3 mm).

Comment

Mammography generators need to supply 3 kW of power for imaging, which is much lower than in conventional radiography and CT (100 kW). Higher power levels cannot be tolerated by the very small focal spots used in mammography (0.3 mm) because these would melt the target. The principal drawback of the low power used in mammography is long exposure times (~1 second), which are much longer than in abdominal radiography (50 ms) and chest radiography (5 ms).

Mammography is unusual in that low x-ray tube voltages (~30 kV) are typically used (Table 2.1). These low voltages ensure that higher energy photons (>30 keV) are not produced. This is important because higher energy photons in mammography would markedly reduce lesion contrast. These low energies also increase contrast of microcalcifications due to increased photoelectric interactions. User control of beam energy in mammography also differs from other modalities. In conventional radiography, the x-ray tube voltage is modified to change the x-ray beam energy. However, in mammography the voltage has only a relatively modest effect on beam energy, which is instead primarily determined by the choice of k-edge filter.

TABLE 2.1 COMPARISON OF TYPICAL TECHNIQUES IN CONTACT MAMMOGRAPHY, ABDOMINAL RADIOGRAPHY, AND CHEST RADIOGRAPHY

Parameter	Contact Mammography	Abdominal Radiography	Chest Radiography
Voltage (kV)	30	80	120
Power (kW)	3	80	100
Exposure time (s)	1	0.05	0.005
Relative output (mAs)	100	20	1

Exposure times

Chest x-ray 0.005 second

10x

Abdominal x-ray 0.05 second

200x

Mammogram 1 second

Fig. 2.7 The exposure time of a typical chest x-ray is 0.005 seconds, with an abdominal x-ray taking 20 times longer (0.05 seconds) and a mammogram taking 1000 times longer (1 second).

Tube currents in mammography are typically 100 mA and are limited by the power that can be tolerated by the small focal spot size. This tube current is much larger than used in fluoroscopy (~3 mA) but markedly lower than the several hundred milliamperes that would be used in radiography and CT. Exposure time in mammography is large (~1 second), which results in approximately 100 mAs typically used for each view. This is much higher than in chest radiography (1 mAs) and abdominal radiography (20 mAs) (Fig. 2.7). The long exposure time is required to ensure a satisfactory exposure at the image receptor (100 µGy) that is substantially higher than in any other radiographic or fluoroscopic imaging modality.

References

Bushberg JT, Seibert JA, Leidholdt Jr EM, Boone JM. Mammography. In: *The Essential Physics of Medical Imaging*. 3rd ed. Philadelphia: Wolters Kluwer; 2012:238–281.

Huda W. Mammography. In: *Review of Radiological Physics*. 4th ed. Philadelphia: Wolters Kluwer; 2016:106–107.

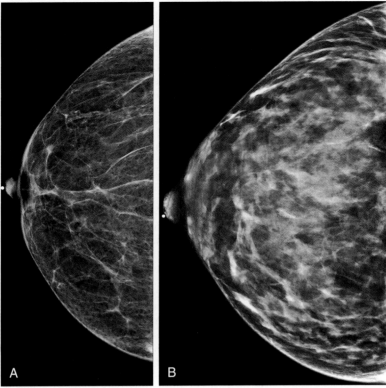

A   B

Fig. 2.8

1. Which filter would most likely be used to image a 50-mm compressed breast (10% glandularity) on an x-ray tube with a molybdenum target?
   A. Aluminum
   B. Molybdenum
   C. Rhodium
   D. Silver

2. Which filter would most likely be used to image an 80-mm compressed breast (20% glandularity) on an x-ray tube with a tungsten (W) target?
   A. Aluminum
   B. Molybdenum
   C. Rhodium
   D. Silver

3. What is the most important determinant of x-ray beam quality in digital mammography?
   A. Tube voltage
   B. Tube output
   C. K-edge filter
   D. A, B, and C

4. When mammography is performed using automatic exposure control (AEC), which is most likely reduced when x-ray beam quality is increased?
   A. Exposure time
   B. Radiation dose
   C. A and B
   D. Neither A nor B

CASE 2.4

X-Ray Beam Quality

Fig. 2.8 Craniocaudal view of a thin, predominantly fatty breast using a rhodium filter/tungsten target (A) and craniocaudal view of a thick, heterogeneously dense breast using a silver filter/tungsten target (B).

1. **B.** A 50-mm-thick compressed breast with 10% glandularity is much less attenuating than average and would use a low beam quality (molybdenum target + molybdenum filter), whereas for a denser breast we would increase the beam quality by replacing the molybdenum filter with a rhodium k-edge filter. Rhodium transmits higher energies because its k-edge energy (23 keV) is higher than that of molybdenum (20 keV).

2. **D.** An 80-mm-thick compressed breast with 20% glandularity is much more attenuating than an average breast and thus uses a higher beam quality (tungsten target + silver filter), whereas a markedly thinner breast would reduce the beam quality by replacing the silver k-edge filter with a rhodium k-edge filter. Silver transmits higher energies because its k-edge energy (26 keV) is higher than that of rhodium (23 keV).

3. **C.** The k-edge filter is the most important determinant of x-ray beam quality in digital mammography and more important than x-ray tube voltage (see Questions 1 and 2). This is different from conventional radiography, where the x-ray tube voltage is generally the most important determinant of x-ray beam quality. Tube output (mAs) does not affect x-ray beam quality.

4. **C.** When beam quality increases, breast penetration also increases. This reduces both exposure times and breast radiation doses because it will require less incident radiation to achieve the required radiation intensity at the image receptor.

Comment

Older x-ray tubes used molybdenum targets with molybdenum or rhodium k-edge filters. Thinner breasts (<60 mm) would be imaged using a molybdenum filter (k-edge 20 keV), and thicker breasts would be imaged using a rhodium filter (k-edge 23 keV). The higher k-edge energy of rhodium results in a higher energy x-ray beam that increases beam penetration. The use of a molybdenum target results in a spectrum that has a substantial contribution from characteristic x-rays (17 and 19 keV). Contrast improvement using new image-processing algorithms means that these energies are now considered to be lower than the optimal value for clinical use.

New mammography x-ray tubes use tungsten targets with rhodium or silver k-edge filters (Fig. 2.9). Thinner breasts (<60 mm) would be imaged using a rhodium filter (k-edge 23 keV), and thicker breasts would be imaged using a silver filter (k-edge 26 keV). The higher k-edge energy of silver results in a higher-energy x-ray beam that increases patient penetration. Tungsten targets produce a spectrum with higher average energies that are more penetrating because of the use of these rhodium and silver filters. Higher average energies reduce patient doses when operated with AEC systems and also reduce exposure times, minimizing motion blur. The loss of contrast normally associated with increased beam energy is now offset by the use of sophisticated vendor-specific image processing.

Beam quality in mammography is typically an order-of-magnitude lower than in conventional radiography, where the HVL of an x-ray beam generated at 80 kV with 3 mm total aluminum filtration is approximately 3 mm aluminum. In 2000, when most facilities operated using screen-film systems using 25 kV and molybdenum/molybdenum target filter combinations, the beam quality was approximately 0.25 mm aluminum. A mammography system in 2010 would be operated at a higher x-ray tube voltage (30 kV) and have an HVL of 0.3 mm aluminum. Modern tungsten targets operated at 30 kV would have an HVL of 0.5 mm with a rhodium k-edge filter, which increases to 0.6 mm aluminum with a silver k-edge filter.

References

Huda W. Mammography. In: *Review of Radiological Physics.* 4th ed. Philadelphia: Wolters Kluwer; 2016:106–107.

Radiological Society of North America. RSNA/AAPM physics modules. Projection x-ray imaging: mammography image quality and dose. https://www.rsna.org/Physics-Modules/. Accessed March 1, 2018.

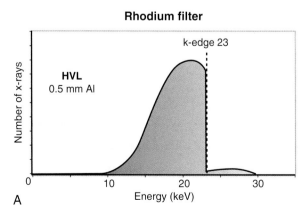

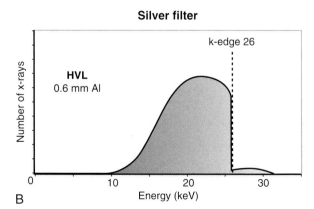

Fig. 2.9 (A) The spectrum from a modern tungsten target and a rhodium k-edge filter, which results in an x-ray beam with a half-value layer *(HVL)* of approximately 0.5 mm aluminum *(Al)* and which would be used on less attenuating breasts. (B) The spectrum from a modern tungsten target and a silver k-edge filter, which results in an x-ray beam with an HVL of approximately 0.6 mm Al and which would be used on the more attenuating breasts.

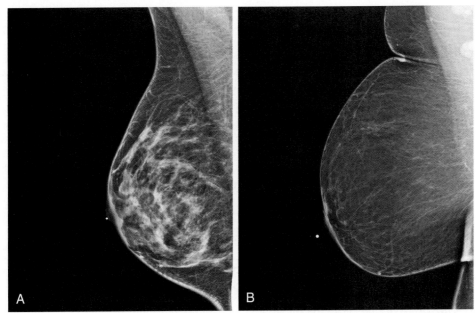

Fig. 2.10

1. What is the average compressed breast thickness (mm) in the United States?
   A. 20
   B. 40
   C. 60
   D. 80

2. What is the average breast glandularity (%) in the United States?
   A. 5
   B. 15
   C. 25
   D. 45

3. What is increased when compression is applied in mammography?
   A. Scattered photons
   B. Radiation dose
   C. Tissue overlap
   D. Radiation penetration

4. What type of blur is affected by the use of breast compression?
   A. Focal
   B. Motion
   C. Detector
   D. A and B

CASE 2.5

Compression and Breast Composition

Fig. 2.10 Screening mammogram mediolateral oblique views of women with a heterogeneously dense breast (A) and fatty breast (B).

1. **C.** The average compressed breast thickness in the United States is approximately 60 mm, with most breasts having a thickness between 40 and 80 mm. This is substantially greater than the 42-mm thickness of the breast phantom used by medical physicists to estimate the mean glandular dose for each clinical imaging system, as required by the Mammography Quality Standards Act (MQSA).

2. **B.** The average breast glandularity in the United States is approximately 15%. Most breasts have a glandularity between 5% and 35%. This is substantially less than the 50% glandularity of the breast phantom used by medical physicists to estimate the mean glandular dose for each clinical imaging system, as required by the MQSA.

3. **D.** Compression is used to reduce the breast thickness and make it approximately uniform thickness. This results in a shorter path for radiation in tissue and thus will decrease scattered photons and also reduce the radiation dose. Tissue overlap is minimized when the tissues are spread out, which markedly improves the visibility of any lesions. Because the breast thickness is reduced, radiation penetration can increase.

4. **D.** Focal blur is reduced with compression as geometric magnification of tissues will be reduced. Motion blur is also reduced when the breast is immobilized by a compression system. Detector blur refers to the detector thickness and pixel size, which is not affected by the amount of compression being applied.

Comment

A compression force of up to 40 lb is typically used in mammography. This compression results in a more uniform thickness, which has several benefits. First, penetration is increased and patient doses are reduced because the thicker regions of an uncompressed breast are flattened. Second, there will be less scattered radiation reducing image contrast. Third, there is less overlap of breast tissues because they are now spread out, which reduces anatomical noise. Subtle findings such as architectural distortion are unmasked and better separated from normal glandular tissue and Cooper ligaments. Finally, the reduction in breast thickness also reduces exposure times, which minimizes the likelihood of motion blur.

The average compressed breast thickness is approximately 60 mm (Fig. 2.11). For the typical patient, 10 mm of tissue will attenuate half of an x-ray beam so that the transmission through a compressed breast is relatively low ($1/2^6$, or around 1%). Increasing the breast thickness by 10 mm will require a doubling of the radiation used to obtain a satisfactory mammogram, and therefore a nominal doubling of the patient dose, when beam quality is kept constant. Average breast glandularity is 15%, and it is this tissue that is of most concern when determining patient radiation dose and cancer induction risk. It is of interest to note that the phantom used for MQSA testing and to estimate breast doses has a thickness of 42 mm and glandularity of 50% (see Case 2.9). However, the breast phantom is a standard used to compare different mammography systems, not to estimate patient doses.

As patients age, fibroglandular breast tissue slowly involutes. This results in an increase in the fatty composition of the breasts, which makes detection of soft tissue masses and focal asymmetries easier. Fatty breasts are also more compressible, again improving detection of subtle abnormalities. When breasts are dense (i.e., high percentage of glandularity), detection is more difficult using mammography. Several states have enacted legislation requiring a note on breast density be included in the patient record. In the case of very dense breasts, tomosynthesis or MRI may be clinically indicated to improve lesion detectability.

References

Huda W. Mammography. In: *Review of Radiological Physics*. 4th ed. Philadelphia: Wolters Kluwer; 2016:108–113.

Yaffe MJ, Boone JM, Packard N, et al. The myth of the 50-50 breast. *Med Phys*. 2009;36(12):5437–5443.

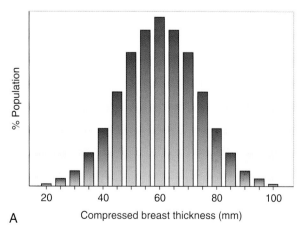

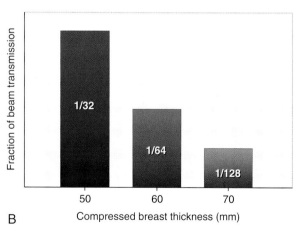

Fig. 2.11 (A) Representative compressed breast thickness values for women undergoing mammography. The average breast thickness is 60 mm, which corresponds to approximately six tissue half-value layers. (B) Transmission through varying thickness values of breast tissue based on a tissue half-value layer of 1 cm, showing that just over 1% of the incident radiation is transmitted through an average sized breast (6 cm).

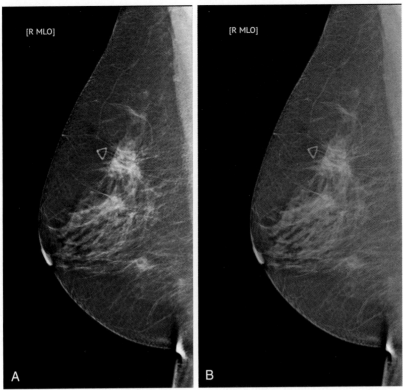

Fig. 2.12

1. What is the typical grid ratio used in screening mammography?
   A. 1:1
   B. 2:1
   C. 5:1
   D. 10:1

2. What detector has the highest spatial resolution in mammography?
   A. Scintillator (cesium iodide)
   B. Photostimulable phosphor (BaFBr)
   C. Photoconductor (selenium)

3. How much radiation (μGy) at the image receptor is needed to obtain a satisfactory mammogram?
   A. 0.01
   B. 1
   C. 10
   D. 100

4. What is the minimum number of megapixels (MP) on a diagnostic workstation used for clinical mammography?
   A. 1
   B. 3
   C. 5
   D. 15

CASE 2.6

Grids, Detectors, and Image Display

Fig. 2.12 Mediolateral oblique *(MLO)* view of spiculated masses in the right breast of a patient with a palpable lump. A was obtained with a grid, and B was simulated without a grid. Note the decreased contrast in B due to scatter.

1. **C.** Most mammography systems make use of a 4:1 or 5:1 scatter removal grid, which is a lower grid ratio than used in conventional radiography (10:1). Grids are used in all contact mammography where the addition of a grid markedly improves image quality (contrast) by removing scatter but results in a doubling of the patient dose for examinations performed using an AEC system.

2. **C.** Photoconductors have the best spatial resolution because the charge produced when x-rays are absorbed does not spread out during the charge collection process (Fig. 2.13). Resolution of scintillators is degraded by the spreading of the light that is produced when x-rays are absorbed. Photostimulable phosphors have the worst resolution because the stimulating laser light is scattered into adjacent pixels.

3. **D.** In radiography, the amount of radiation at the image receptor is approximately 3 μGy. In mammography, the radiation intensity is much higher, at 100 μGy or so, which results in much less mottle, and thereby improves the lesion contrast-to-noise ratio (CNR).

4. **C.** Most mammography workstations have 5 MP, which is higher than in standard workstations (3 MP). Even so, a typical mammogram has up to 15 million pixels so that zoom functions have to be used to achieve the native image resolution.

Comment

All current contact (i.e., nonmagnification mammographic) examinations make use of a scatter removal grid. Scatter in mammography is lower than in standard radiography because of the relatively thin body part thickness, small field of view, and low beam energy, which causes more photoelectric interactions than Compton scatter in soft tissues. Because there is less scatter in mammography, lower grid ratios are used than in conventional radiography (5:1 vs. 10:1). Bucky factors in mammography are also lower (2:1) than in abdominal radiography (5:1), which means that when a mammogram is performed with the grid removed in a system with AEC, the breast dose would be halved.

Scintillators based on cesium iodide can be used as the detector in mammography (see Fig. 2.13). These detectors are very reliable and are "fast" so that multiple exposures can be performed in a short time. Amorphous selenium photoconductors are also used in many commercial mammography systems. These detectors have high absorption efficiencies at the low photon energies used in mammography and also have excellent spatial resolution. Photostimulable phosphors (e.g., BaFBr) do not perform as well as scintillators and photoconductors in mammography, and there is now evidence that these detectors may result in a lower clinical diagnostic performance. Pixels sizes in mammography generally range between 50 and 100 μm (~7 lp/mm), which is necessary to visualize and characterize microcalcifications, which can be as small as a couple of hundred microns in diameter.

A 24 × 30 cm digital detector with 75 μm pixels will consist of just over 12 million individual pixels. Because 2 bytes are required to store each individual pixel value, the data content of a mammogram is 25 MB. A screening mammogram with two views of each breast (craniocaudal [CC] and mediolateral oblique [MLO]) would require a total storage capacity of 100 MB, or the equivalent of 10 chest x-rays. Mammograms are currently displayed on 5 or 8 MP monitors, which have markedly fewer monitor pixels than the number of pixels in each mammogram. As a result, image zoom must be used to see all of the detail in any mammogram. The American College of Radiology (ACR) requires display monitors in mammography to have a high luminance (>420 cd/m²) and for images to be viewed in darkened rooms (e.g., <50 lux).

References

American College of Radiology. ACR–AAPM–SIIM practice parameter for determinants of image quality in digital mammography (resolution 42), revised 2017. https://www.acr.org/-/media/ACR/Files/Practice-Parameters/dig-mamo.pdf?la=en. Accessed March 1, 2018.

Huda W. Mammography. In: *Review of Radiological Physics*. 4th ed. Philadelphia: Wolters Kluwer; 2016:108–113.

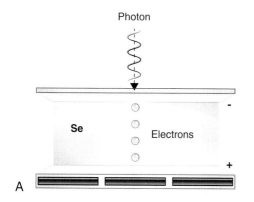
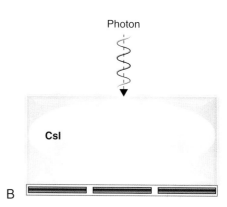

Fig. 2.13 (A) A selenium *(Se)* photoconductor that directly measures the charge produced when x-rays interact directly. (B) A cesium iodide *(CsI)* scintillator converts a fraction of the absorbed x-ray energy into light that is subsequently converted into charge (i.e., indirect). Photoconductors have superior resolution because the charge does not disperse in the way light does in a scintillator.

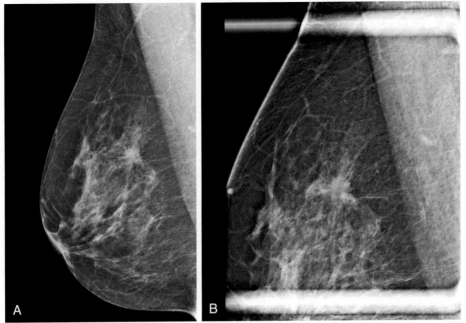

Fig. 2.14

1. How are focal spot sizes and tube output (mAs) adjusted when switching to magnification mammography?
   A. Decreased and decreased
   B. Decreased and increased
   C. Increased and decreased
   D. Increased and increased

2. How is the grid ratio modified when performing magnification mammography?
   A. Increased
   B. Unchanged
   C. Reduced
   D. Grid removed

3. Which is likely true of current breast tomosynthesis?
   A. Fixed x-ray tube angle; one image
   B. Fixed x-ray tube angle; many images
   C. Varied x-ray tube angle; one image per angle
   D. Varied x-ray tube angle; many images per angle

4. Which is likely improved the most when replacing contact mammography with digital breast tomosynthesis (DBT)?
   A. Resolution
   B. Contrast
   C. Mottle
   D. Dose

CASE 2.7

Magnification Mammography and Breast Tomosynthesis

Fig. 2.14 Mediolateral oblique view (A) of the right breast and spot magnification view (B) depicting a spiculated mass in the upper outer quadrant with associated architectural distortion.

1. **A.** In magnification mammography, the focal spot is reduced from 0.3 to 0.1 mm to reduce focal spot blurring. The tube current that can be tolerated by the small focal spot is only a quarter of that tolerated by the large focal spot and is therefore reduced from 100 to 25 mA. Because there is no grid being used, there is no longer a 30% loss of primary photons by the grid, so techniques (mAs) can be reduced by approximately 30%.

2. **D.** Magnification mammography is performed by moving the breast closer to the x-ray tube, thereby creating an air gap between the compressed breast and the image receptor. The air gap effectively removes most of the scatter and eliminates the need for a grid.

3. **C.** In breast tomosynthesis, a typical system would acquire 15 images during an x-ray tube angular rotation of approximately 15 degrees. Commercial systems are available that offer 25 images over an x-ray tube movement of 50 degrees *or* 9 images over an x-ray tube movement of 25 degrees. Only one image is ever obtained at any given x-ray tube angle location.

4. **B.** The principal benefit of tomography is the removal of overlying tissues so that the lesion contrast (lesion vs. surrounding tissues) is markedly improved. In-plane resolution will be comparable to contact mammography, as will mottle, because a similar number of photons (dose) is generally used to generate the tomosynthesis images.

Comment

Magnification mammography is performed to obtain magnified views of suspicious areas of the breast. This is achieved by placing the breast on an elevated stage above the detector (closer to the x-ray tube source). In magnification mammography, the source-to-image receptor distance (SID) stays the same (65 cm) and the source-to-object distance (SOD) is halved to approximately 32 cm. The geometric magnification is thus approximately 2 (SID/SOD), and each pixel in the contact mammography now has four pixels in the magnified image, improving lesion visibility. A small focal spot (0.1 mm) must be used because the 0.3-mm focal spot results in focal spot blur.

There is a substantial air gap in magnification mammography (~32 cm), which ensures that most of the scattered photons generated in the breast will miss the image receptor. As a result, there is no need to use a scatter removal grid. When the grid is removed, all of the primary radiation transmitted by the breast reaches the detector. Most of the radiation incident on the breast is absorbed in the breast, with only a very small percentage transmitted to create the image. Because magnification mammography uses 30% less radiation than contact mammography, mean glandular doses will be 30% lower.

Breast tomosynthesis creates tomographic images using a conventional digital mammography imaging chain. X-ray tubes typically rotate in a 15-degree arc around the breast, with projection images obtained every degree (Fig. 2.15). A total of 15 projection radiographs are acquired, with examination times of approximately 5 seconds. Compression is required to help minimize motion artifacts. Acquired image data are processed using a type of back projection to yield approximately 45 images, each with a slice thickness of approximately 1 mm. Radiation dose in DBT is comparable to digital contact mammography.

Reports have claimed that DBT screening can detect 40% more invasive cancers, with reductions in patient recall rate of approximately 15%. Some systems offer the ability to acquire only DBT images and then reconstruct a synthetic contact mammogram. The advantage of using this synthetic contact mammogram is a reduction in patient dose by eliminating the radiation associated with an additional contact mammogram. DBT currently appears more promising than dedicated breast CT systems because the latter has a much lower in-plane resolution.

References

Huda W. Mammography. In: *Review of Radiological Physics*. 4th ed. Philadelphia: Wolters Kluwer; 2016:110–113.

Roth RG, Maidment ADA, Weinstein SP, et al. Digital breast tomosynthesis: lessons learned from early clinical implementation. *RadioGraphics*. 2014;34(4):E89–E102.

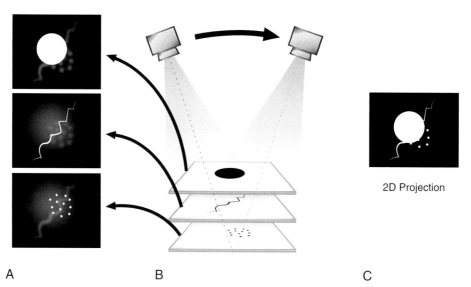

A   B   C

2D Projection

Fig. 2.15 Obtaining a number of projection images at different x-ray tube angles (B) enables tomographic slices to be generated (A) that depict a mass in the upper slice, a fibril in the middle slice, and a cluster of microcalcifications in the lower slice. These lesions have markedly superior lesion contrast than a conventional projection image (C).

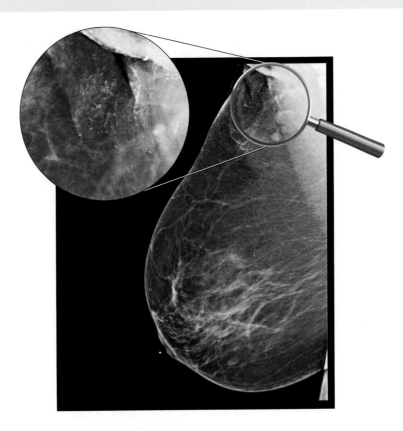

Fig. 2.16

1. What is the most likely etiology of the axillary hyperdensities depicted in the case above?
   A. Treated lymphoma
   B. Ductal carcinoma in situ
   C. Deodorant artifact
   D. Dermatomyositis

2. How does the resolution of digital mammography compare with that of the human eye at a normal viewing distance of 25 cm (5 lp/mm)?
   A. Superior
   B. Similar
   C. Worse

3. What happens to the subject contrast of a lesion when the k-edge energy of a mammography filter is increased?
   A. Increased
   B. Unchanged
   C. Reduced

4. How much more radiation is used to generate a mammogram relative to a single fluoroscopy frame?
   A. 10
   B. 100
   C. 1000
   D. 10,000

CASE 2.8

Image Quality and Artifacts

Fig. 2.16 Mediolateral oblique view of the right breast demonstrating antiperspirant artifact.

1. **C.** Calcifications and focal hyperdensities are commonly encountered in mammography, but not all of them are located within the body. Antiperspirant artifact is commonly seen in the axilla and will disappear when the skin is cleaned. This artifact is not caused by calcium, but rather by the presence of additives in the deodorant such as aluminum. Although antiperspirant artifact usually has a distinctive appearance, an alternative etiology of axillary hyperdensities should be considered if there are other concerning imaging characteristics.

2. **A.** The human eye can resolve approximately 5 lp/mm at a normal viewing distance (~25 cm). Digital mammography is generally limited by the pixel size, which is approximately 70 μm on most commercial systems, which translates into 7 lp/mm. Screen-film systems could achieve much higher resolution (15 lp/mm) because there was no loss of resolution due to detector pixels.

3. **C.** Increasing the k-edge energy of a filter in digital mammography will increase the average energy and make the beam "more penetrating." In general, more penetrating beams will reduce contrast, as well as the corresponding dose when an AEC system is used. Clinical systems are increasing their beam quality (penetrating power) to reduce patient doses because these can offset the loss of contrast by sophisticated image processing algorithms.

4. **D.** The radiation intensity at the image receptor in mammography is generally 100 μGy. In fluoroscopy, the radiation intensity for a single frame is extremely low (~0.01 μGy), which is why these noisy images are seldom diagnostic quality. Accordingly, a mammography frame uses approximately 10,000 times more radiation than a fluoroscopy frame.

Comment

The limiting resolution of the human eye at a normal reading distance of 25 cm (arm's length) is only 5 lp/mm. The limiting spatial resolution of screen film was 15 to 20 lp/mm, which is why mammographers always used a magnifying glass when inspecting film images. By contrast, digital mammography is limited by the pixel size of the detector and is typically 7 lp/mm. To maintain this resolution, focal spot blur must be reduced by using a small (0.3 mm) focal spot size and motion blur minimized by the use of a compression paddle that immobilizes the breast. Regulations required screen-film systems to achieve approximately 12 lp/mm; however, digital systems are currently only required to achieve the manufacturer's specified resolution. The fact that digital resolution is markedly inferior to that of screen film illustrates that this aspect of image quality in mammography is not as important as the ability to improve lesion CNR through image processing.

Lesion contrast is very low when a tumor is embedded in a fibroglandular tissue background. Scatter from the patient is partially responsible for the low contrast. The use of a scatter removal grid (or air gap in magnification mammography) plays an important role in maximizing lesion contrast by removing scatter. Lesion contrast always increases as photon energy is reduced, but there is a limit on the lowest energy that can be used because the radiation must penetrate through the patient to reach the detector. With a rhodium filter, the average energy will be less than the 23 keV k-edge, with an HVL of 0.5 mm aluminum. When a silver filter is used, the average energy will be less than the 26 keV k-edge, with a correspondingly higher HVL of 0.6 mm aluminum.

All mammograms are performed using AEC, where the radiation intensity at the image receptor is kept constant irrespective of the size and composition of the compressed breast. The value that is currently used for the receptor air kerma in digital mammography is 100 μGy, which has not changed significantly over the past 30 years. This value is much greater than in radiography, where a standard chest x-ray will have an image receptor air kerma of approximately 3 μGy, or 30 times less than in mammography. Increasing air kerma reduces image mottle, and

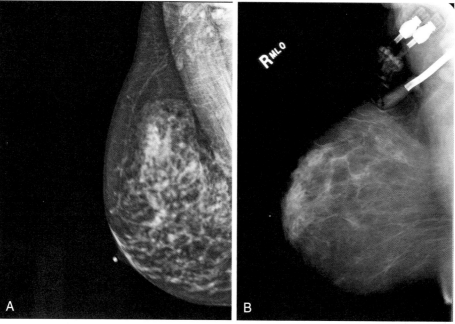

Fig. 2.17 (A) Motion blur reduces the resolution of the image. This may result in missed microcalcifications or architectural distortion. (B) The presence of an implanted catheter confuses image processing and results in a loss of the breast skin line and suboptimal contrast in the parenchyma.

mammograms have the lowest amount of quantum mottle in all of projection radiographic imaging by far.

Similar to conventional projection radiography, mammography is subject to artifact. Artifacts common to digital detectors, such as dead pixels and debris on the detector, may also occur in mammography. Such artifacts are arguably more dangerous in mammography because they could mimic calcifications or architectural distortion. Unlike radiography, where typical exposure times are 1 to 10 ms, the typical exposure time in mammography is greater than 500 ms. This allows more time for the patient to move, and the resulting blur may distort the shape of microcalcifications or render them invisible (Fig. 2.17). Foreign objects may also cause image artifacts. Lotion or deodorant on the skin may mimic architectural distortion or microcalcifications, respectively. Furthermore, implants such as the Permcath in Fig. 2.17B may confuse image processing algorithms, resulting in suboptimal contrast.

References

Ayyala RS, Chorlton M, Behrman RH, et al. Digital mammographic artifacts on full-field systems: what are they and how do I fix them? *Radiographics.* 2008;28(7):1999-2008.

Huda W. Mammography. In: *Review of Radiological Physics.* 4th ed. Philadelphia: Wolters Kluwer; 2016:105-114.

Fig. 2.18

1. What does the MQSA require of mammographic imaging facilities?
   A. FDA certification
   B. ABR accreditation
   C. Certificate of need
   D. A, B, and C

2. Who is ultimately responsible for quality assurance (QA) and quality control (QC) according to the MQSA?
   A. Radiology technologist
   B. Medical physicist
   C. Radiology manager
   D. Interpreting physician

3. What is the most likely total number of masses, fibers, and microcalcification clusters visible on an accreditation phantom image?
   A. 6
   B. 9
   C. 12
   D. 16

4. What is the current MQSA regulatory patient dose limit (mGy) for a four-view mammogram?
   A. 3
   D. 6
   C. 12
   D. No limit

CASE 2.9

Mammography Quality Standards Act

Fig. 2.18 Mammography phantom used by the American College of Radiology and Mammography Quality Standards Act.

1. **A.** MQSA requires that all facilities that offer mammography are certified by the FDA. To obtain certification requires accreditation by an approved body such as the ACR or by one of the several states that offer this service (e.g., California and Texas).

2. **D.** The MQSA refers to an interpreting physician as opposed to a radiologist, and this individual is ultimately responsible for all aspects of the MQSA regulations. This responsibility includes the appointment of a technologist who is responsible for QC and a medical physicist who is responsible for testing the equipment and overseeing the QC program.

3. **C.** A phantom image will generally depict five or six fibers, three or four groups of microcalcification clusters, and four or five masses. A typical image would show 12 or so objects. Note that a perfect score in the accreditation phantom is 16 and that 9 would fail the image quality test.

4. **D.** MQSA requires that the dose from a single view to a standard phantom (4.2 cm thick with 50% glandularity) be less than 3 mGy for a contact mammogram with a scatter removal grid. There is no limit to the dose to any patient, and thick/dense breasts will most likely need to receive doses well in excess of 3 mGy to generate a clinically acceptable mammogram.

Comment

The MQSA was passed in 1992 and requires all 10,000 mammography facilities in the United States to be accredited and certified every 3 years. Accreditation is provided by the ACR, as well as individual states (e.g., California, Texas, and Iowa). The FDA is responsible for enforcing MQSA, but this task is often contracted to state regulatory agencies. For screen-film systems, the MQSA requires a limiting resolution of 12 lp/mm. The MQSA requires each facility to achieve the vendor specifications in digital mammography, which are typically 7 lp/mm. Failure of any MQSA requirement during QC testing must be assessed and corrected. Facilities must correct QC failures in a time frame ranging from immediate (e.g., failed phantom image quality test) to within 30 days (e.g., failed AEC performance test).

The ACR provides explicit accreditation guidelines for radiologists, technologists, and medical physicists. MQSA requires the "interpreting physician" (i.e., radiologist) to take charge of all aspects of clinical mammography. The ACR phantom used to assess image quality contains various-sized fibers (6), speck groups (5), and masses (5). Images of the accreditation phantom are acquired weekly by technologists and annually by a medical physicist. The images are acquired using acquisition parameters (e.g., tube voltage and beam filtration) that would be used

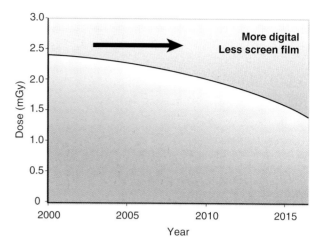

Fig. 2.19 Demonstration of the steady decrease in dose from just below 2.5 mGy in 2000 to just below 1.5 mGy in 2015. The decrease is primarily due to the use of progressively increased beam energies associated with the shift from screen-film systems in 2000 to the digital systems currently used that use tungsten targets.

clinically. To pass, the phantom image must show a minimum number of fibers, speck groups, and masses and demonstrate few or no artifacts. This minimum passing image quality score is currently set by each digital mammography system vendor and was four fibers, three speck groups, and three masses for screen film (i.e., 10 lesions fully visualized).

The ACR requires an average glandular dose (AGD) for a 4.2-cm thick breast phantom (50% glandularity) below 3 mGy per image (with grid). The MQSA dose limit applies only to the standard phantom (i.e., not to any patient). The benefit of a standard phantom, even though it does not accurately reflect the anatomy of an average patient, is that it enables facilities to accurately intercompare doses delivered by their mammography systems. The AGD in the United States has continued to decline over the past few decades due to the shift toward full-field digital mammography, which uses higher beam energies (Fig. 2.19). There is little doubt that radiation doses in all US facilities are consistently less than 3 mGy for the ACR phantom and that the differences between facilities will be relatively modest. This state of affairs differs from other modalities, where different facilities have dose settings that can differ by more than an order of magnitude when imaging the same patient.

References

Huda W. Image quality. In: *Review of Radiological Physics*. 4th ed. Philadelphia: Wolters Kluwer; 2016:33–40.

Radiological Society of North America. RSNA/AAPM physics modules. Projection x-ray imaging: mammography image quality and dose. https://www.rsna.org/Physics-Modules/. Accessed March 1, 2018.

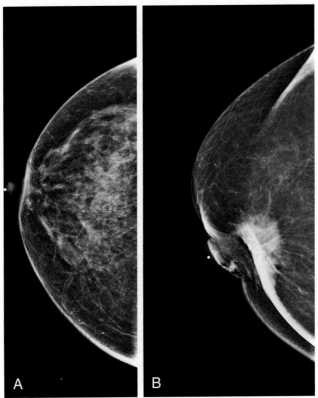

Fig. 2.20

1. What is the most likely AGD in milligray for an 80-mm-thick breast with a 30% glandularity?
   A. 1
   B. 2
   C. 3
   D. >3

2. What is the most likely AGD (mGy) for a screening mammogram (CC and MLO views for each breast) for an average patient (60-mm-thick breast and 15% glandularity)?
   A. 1.5
   B. 3
   C. 6
   D. 10

3. How many fatal breast cancers would be expected when 1 million 40-year-old women undergo a screening mammogram?
   A. 1
   B. 10
   C. 100
   D. 1000

4. What is the most likely reduction (%) in breast cancer mortality that is believed to be achieved by screening mammography?
   A. 10
   B. 20
   C. 50
   D. 90

CASE 2.10

Dose and Risk

Fig. 2.20 Negative screening mammogram (A) and subareolar mass with associated nipple retraction and skin thickening (B).

1. **D.** The dose to a woman undergoing a contact mammogram with a thick and dense breast is most likely to be well in excess of 3 mGy. Note the MQSA limit of 3 mGy for a mammogram applies to a standard phantom (4 cm thick + 50% glandularity) and is not directly applicable to patients.

2. **B.** Although the maximum AGD dose to the accreditation phantom is 3 mGy per view, the AGD to an average size breast is typically between 1 and 2 mGy. This means that each breast receives between 2 and 4 mGy. By definition, if each breast receives approximately 3 mGy, the AGD for the exam will be 3 mGy.

3. **B.** The fatal cancer risk factor for 40-year-old women is approximately 3 per million per mGy, so with an AGD of 3 mGy, the risk is approximately 10 per million. The uncertainty in this risk factor is substantial because of the absence of epidemiologic data at such low doses.

4. **B.** Clinical trials indicate that breast cancer mortality is generally reduced by approximately 20% or so when screening of asymptomatic patients is introduced. It is noteworthy that this number has a much firmer scientific basis than current radiation risk estimates but is also subject to some uncertainty.

Comment

Average glandular dose, also called mean glandular dose, is calculated by medical physicists who take into account x-ray technique (i.e., entrance K_{air} and HVL). Breast doses also depend on patient breast characteristics (i.e., thickness and glandularity). Vendors can estimate the AGD for each mammogram and then store these values in the image Digital Imaging and Communications in Medicine (DICOM) header so they are available for display on most picture archiving and communication system (PACS) workstations.

The AGD for a 60-mm-thick 15% glandular breast is currently between 1 and 2 mGy per view and is markedly lower than for screen-film systems. The reduction in patient dose is attributable to the use of higher beam energies (see previous cases). For a two-view screening mammogram, each breast will receive between 2 and 4 mGy, and the patient AGD is thus between 2 and 4 mGy. Note that left and right breast doses are *never* added together, because it is the AGD that determines the corresponding carcinogenic radiation risk (Table 2.2).

Tissue HVL at the low photon energies encountered in mammography is typically 1 cm. At a constant beam quality, increasing the compressed thickness by 10 mm would likely double the AGD when the examinations are performed using an AEC system. However, beam quality is always increased in larger and denser breast, which limits dose increases in thicker breasts. For example, on a tungsten target, a rhodium filter would likely be used for thinner breasts and a silver filter for thicker and/or denser breasts. Even with the use of an increased beam quality, as breast thickness and density increase, so does AGD. For an 80-mm dense breast, imaged at the highest beam quality, the AGD would almost certainly be higher than 3 mGy per view.

Radiation risk *estimates* in screening mammography are extremely low and should not deter women from having this diagnostic test. If 1,000,000 middle-aged women receive 3 mGy from a mammogram, the estimated number of induced cancers will be less than 45 and the estimated number of fatal cancers will be less than 12 (see Table 2.2). It has been estimated that radiation risks from screening mammograms are equivalent to the risk of dying in an accident when traveling approximately 1000 miles by car. Patients, especially those hesitant to undergo screening due to radiation, should always be encouraged to consider the benefits of screening mammography in comparison to the relatively low risk of radiation (Fig. 2.21).

References

Hall EJ, Giaccia AJ. Radiation carcinogenesis. In: *Radiobiology for the Radiologist*. 7th ed. Philadelphia: Wolters Kluwer; 2012.

Huda W. Mammography. In: *Review of Radiological Physics*. 4th ed. Philadelphia: Wolters Kluwer; 2016:105, 114.

TABLE 2.2 RADIATION CANCER RISKS PER MILLIGRAY (AVERAGE GLANDULAR DOSE) PER MILLION EXPOSED INDIVIDUALS BASED ON DATA PRESENTED IN BIOLOGIC EFFECTS OF IONIZING RADIATION (BEIR) VII

Age	Breast Cancer Incidence	Breast Cancer Fatality
40	14	3.5
50	7	1.9
60	3	0.9

Breast Cancer Screening

Screening **1 million** women	Benefit of screening is an expected **20% reduction in mortality**	Breast doses would be **3 mGy** for a four-view screening exam	In **1 million** 40-year-old women, the number of fatal radiation-induced breast cancers is **about 10** (less as age increases)	Uncertainties in breast cancer risks are *very large*; epidemiological data only exists at doses >1000 mGy	Benefits of screening are measured in clinical trials and have been established to a *much greater* degree than radiation risks
Identifies **4000** breast cancers **25%** will be fatal (1000 lives)					

Fig. 2.21 Current radiation risk estimates from screening mammography are much lower than the corresponding benefit in terms of lives saved. This benefit-to-risk estimate (1000 vs. 10) is likely very conservative given that benefits are better known than the corresponding radiation risks.

CASE 3.1

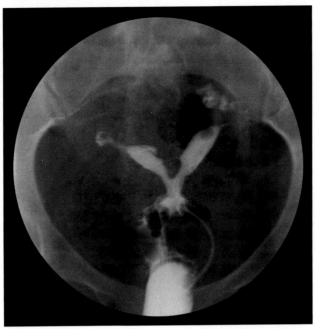

Fig. 3.1

1. Which of the following is most likely used as the x-ray detector in an image intensifier (II)?
   A. Cesium iodide (CsI)
   B. Sodium iodide (NaI)
   C. Amorphous selenium (Se)
   D. Barium fluorobromide (BaFBr)

2. What does a photocathode initially absorb and subsequently emit?
   A. X-rays and light
   B. X-rays and electrons
   C. Electrons and light
   D. Light and electrons

3. If the input diameter of an II is 10 inches (25 cm) and the output phosphor is 1 inch (2.5 cm), what is the minification gain?
   A. 3
   B. 10
   C. 30
   D. 100

4. What is the most likely brightness gain of an II compared with a simple scintillator screen?
   A. 5
   B. 50
   C. 500
   D. 5000

CASE 3.1

Image Intensifiers

Fig. 3.1 Hysterosalpingogram in a patient with a bicornuate uterus.

1. **A.** Cesium iodide (CsI) is a scintillator that is used in virtually all IIs. Sodium iodide is a scintillator used in gamma cameras. Amorphous selenium is a poor absorber of x-rays because of its low k-edge (13 keV), and photostimulable phosphors are not suitable for an II because of the lengthy "readout" mechanism.

2. **D.** An II "works" by first absorbing x-rays that produce light. The light is absorbed by a photocathode that emits low-energy electrons. These photoelectrons are accelerated to high energies and focused onto an output scintillator, producing a very large number of light photons from each energetic electron.

3. **D.** The minification gain is the ratio of the area of the input scintillator (CsI) to the output phosphor (zinc cadmium sulfide [ZnCdS]). The area of the input phosphor (diameter, 10 inches) is 100 times larger than the output phosphor because the area is proportional to the square of the diameter. Squeezing the light into an area 100 times smaller clearly makes the image 100 times brighter (i.e., minification gain is 100).

4. **D.** The output light at an II is approximately 5000 times brighter than the initial brightness at the input scintillator. The output image is brighter because of the flux gain and the minification gain. For an average intensifier, the flux gain is approximately 50 and the minification gain is approximately 100. The brightness gain is the product of the flux gain and the minification gain, or approximately 5000.

Comment

IIs convert a pattern of incident x-rays into a bright light image. An II consists of an evacuated envelope made of glass or aluminum, which contains an input phosphor, photocathode, electrostatic focusing lenses, accelerating anodes, and output phosphor. Input phosphors, typically made of CsI, absorb x-rays and emit light photons, which has excellent x-ray absorption properties with a k-edge of approximately 30 keV that matches the average energy of most x-ray beams. The light image produced by the input phosphor is too faint to be usable, and methods are applied to intensify the beam to make it brighter at the output phosphor.

Some of these emitted light photons are absorbed by a photocathode, resulting in the emission of low-energy electrons called photoelectrons. These low-energy electrons are accelerated using a high voltage applied across the II tube and focused onto the output phosphor. When these energetic electrons are absorbed by the output phosphor, they emit a very large number of light photons (Fig. 3.2). The intensification factor is known as brightness gain and is the product of flux and minification gain (Fig. 3.3). In practice, the overall brightness gain is approximately 5000 because a typical flux gain is approximately 50 and a typical minification gain is approximately 100.

Flux gain is the number of light photons emitted at the output phosphor for each photon emitted at the input phosphor, typically approximately 50. This increase occurs because light at the input phosphor produces photoelectrons that are accelerated to high energies. These electrons produce many more light photons when they hit the output phosphor and convert their kinetic energy into light. Flux gain is a fixed property of the II system and cannot be adjusted by the operator.

Minification gain is the increase in image brightness resulting from reduction in image size from the input phosphor to the output phosphor. The output phosphor diameter is approximately 2.5 cm, which corresponds to the size of the camera that views the II output. A typical II has a 25-cm diameter, with an input area 100 times greater than the output area (2.5-cm diameter), resulting in a minification gain of approximately 100. It is important to note that reducing the area of the II input that is exposed to x-rays, as is done in electronic magnification, will reduce minification gain and the corresponding brightness gain.

References

Huda W. Fluoroscopy. In: *Review of Radiological Physics*, 4th ed. Philadelphia: Wolters Kluwer; 2016:119–121.

Schueler BA. The AAPM/RSNA physics tutorial for residents: general overview of fluoroscopic imaging. *Radiographics*. 2000;20:1115–1126.

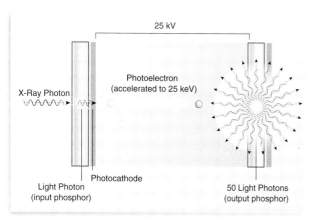

Fig. 3.2 An input phosphor *(left)* that absorbs x-rays and emits light. This light is absorbed by the photocathode that subsequently emits low-energy light photons. Each light photon produced by the input phosphor will produce 50 light photons at the output phosphor as a result of the electrons being accelerated to a high energy (flux gain).

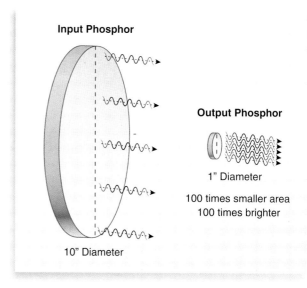

Fig. 3.3 When the light produced by a large area (10-inch diameter) input phosphor is squeezed into a 1-inch diameter output phosphor, there is a 100-fold increase in image brightness (brightness gain).

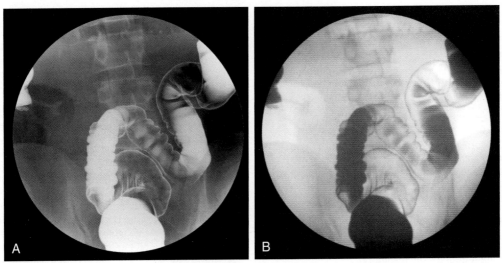

Fig. 3.4

1. How many "lines" are most likely used to generate a standard fluoroscopy frame?
   A. 500
   B. 1000
   C. 2000
   D. 4000

2. How many "lines" are most likely used to generate a photospot image?
   A. 500
   B. 1000
   C. 2000
   D. 4000

3. What TV mode is likely to have the least motion artifacts?
   A. Interlaced
   B. Progressive
   C. Comparable

4. How will fluoroscopy patient doses be affected when a TV camera is replaced by a solid-state camcorder?
   A. Increased
   B. Remain constant
   C. Reduced

CASE 3.2

Television Systems

Fig. 3.4 Photospot image from a barium enema (A) obtained using 1000 lines of resolution compared with a fluoroscopic image obtained with 500 lines of resolution (B). Note the simulated scan lines in B, which are more apparent when displaying a moving area in interlaced mode on a TV system.

1. **A.** Standard fluoroscopy is not diagnostic and is typically used to identify the patient anatomy so that diagnostic photospot images can be obtained. As such, TV systems during fluoroscopy can be operated in a standard 500-line mode, which offers an adequate level of resolution.

2. **B.** Photospot images are always of diagnostic quality and so need improved spatial resolution performance compared with fluoroscopy. The TV cameras are switched to a 1000-line mode, which doubles the resolution.

3. **B.** In interlaced mode, first the odd lines are read out (and displayed), followed by the even lines. If there is motion between the two sets of readouts (~16 ms or half a frame), artifacts will be seen. In progressive mode, the lines are read out sequentially to generate a whole frame in approximately 33 ms and motion artifacts will be reduced compared with interlaced mode.

4. **B.** Virtually all x-ray imaging is quantum mottle limited, which means that the number of x-rays absorbed in the detector determines the amount of noise in the image (as opposed to electronic or thermal noise). Replacing any component, such as the TV camera, will not affect the amount of mottle in the resultant image.

Comment

The bright image generated at the output phosphor can be viewed on a monitor using a TV system. A TV camera is pointed at the II output phosphor. Light is collected by the camera, which generates an electronic signal (video) that is fed into a display monitor. Spatial resolution increases as the number of lines used to generate the image increases and as the field of view decreases. In general, fluoroscopy images are not diagnostic, so 500 lines giving a spatial resolution of approximately 1 line pair per mm (lp/mm) is acceptable. Photospot images are diagnostic images acquired through the fluoroscopic imaging chain. They use 1000 lines to attain a spatial resolution of approximately 2 lp/mm.

TV systems build up images in a series of horizontal lines using one of two raster scanning modes known as interlaced and progressive (Fig. 3.5). When raster scanning is progressive, each line in the image (frame) is read sequentially. When raster scanning is interlaced, first the odd lines are read (and combined with the previous frame's even lines), followed by even lines being read (combined with the previous frame's odd lines). Interlacing halves the number of lines updated for each frame and doubles the frame rate to prevent image flicker that can cause viewer fatigue, headaches, and misdiagnosis. Although interlaced mode minimizes the appearance of flicker, it will create motion artifacts when viewing rapidly moving objects, because there will be a mismatch between the even and odd fields.

Standard definition (SD) uses 480 lines, and high definition (HD) is typically operated using either 1080 lines (interlaced) or 720 lines (progressive). TV systems generally display images at 30 frames per second, with each frame taking 1/30 of a second, in both interlaced and progressive modes of operation. In modern fluoroscopy systems, the TV camera is replaced by a solid-state detector such as a charge-coupled device (CCD) because these are cheaper and more reliable. CCDs and TV cameras result in similar levels of mottle because fluoroscopy imaging is quantum mottle limited. Reducing fluoroscopy mottle using technical means, as opposed to image processing, requires using more photons, which will inevitably increase patient dose.

References

Huda W. Fluoroscopy. In: *Review of Radiological Physics*. 4th ed. Philadelphia: Wolters Kluwer; 2016:120–121.

Radiological Society of North America. RSNA/AAPM physics modules. Fluoroscopy: fluoroscopy systems. https://www.rsna.org/Physics-Modules/. Accessed March 1, 2018.

Interlaced   Progressive

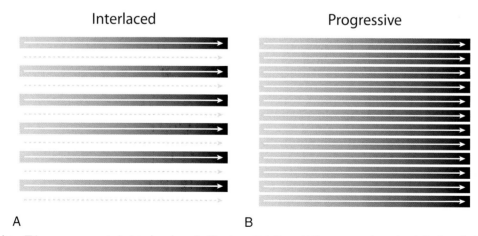

A   B

Fig. 3.5 A shows how TV cameras operate in interlaced mode (I), where first the odd lines are read out (and displayed), followed by the even lines. B shows progressive mode (P), where all lines are read out (and displayed) sequentially.

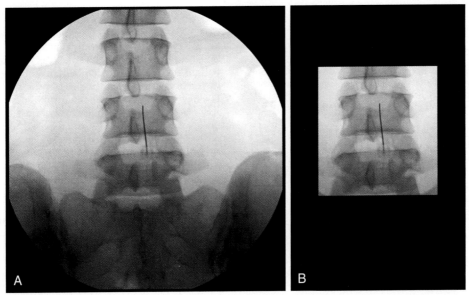

Fig. 3.6

1. How will increased collimation during fluoroscopy affect kerma area product (KAP) and lesion contrast-to-noise ratio (CNR)?
   A. ↑ KAP; ↑ CNR
   B. ↑ KAP; ↓ CNR
   C. ↓ KAP; ↑ CNR
   D. ↓ KAP; ↓ CNR

2. How much of an incident x-ray beam (%) is most likely attenuated by a 2-mm aluminum fluoroscopy table?
   A. 1
   B. 3
   C. 10
   D. 30

3. If the radiation incident on a fluoroscopy patient is 100%, what is the scatter intensity (%) to the operator at a distance of 1 m?
   A. 0.01
   B. 0.1
   C. 1
   D. 10

4. What is the minimum distance (cm) from the patient to the focal spot on a mobile C-arm fluoroscopy system?
   A. 20
   B. 30
   C. 38
   D. No limit

CASE 3.3

Imaging Chain

Fig. 3.6 Lumbar puncture under fluoroscopic guidance without collimation (A) and with collimation (B). Using collimation decreases the total amount of radiation incident on the patient, reduces operator scatter, and also improves image contrast by reducing scatter at the image receptor.

1. **C.** When collimation is increased, the area of the x-ray beam is reduced, but the radiation intensity (entrance air kerma) remains the same. As a result, the peak skin dose will not be affected, but the KAP will be reduced. Because less of the patient is irradiated, scatter from the patient's body will be reduced, which increases the contrast. Collimation is thus a win-win-win scenario because patient dose is reduced, operator dose is reduced (less scatter), and image quality (contrast) gets better.

2. **D.** The US Food and Drug Administration (FDA) requires that fluoroscopy tables sold in the United States have tables with an aluminum equivalent thickness of 2 mm or less. A thickness of 2 mm will attenuate 30%, which is important because only the operator knows if the x-ray beam is being attenuated by a table. For lateral projections, the patient is subject to the full displayed radiation intensities, whereas for under-the-table x-ray tube positions, a third of the displayed radiation intensity will have been absorbed by the table, not the patient.

3. **B.** At 1 m from the patient, the scattered radiation intensity is approximately 0.1% of the radiation intensity incident on the patient. A typical photospot image will have approximately 1 mGy incident on an average patient, so the dose at 1 m will be 1 μGy. During fluoroscopy, approximately 10 mGy/min will be incident on an average patient, so the dose rate at 1 m will be 10 μGy/min. One important lesson is that the operator dose will always be directly proportional to the patient dose.

4. **B.** When the distance to the focal spot is reduced from 100 to 10 cm, the radiation intensity increases by a factor of 100 because of the inverse square law. To minimize such accidents, the FDA requires a spacer cone to be installed to ensure the patients cannot get "too close" to the focal spot. The minimum distance is 30 cm on mobile C-arms and 38 cm on fixed fluoroscopy system.

Comment

Fluoroscopy x-ray tubes are similar to those used in radiography but are operated at very low x-ray tube currents (e.g., ~3 mA). Because the power used is very modest (<0.5 kW), x-ray tubes should *never overheat* during fluoroscopy. After an extended period of fluoroscopy, heat deposition and dissipation reach equilibrium before the anode reaches saturation. The low power also means that fluoroscopy can be performed using a small focal spot (0.6 mm) that easily tolerates power loading of 0.5 kW. A spacer cone is used to ensure that the patient cannot get too close to the focal spot, because this could dramatically increase skin doses due to the inverse square law. FDA requires the minimum focus-to-skin distance on fixed units to be 38 and 30 cm on mobile systems.

Adult fluoroscopy systems would typically use a total filtration of approximately 3 mm aluminum. At 80 kV, the resultant x-ray beam quality will be 3-mm aluminum half-value layer (HVL), which is just above the current regulatory minimum. However, dedicated pediatric fluoroscopy systems may have an additional 0.1- or 0.2-mm copper (Cu) filtration. Increased beam quality achieved by the use of copper filters reduces pediatric patient doses because the x-ray beams are more penetrating

and so require less radiation when automatic exposure control (AEC) is used. The loss of photons from added copper filters is acceptable because anatomy thickness is generally small in pediatric imaging but may not be in adult imaging, where patient attenuation is much higher. Collimators can be adjusted by the operator to limit the size of the x-ray beam to that which is clinically required. Collimation reduces patient dose. It also reduces scatter, which improves contrast and reduces operator dose. Scatter radiation is also reduced using grids, with grid ratios similar to those used in radiography (i.e., 10:1).

Radiography/fluoroscopy (R/F) units are used to perform gastrointestinal (GI) studies that visualize barium administered to patients (Fig. 3.7). An R/F unit generally has the x-ray tube under the table and uses lead drapes to reduce operator scatter. Tables on R/F units typically rotate 30 degrees cranially and 90 degrees caudally. Fluoroscopy tables are required by the FDA to be less than approximately 2-mm aluminum equivalence. R/F tables attenuate approximately a third of the incident x-ray beam since 3 mm aluminum attenuates 50% (HVL). Portable fluoroscopy systems are C-arm devices often used outside of radiology departments (e.g., in operating rooms). Mobile C-arms are used clinically by nonradiologists but operated by certified radiology technologists. Dedicated systems for performing urological examinations are similar to R/F units except that the x-ray tube is located above the patient. Overhead x-ray tubes are used to minimize magnification of the kidneys. An overhead tube increases operator doses from radiation that is scattered from the patient, especially to the eye lens.

References

Bushberg JT, Seibert JA, Leidholdt Jr EM, Boone JM. Fluoroscopy. In: *The Essential Physics of Medical Imaging*. 3rd ed. Philadelphia: Wolters Kluwer; 2012:282–283.

Huda W. Fluoroscopy. In: *Review of Radiological Physics*. 4th ed. Philadelphia: Wolters Kluwer; 2016:119–121.

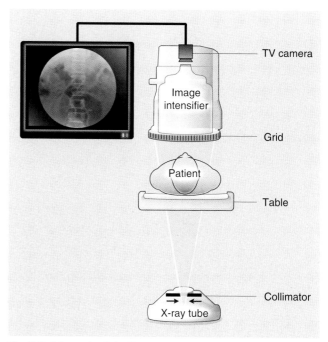

Fig. 3.7 Schematic of the fluoroscopic imaging chain typical for gastrointestinal/genitourinary examinations. An x-ray tube is used with adjustable collimators to limit the x-ray beam to the area of clinical interest. Grids are used to minimize the amount of scatter reaching the image intensifier (II). The resultant image on the II output phosphor is captured by a TV camera and displayed on a display monitor.

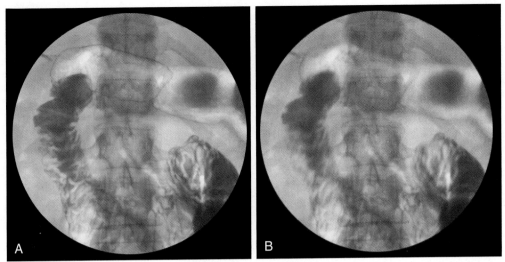

Fig. 3.8

1. What is the x-ray tube technique (mAs) required to generate a single fluoroscopy frame?
   A. 0.01
   B. 0.1
   C. 1
   D. 10

2. How many frames are most likely generated in 1 minute of fluoroscopy?
   A. 10
   B. 100
   C. 1000
   D. 10,000

3. What is the principal benefit of performing pulsed fluoroscopy (30 frames per second) compared with continuous fluoroscopy?
   A. Sharper images
   B. Reduced mottle
   C. Increased contrast
   D. Improved CNR

4. What is the most likely patient dose reduction if the frame rate in pulsed fluoroscopy is reduced from 30 to 15 frames per second (%)?
   A. <50
   B. 50
   C. >50

CASE 3.4

Pulsed Fluoroscopy and Fluoroscopy Times

Fig. 3.8 Upper gastrointestinal examination using pulsed fluoroscopy (A) and continuous fluoroscopy (B). Note the motion blur in B due to bowel peristalsis.

1. **B.** A typical tube current is between 1 and 5 mA and images at 30 frames per second, which takes 33 ms to generate each image. The x-ray tube output is thus approximately 3 mA × 0.03 ms, or 0.1 mAs. A fluoroscopy frame likely has too much mottle to be of diagnostic value, whereas radiographs are always diagnostic because their mottle is approximately 10 times lower.

2. **C.** Modern fluoroscopy systems are generally pulsed and are likely operated at 15 frames per second (vs. 30 frames per second in continuous fluoroscopy). In 60 seconds there are nearly 1000 images acquired (15 × 60), where each image would be displayed twice (i.e., 30 frames per second) to reduce perceived flicker.

3. **A.** In continuous fluoroscopy, the x-ray tube current is on for all 33 ms using a low current of approximately 3 mA, so the total output is approximately 0.1 mAs. In pulsed fluoroscopy, the frame rate displayed on the screen is the same but exposure time is reduced (e.g., ⅒). Reduced exposure time reduces motion blur.

4. **A.** Psychophysics experiments have shown that to maintain perceived noise in dynamic imaging requires the dose per frame to be increased when the frame rate is reduced. In practice, reducing the pulsed rate from 30 to 15 frames per second will reduce patient doses by only 25% to keep perceived noise the same.

Comment

In normal fluoroscopy, the tube current flows continuously and an image is read out by the TV camera every 1/30 of a second (33 ms). These images are displayed on a monitor at a rate of 30 frames per second. In each minute of fluoroscopy, nearly 2000 images will have been created. There are no regulatory limits on the total exposure times, but a 5-minute alarm is used to alert the operator to the amount of time that has passed.

Operators are familiar with the total fluoroscopy time, but it is also important to note that patients are also exposed from photospot and spot images associated with most fluoroscopy-guided examinations.

In 1/30 second, the total x-ray production for a 3-mA tube current is approximately 0.1 mAs. If a pulse lasting one-tenth of this normal frame time is used (i.e., 3.3 ms) that has a current 10 times higher (i.e., 30 mA), the total number of x-rays produced remains the same (i.e., 0.1 mAs). When pulsed mode is used, patient motion is substantially reduced because less patient motion occurs during the pulse (3.3 ms) than would occur during the entire frame (33 ms). Each image created in 3.3 ms is displayed for the entire 33-ms frame before being updated by during the next frame. Replacing continuous with pulsed fluoroscopy (30 pulses per second) results in sharper images but does not affect patient dose. For diagnostic tasks that require excellent temporal resolution (i.e., modified barium swallow study), high pulse rates of 30 frames per second are essential.

Pulsed fluoroscopy will cause motion to appear less smooth (i.e., decrease temporal resolution) as the frame rate is decreased. Displaying images on the TV screen at 15 frames per second would result in image flicker, thus each image is displayed twice to simulate 30 frames per second and minimize flicker. Pulsed fluoroscopy will also reduce patient dose compared with continuous fluoroscopy when frames are acquired at a rate that is less than 30 frames per second. Operators should be encouraged to reduce the pulse rate for simple diagnostic tasks such as detecting a coin swallowed by a child.

Pulsed fluoroscopy performed at frame rates less than 30 frames per second generally uses a *higher dose per frame* (Fig. 3.9). This increase in dose per frame will help to reduce the perceived level of random noise and thereby maintain diagnostic performance. Fluoroscopy performed at 15 acquired frames per second (which increases the radiation per frame by 50%) would reduce patient doses by 25%, instead of the 50% that would be achieved if a constant radiation per frame were used.

References

Balter S. Fluoroscopic frame rates: not only dose. *AJR Am J Roentgenol.* 2014;203:W234–W236.

Huda W. Fluoroscopy. In: *Review of Radiological Physics.* 4th ed. Philadelphia: Wolters Kluwer; 2016:124–126.

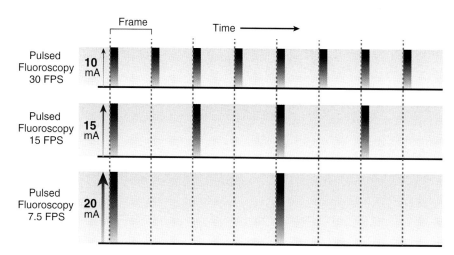

Fig. 3.9 The upper image shows pulsed fluoroscopy performed at 30 frames per second using a tube current of 10 mA. When the pulse rate is reduced, the tube current increases to maintain diagnostic performance. At 15 frames per second, the patient dose savings will be 25% (not 50%), and at 7.5 frames per second, the patient dose savings will be 50% (not 75%).

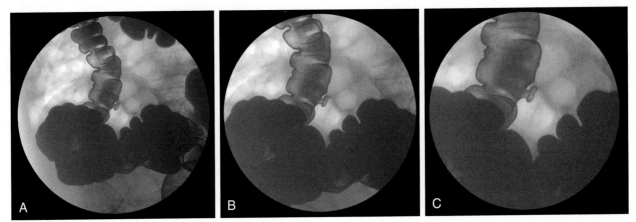

Fig. 3.10

1. What percentage (%) of the 36-cm II input area is exposed to x-rays in electronic magnification (Mag 2) when the input diameter is reduced to 18 cm?
   A. 90
   B. 50
   C. 25
   D. 10

2. How is the minification gain changed in electronic magnification (Mag 1) when the exposed scintillator diameter is reduced from 10 to 7 inches?
   A. 2
   B. 4
   C. 1/2
   D. 1/4

3. Which is most likely kept constant in electronic magnification when compared with normal fluoroscopy mode?
   A. KAP
   B. Entrance air kerma
   C. A and B
   D. Neither A nor B

4. Which will likely result in substantial changes (e.g., >10%) to air kerma rates (mGy/min) during fluoroscopic imaging?
   A. Electronic magnification
   B. X-ray beam collimation
   C. A and B
   D. Neither A nor B

CASE 3.5

Electronic Magnification

Fig. 3.10 No magnification (A), Mag 1 (B), and Mag 2 (C) demonstrating a colonic diverticulum.

1. **C.** A 36-cm II (normal mode) also has two additional electronic magnification modes, Mag 1 and Mag 2. In normal mode, the exposed area is $\pi (18)^2$ square cm. In Mag 2 mode, the II diameter that is exposed to x-rays is reduced to 18 cm, so the area is one-quarter of that normal mode (i.e., $\pi [9]^2$).

2. **C.** The minification gain is directly proportional to the area of the input phosphor that is directly exposed to x-rays. Compared to normal mode, the exposed area in Mag 1 mode is ½ and in Mag 2 mode is ¼ (see Question 1 earlier).

3. **A.** In Mag 1 mode, the exposed area (beam area) is halved, but the corresponding air kerma used is doubled by the automatic brightness control (ABC) system that operates only when switching between electronic Mag modes. This increase in air kerma is achieved by increasing the fluoroscopy tube current. Because the area is halved in Mag 1 mode and the air kerma is doubled, the KAP remains the same.

4. **A.** In electronic magnification, the ABC system is active and will increase air kerma values substantially in Mag 1 and Mag 2 modes. Accordingly, the peak skin dose will increase. When the x-ray beam is collimated, the air kerma stays essentially the same and the peak skin dose is unaffected. However, in collimation mode, the area is decreased so that the resultant KAP value (and corresponding stochastic cancer risk) will be reduced.

Comment

Electronic magnification irradiates a smaller diameter (area) of the II scintillator but maintains the same output image size. Consider an II with a 36-cm (14-inch) diameter and that has three electronic magnification modes (Fig. 3.11). In Mag 1 mode, the exposed area has a diameter that is reduced to 25 cm (10 inches), and this area is halved in comparison to normal mode. In Mag 2 mode, the exposed area has a diameter that is reduced to 18 cm (7 inches) (i.e., 50% of normal) and the exposed area is down to a quarter. In Mag 3 mode, the exposed area has a diameter that is reduced to 13 cm (5 inches) and this area is down to ⅛ of the normal mode.

Minification gain is directly proportional to the input area of the scintillator that is exposed to radiation; thus this will be reduced when electronic magnification modes are selected (e.g., Mag 1). To compensate for this drop in minification gain, the fluoroscopy ABC automatically increases radiation incident on the II in fluoroscopy to maintain a constant brightness at the II output. The ABC is normally active only during electronic magnification and is not active during collimation. Because switching from normal to Mag 1 mode normally halves the exposed II area and thereby halves the minification gain, the ABC mode will double radiation intensity that is used. Similarly, the incident radiation intensity used in Mag 2 quadruples relative to normal mode and in Mag 3 mode would increase by a factor of eight.

The limiting resolution of an II is approximately 5 lp/mm when the output phosphor is viewed directly (i.e., removing TV). When the II is viewed through a TV camera, the limiting resolution is reduced. Limiting resolution in fluoroscopy is approximately 1 lp/mm with a 500-line TV, which can be doubled by using a 1000-line TV system. Fluoroscopy resolution can be improved using electronic magnification so that when the II field of view is halved, the spatial resolution will be doubled.

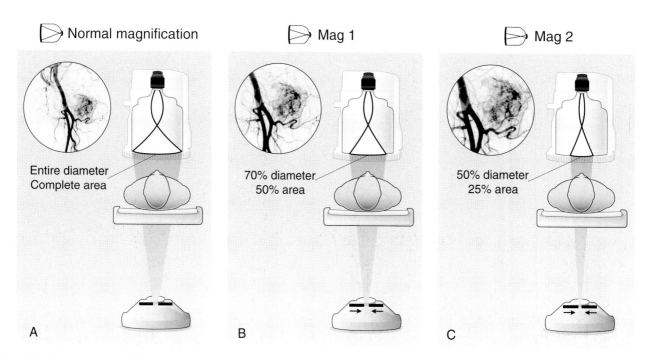

Normal magnification — Entire diameter / Complete area

Mag 1 — 70% diameter / 50% area

Mag 2 — 50% diameter / 25% area

A B C

Fig. 3.11 A) An image intensifier operated in normal mode where the full area of the input scintillator is exposed to x-rays. (B) In Mag 1 mode, the input area is reduced to one-half so that the incident air kerma is doubled by the automatic brightness control (ABC) to maintain constant output image brightness. (C) Mag 2 mode where the input area is reduced to one-quarter and the incident air kerma is quadrupled by the ABC.

Resolution in photospot imaging through an II is 2 lp/mm, mainly determined by the 1000 × 1000 matrix used (1000-line TV system).

Use of electronic magnification does not affect the KAP because the increase in radiation intensity will be offset by the corresponding decrease in x-ray beam area. Consequently, patient stochastic risks are not increased when switching electronic magnification because the total radiation incident on the patient is not changed. When collimation is applied, the radiation intensity does not substantially change but the exposed area is reduced. As a result, skin doses are unchanged while stochastic radiation risks are reduced. Scatter is also reduced, resulting in lower doses to the operator and slightly improving image contrast because less scatter hits the receptor.

References

Huda W. Fluoroscopy. In: *Review of Radiological Physics.* 4th ed. Philadelphia: Wolters Kluwer; 2016:123–125.

Nickoloff EL. *Radiology Review: Radiologic Physics.* Philadelphia: Elsevier; 2006.

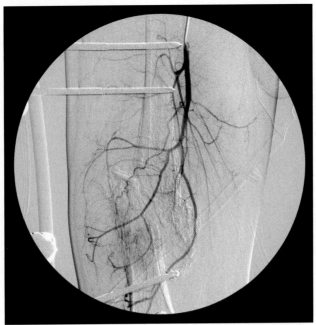

Fig. 3.12

1. How large is the file to store a single fluoroscopy frame (MB)?
   A. 0.1
   B. 1
   C. 10
   D. 100

2. Who requires fluoroscopy systems sold in the United States to have last image hold (LIH) capability?
   A. US Food and Drug Administration (FDA)
   B. Nuclear Regulatory Commission (NRC)
   C. Conference of Radiation Control Program Directors (CRCPD)
   D. National Council on Radiation Protection and Measurements (NCRP)

3. Tube currents in fluoroscopy are typically what percentage of those in radiography (%)?
   A. 1
   B. 3
   C. 10
   D. 30

4. What is the most likely x-ray tube voltage for a fluoroscopy examination performed after the administration of iodine and barium contrast media?
   A. 70 and 70
   B. 70 and 110
   C. 110 and 70
   D. 110 and 110

CASE 3.6

Digital Fluoroscopy and Fluoroscopy Techniques

Fig. 3.12 Digital subtraction angiogram of the right profunda femoris artery during intra-arterial chemotherapy infusion for a patient with pathologic fracture due to osteosarcoma.

1. **B.** A fluoroscopy image contains 500 to 1000 lines, and each line consists of 500 to 1000 pixels, so the total number of pixels is between 250,000 and 1,000,000. Each pixel is 2 bytes (16 bits) in order to encode the range of possible gray levels seen in the image. Therefore there are 250,000 to 1,000,000 pixels × 2 bytes each, which corresponds to a file size between 0.5 and 2 MB per frame.

2. **A.** The FDA requires fluoroscopy systems to have features such as LIH if these are sold in the United States. The NRC regulates the use of radioactive materials, not diagnostic x-rays. States regulate the use (not sale) of x-ray–emitting devices, and the 50 states coordinate their regulatory activities through an annual meeting of the CRCPD. The NCRP is the most important radiation safety advisory body (*i.e., not regulatory*) in the United States.

3. **A.** Fluoroscopy images are generated at average tube currents of approximately 3 mA, which is 100 times lower than radiography. One consequence of this is that fluoroscopy frames have much higher (10 times) mottle and are not diagnostic.

4. **B.** Iodine has a k-edge of 33 keV, so 70 kV (which is the maximum energy) will produce an x-ray beam with an average energy that maximizes iodine attenuation (and thus visibility). A 70-kV beam used for barium contrast would not penetrate contrast material and thus would look uniformly white. A 110-kV beam is essential to achieve some penetration, thereby providing the required differentiation between different parts of the GI tract.

Comment

Digital fluoroscopy requires the analog voltage signal from a TV camera be digitized in real time by use of an analog-to-digital converter (ADC). Digitized TV images may be stored in a computer and subsequently processed and/or displayed in real time. A 500-line TV frame contains approximately a quarter of a million pixels and requires 2 bytes per pixel. A fluoroscopy frame thus needs 0.5 MB, and 5 minutes of fluoroscopy acquired at 30 frames per second would thus require a storage space of 4.5 GB (9000 images), which is why all images are not typically stored. When images are acquired by a computer, LIH software permits the visualization of the last image after an x-ray beam has been switched off (Fig. 3.13). LIH is an FDA regulatory requirement for all fluoroscopes sold in the United States after June 2006.

Digitization of fluoroscopy will not reduce dose per se but does permit the use of image processing techniques to improve image quality. Temporal filtering, which averages successively acquired frames, can be performed in real time and will reduce the random noise. However, temporal filtering may also introduce noticeable lag of dynamic objects, which increases blur due to motion. In medical imaging, there is generally a price to be paid when one parameter (noise) improves in that another imaging aspect (lag) may get worse. Availability of digital data can improve image quality by image processing, which may enable techniques to be adjusted to reduce patient doses without adversely affecting diagnostic performance.

Fluoroscopy is always performed at low average x-ray tube currents (e.g., ~3 mA), irrespective of whether an iodinated or barium contrast agent is administered. Iodinated contrast agents use tube voltages of approximately 70 kV because this results in an average x-ray photon energy that matches the iodine k-edge energy of 33 keV. A voltage of 70 kV will maximize absorption by iodine due to the photoelectric effect and generate the best achievable visibility of the vasculature.

Unlike angiography, barium is readily visible in the GI tract because the mass of contrast material administered to patients is much higher. In GI studies, high voltages (>100 kV) are used to penetrate barium. High voltages, which result in increased penetration, allow the barium distribution within the GI tract to be visualized. Higher voltages also result in lower patient dose. A lower voltage would simply exhibit a white image in areas with barium, with no subtle shades of gray that can offer valuable diagnostic information.

References

Bushberg JT, Seibert JA, Leidholdt Jr EM, Boone JM. Fluoroscopy. In: *The Essential Physics of Medical Imaging*. 3rd ed. Philadelphia: Wolters Kluwer; 2012:298–301.

Huda W. Fluoroscopy. In: *Review of Radiological Physics*. 4th ed. Philadelphia: Wolters Kluwer; 2016:124–127.

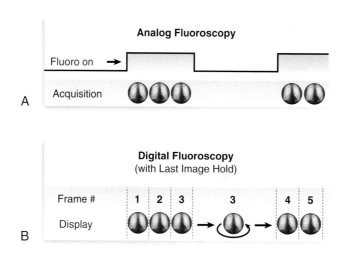

Fig. 3.13 In analog fluoroscopy (A), the image disappears when the x-ray beam is switched off, and the only way to obtain an image is by irradiating the patient. In digital fluoroscopy (B), images are stored in a computer and can be displayed even if the x-ray beam is switched off.

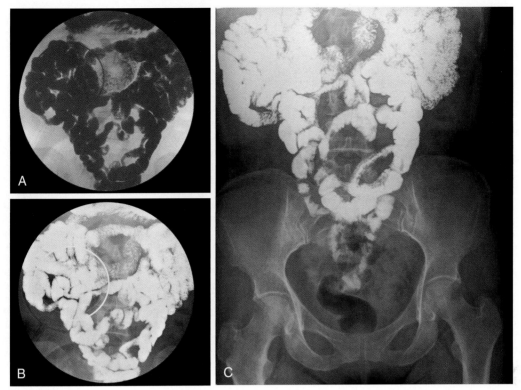

Fig. 3.14

1. Why are fluoroscopy images generally not deemed to be of diagnostic quality?
   A. High mottle
   B. Low contrast
   C. Poor resolution
   D. Elevated artifacts

2. Which dose mode would be most appropriate to use when performing fluoroscopy of a low-contrast lesion?
   A. Low
   B. Normal
   C. High

3. Which of the following digital images obtained during a fluoroscopy examination (40-cm diameter II) would most likely have the best spatial resolution?
   A. LIH
   B. Photospot image
   C. Spot radiograph
   D. All comparable

4. What image artifact is characterized by signal from the previous image still present in the current image?
   A. Pincushion
   B. Lag
   C. Dead pixel
   D. Mottle

CASE 3.7

Fluoroscopic and Diagnostic Images

Fig. 3.14 Small bowel follow-through showing the three different types of image acquisition that can be used during a fluoroscopic examination: fluoroscopic image with last image hold (A), photospot (B), and spot radiograph (C).

1. **A.** The radiation intensity used to create a fluoroscopy frame is approximately 0.1 mAs, which is 100 times less than in conventional radiography (~10 mAs). As a result, the mottle in fluoroscopy is approximately 10 times higher than in radiography (noise is inversely proportional to the square root of the radiation intensity).

2. **C.** Fluoroscopy systems offer choices of "low," "normal," and "high" radiation intensities. High mode will use approximately 50% more than in normal mode and should always be selected when low-contrast lesions are being imaged, because lesion visibility can be increased when mottle is reduced.

3. **C.** The best resolution occurs with the smallest pixel size, which in turn requires the largest matrix size. Smallest pixels are with a spot radiograph because the matrix size is 2000 × 2500. The largest pixels are for an LIH because the matrix size is 500 × 500. A photospot image would likely have a matrix size of 1000 × 1000, so the resolution is intermediate between a spot radiograph and the LIH.

4. **B.** Lag occurs when the signal from a prior image is still present while a new image is being acquired. Consequently, this occurs during fluoroscopy when new images are continuously being created and displayed.

Comment

Fluoroscopy involves viewing dynamic images in real time. Because the number of x-ray photons used to create a fluoroscopy frame is very low, such images are rarely of diagnostic image quality. Exposure times in radiography are typically very short (<0.1 seconds), whereas in fluoroscopy exposure times are generally measured in minutes. The number of acquired images in radiography is generally low and consists of a few images. In fluoroscopy, it is common to acquire thousands or tens of thousands of images (or even more). Although the patient dose per fluoroscopy frame is relatively low, the large number of acquired frames can result in high patient doses.

Fluoroscopy systems generally offer choices of "low," "normal," and "high" radiation intensities for use during fluoroscopy. The radiation intensity in a low-dose mode would likely be 50% of that offered in normal mode. Low-dose mode is typically used when misdiagnosis due to mottle is less of a concern. For example, when a child has swallowed a coin (i.e., high contrast lesion) and the diagnostic task is to locate this coin, mottle is unlikely to prevent detection. Radiation intensity in high mode is approximately 50% higher than in normal mode and would be selected when low-contrast lesions are being imaged. When the inherent contrast of a lesion is low, lesion visibility can be increased if the amount of mottle is reduced. This may be accomplished by increasing the number of photons or through image processing such as temporal filtering.

Diagnostic images for fluoroscopy-guided GI and genitourinary (GU) examinations consist of spot and photospot images. Spot images are conventional radiographs, usually obtained by introducing an overhead x-ray tube to expose a digital detector in the table Bucky. Photospot images are diagnostic quality images that are obtained through the fluoroscopic imaging chain. Photospot images use x-ray tube currents that are increased from a few mA during fluoroscopy to several hundred mA. Increasing the number of x-ray photons a hundredfold in a photospot image makes it diagnostic (10 times less noise) (Fig. 3.15). Photospot imaging would always use a large focal spot (1.2 mm) and use 1000-line TV mode, doubling the spatial resolution compared with fluoroscopy. A photospot has approximately a million pixels, requires 2 bytes per pixel, and results in a total data content of approximately 2 MB (Table 3.1).

References

Huda W. Fluoroscopy. In: *Review of Radiological Physics*. 4th ed. Philadelphia: Wolters Kluwer; 2016:122–123.

Nickoloff EL. AAPM/RSNA physics tutorial for residents: physics of flat-panel fluoroscopy systems: survey of modern fluoroscopy imaging: flat-panel detectors versus IIs and more. *Radiographics*. 2011;31:591–602.

TABLE 3.1 COMPARING LAST IMAGE HOLD, PHOTOSPOT, AND SPOT IMAGE PROPERTIES ON AN IMAGE INTENSIFIER–BASED FLUOROSCOPY SYSTEM

Parameter	Last Image Hold	Photospot	Spot
Matrix (k)	0.5 × 0.5	1 × 1	2.5 × 2
Number of pixels (M)	0.25	1	5
Bytes per pixel (B)	1 or 2	2	2
Image size (MB)	0.25 or 0.5	2	10
Nominal resolution (line pairs/mm)	1	2	3

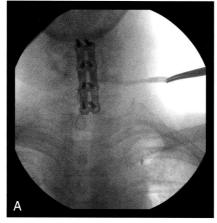 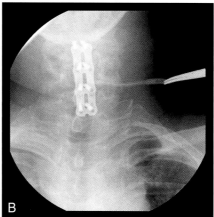

Fig. 3.15 Fluoroscopy frame (A) compared with a photospot image (B) during foreign body retrieval in the neck demonstrates the superior quality of photospot imaging. The retained surgical drain fragment is clearly more visible in the photospot image (B), which is a result of a markedly lower level of mottle.

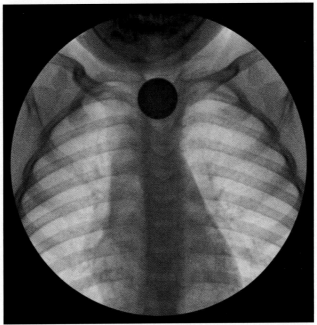

Fig. 3.16

1. What is the most likely entrance air kerma rate (mGy/min) when a 23-cm thick adult patient undergoes fluoroscopy?
   A. 0.1
   B. 1
   C. 10
   D. 100

2. Which is most likely activated when patient attenuation changes during a fluoroscopy examination performed using an II-based system?
   A. ABC
   B. AEC
   C. A and B
   D. Neither A nor B

3. Which are expected to be significantly affected (i.e., >10%) when x-ray beam area is manually reduced (i.e., collimated) to 50% by the radiologist?
   A. kV
   B. mA
   C. A and B
   D. Neither A or B

4. When patient attenuation increases and operators choose to increase mA more than kV, how does lesion contrast under this scheme compare with when the operator chooses to increase kV more than mA?
   A. Higher
   B. Unchanged
   C. Lower

CASE 3.8

Automatic Exposure Control

Fig. 3.16 Fluoroscopic image of a child's chest following ingestion of a coin.

1. **C.** For an average-sized patient, (~23 cm), the entrance air kerma rate is approximately 10 mGy/min. This value may vary depending on factors such as the pulse rate and the dose level selected. For thin body parts and infants, the entrance exposure rate can be up to an order-of-magnitude lower and for large patients may be up to an order of magnitude higher!

2. **B.** When patient attenuation increases, the AEC system will increase the amount of radiation being used and reduce the amount of radiation used when patient attenuation is reduced. ABC is used only when switching between normal and electronic magnification modes (Mag 1, Mag 2, etc.) that are available on a given II system.

3. **D.** Use of collimation has no significant effect on either the kV or mA used, because the air kerma rate is kept constant during collimation. The only change occurs to the x-ray beam area, which is reduced as collimation increases. AEC operates when patient attenuation changes (not area), and ABC operates when applying electronic magnification using an II-based fluoroscopy system.

4. **A.** Strategy I, sometimes called high-contrast mode, increases kV a bit (and mA a lot), whereas strategy II, sometimes called low-dose mode, increases kV a lot (and mA a bit). Strategy I has more contrast (and higher dose) than would have been achieved using strategy II.

Comment

AEC systems maintain a fixed amount of radiation at the image receptor. Accordingly, the amount of mottle in images generated with AEC systems is "fixed." When an AEC is operational, the CNR of a lesion can be changed only by varying the amount of contrast (i.e., noise is fixed). Selection of a low tube voltage in AEC fluoroscopy will therefore increase lesion contrast and thereby increase lesion CNR, which means the lesion will be easier to see. However, because lower-energy x-ray beams are less penetrating, more radiation will be required (vs. high kV), and patient doses will increase. Conversely, the use of high voltages in AEC fluoroscopy will reduce lesion visibility but will also reduce patient doses.

Imaging a larger patient requires more radiation incident on the patient to maintain a constant kerma at the image receptor and thereby maintain a constant level of image mottle. Increasing patient size causes the AEC system to increase tube voltage (kV) and/or current (mA). One option to achieve more radiation incident on a larger patient would be to simply increase the tube current (mA). Increasing the tube current does not affect beam quality (HVL) and would also maintain image contrast. Another option to achieve more radiation incident on a larger patient and penetration through the patient is to increase the tube voltage (kV). However, increasing the tube voltage reduces contrast. This is particularly noticeable in high Z materials, such as iodinated contrast material in a blood vessel.

Most fluoroscopy AEC systems modify both kV and mA when patient size changes. When a larger patient undergoes fluoroscopy, increasing the kV a lot and increasing the mA a little gives lower patient dose than increasing the kV a little and increasing the mA a lot. However, choice of the latter option (small kV increase + large mA increase) provides better contrast than the alternative (large kV increase + small mA increase). All modern fluoroscopy systems offer several "AEC curves" allowing operators to select optimal techniques for any given diagnostic task (Fig. 3.17). In general, when choosing a specific curve, the operator is deciding whether to try to minimize the dose to the patient or alternatively to try to get as a good an image quality (contrast) as possible.

References

Huda W. Fluoroscopy. In: *Review of Radiological Physics*. 4th ed. Philadelphia: Wolters Kluwer; 2016:124–127.

The Image Gently Alliance. Fluoroscopy imaging—pause and pulse resources. Section II: steps to manage radiation dose during the examination. http://www.imagegently.org/Procedures/Fluoroscopy/Pause-and-Pulse-Resources. Accessed March 1, 2018.

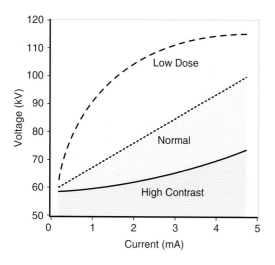

Low Dose

↑↑↑ kV + ↑ mA
Lower Dose
Lower Contrast

High Contrast

↑ kV + ↑↑↑ mA
Higher Dose
Higher Contrast

Fig. 3.17 When imaging a larger patient, one option is to increase the tube voltage a lot and the tube current a little, as shown on the top automatic exposure control (AEC) curve in A ("low dose") and left silhouette (B). Another AEC curve could increase the tube current a lot and the tube voltage a little, as shown on the bottom curve in A ("high contrast") and right silhouette (C). Choosing the low-dose option reduces lesion contrast compared with choosing the high-dose option.

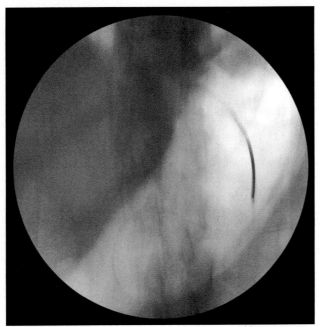

Fig. 3.18

1. Changing which contrast agent characteristic would likely result in the highest image contrast?
   A. Density
   B. Atomic number
   C. A and B similar

2. Ignoring all other factors, which agent would produce the highest vascular contrast if injected?
   A. Iodine ($Z = 53$)
   B. Barium ($Z = 56$)
   C. Gadolinium ($Z = 64$)
   D. A, B, and C equivalent

3. How will the CNR in the fluoroscopy frame likely change when four digital frames are added together?
   A. Quadrupled
   B. Doubled
   C. Halved
   D. Quartered

4. What fluoroscopy image artifact may occur if a mobile C-arm is used in a room next to a magnetic resonance (MR) scanner?
   A. Pincushion
   B. S-distortion
   C. Vignetting
   D. Veiling glare

CASE 3.9

Contrast Media, Quality, and Artifacts

Fig 3.18 Fluoroscopy frame demonstrating motion blur while moving the table position cranially.

1. **B.** Increasing the atomic number of a contrast agent will have a dramatic impact on x-ray absorption because the photoelectric effect is approximately proportional to Z^3 so that doubling the atomic number will increase the photoelectric absorption by a factor of eight. Doubling the density of a material would (only) double the number of x-ray interactions.

2. **C.** Physical density and atomic number (Z) determine how much a substance attenuates radiation. The higher the Z number is, the more photoelectric events occur, which causes a large increase in attenuation. Gadolinium ($Z = 64$) has a higher Z than barium ($Z = 56$) and iodine ($Z = 53$) and thus would give the highest contrast (though it is not used for safety reasons).

3. **B.** When four images are added together, the signal is simply quadrupled because the signal is deterministic. However, because noise is random, the fluctuations would only be doubled (not quadrupled) so that the lesion CNR would be doubled.

4. **B.** A strong magnetic field (either from variation in the Earth or due to proximity to an MR scanner) will deflect the electrons moving in the image receptor. This results in geometric distortion that makes a straight line in the image look like an "S."

Comment

Contrast agents may be used during fluoroscopy to improve image contrast. Common agents include iodine, barium, air, and carbon dioxide. Barium and iodine are positive contrast agents and appear white due to their high Z and correspondingly high attenuation. Iodine contrast is maximized using voltages of 70 kV to take advantage of the increased attenuation from the iodine k-edge. Air and carbon dioxide are negative contrast agents and appear black due to their low density and correspondingly low attenuation.

Detectability of mid-sized abnormalities in fluoroscopy is determined by both the contrast and noise. The low doses used in fluoroscopy cause high noise, which may overwhelm any contrast between a lesion and surrounding tissue (thus rendering it invisible). Contrast agent may be used to boost the lesion contrast and make the signal stand out on a noisy background. Similarly, higher dose modes or photospots can be taken because their higher dose will reduce the noise background, making the lesion more visible.

II-based fluoroscopy is subject to several unique artifacts. The II input is a thin curved surface, but the output phosphor and image are flat. Projecting a curved surface onto a flat output phosphor results in geometric distortions similar to flat maps that try to show the curved surface of the Earth. Pincushion distortion is where straight lines appear curved toward the center due to this geometric distortion (Fig. 3.19). S-distortion is a geometric distortion caused by local magnetic fields, as might occur near an MR suite. These fields push the electrons moving between input and output surfaces causing the artifact.

II-based fluoroscopy units can also encounter artifacts that impact contrast. Veiling glare is a decrease in image contrast caused mostly by scattering of light in the output window. Saturation occurs when the II response has reached its peak value. This causes pixels to appear as a uniform white intensity in the image. Furthermore, this may "bleed" into surrounding image pixels, making them brighter. Saturation is undesirable because saturated areas exhibit no image contrast. Vignetting is darkening of the periphery of the image due to less light being collected in the periphery of the output surface than in the center, caused by the curved input surface being mapped onto the flat output window (see Fig. 3.19).

References

Huda W. Fluoroscopy. In: *Review of Radiological Physics*. 4th ed. Philadelphia: Wolters Kluwer; 2016:127–128.

Nickoloff EL. AAPM/RSNA physics tutorial for residents: physics of flat-panel fluoroscopy systems: survey of modern fluoroscopy imaging: flat-panel detectors versus IIs and more. *Radiographics*. 2011;31:591–602.

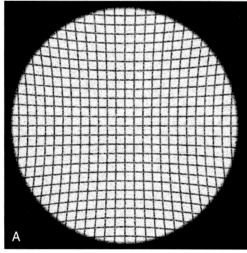
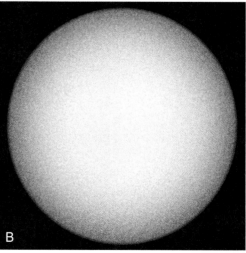

Fig. 3.19 (A) Pincushion distortion that occurs because of the curved nature of the input phosphor, which is used because it can maintain a vacuum while being kept thin. Vignetting is displayed in B where there is a loss of image intensity at the periphery of the image.

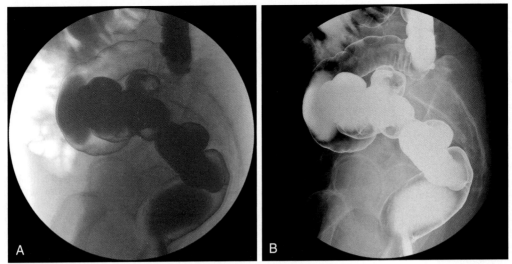

Fig. 3.20

1. What is the most likely KAP (Gy-cm^2) in a GI/GU fluorosco-py-based radiologic examination?
   A. 0.1
   B. 1
   C. 10
   D. 100

2. How much higher is the average amount of radiation used in a GI/GU study compared with an average radiographic examination?
   A. 3
   B. 10
   C. 30
   D. 100

3. How many diagnostic images (photospot) would most like-ly use the same amount of radiation as 1 minute of fluoros-copy?
   A. 3
   B. 10
   C. 30
   D. 100

4. What is the most likely contribution (%) of fluoroscopy to the total amount of radiation used in a GI/GU examination?
   A. 10
   B. 30
   C. 70
   D. 90

CASE 3.10

Kerma Area Product

Fig. 3.20 Lateral views during a barium enema comparing a low-dose fluoroscopic image (A) and a higher-dose photospot image (B).

1. **C.** An average GI/GU examination would most likely use 10 Gy-cm^2 of radiation. One-third of this radiation would likely be a result of fluoroscopy. The remaining two-thirds would be from the photospot and spot diagnostic quality images.

2. **B.** A typical radiographic examination (e.g., two-view skull radiographic study) would likely use approximately 1 Gy-cm^2 of radiation incident on the patient. This is 10 times less than that associated with a typical GI/GU examination. It is reasonable to assume that effective doses and corresponding stochastic (cancer) risks in GI/GU examinations are approximately 10 times higher than in an average radiographic examination.

3. **B.** A typical diagnostic quality photospot image would likely have an incident air kerma of 1 mGy, and the entrance air kerma rate during fluoroscopy would likely be 10 mGy/min. It is appropriate to equate 1 minute of fluoroscopy with approximately 10 diagnostic photospot images.

4. **B.** Experience has shown that fluoroscopy accounts for between 25% and 50% of the total radiation used. Most medical physicists would assume that fluoroscopy contributed ⅓ of the radiation in the absence of more detailed information.

Comment

Entrance air kerma in a normal-sized adult (23 cm tissue) is approximately 10 mGy/min. Similarly, entrance air kerma for diagnostic images of a normal-sized adult will be approximately 1 mGy per photospot image. One minute of standard fluoroscopy most likely results in the same patient dose as approximately 10 photospot images (Fig. 3.21). Patient fluoroscopy entrance air kerma rates increase with increasing patient size. At the same beam quality, increasing the patient thickness by 3 cm would likely double the incident (entrance) air kerma rate.

The typical value of KAP for a GI/GU examination is 10 Gy-cm^2. Fluoroscopy examinations use KAP values that are approximately 10 times higher than the corresponding value in radiography. The variation in average KAP is relatively modest, with a barium swallow using approximately two to three times less than average and a barium enema using two to three times more than average. Factors that can increase the KAP in any fluoroscopic examination include increased patient size and AEC curve selected.

Radiologists and technologists are responsible for selecting both the quality (HVL) and quantity (air kerma and KAP) used in fluoroscopic examinations. Converting these into patient doses and their corresponding risks is the responsibility of the medical physicist. Any organ dose can be related to the incident radiation via a conversion factor that may be measured or calculated by medical physicists. When the incident radiation quantity/quality is combined with dose conversion factors, organ and effective doses may be readily determined.

Operators are close to patients undergoing fluoroscopy and may be exposed to radiation scattered from the patient. It is important to note that operator doses are likely to be directly proportional to patient doses. Operators have their exposures monitored and wear lead aprons. Radiologists, technologists, and ancillary staff are routinely exposed to radiation that is scattered from the patient in fluoroscopy. Operator and staff dose can be kept low by ensuring that proper shielding is used, exposure time is kept to the minimum required, and staff maximize the distance to the entrance surface of the patient.

References

Huda W. Fluoroscopy. In: *Review of Radiological Physics*. 4th ed. Philadelphia: Wolters Kluwer; 2016:127–129.

Radiological Society of North America. RSNA/AAPM physics modules. Fluoroscopy: radiation dose and safety in interventional radiology. https://www.rsna.org/Physics-Modules/ Accessed March 1, 2018.

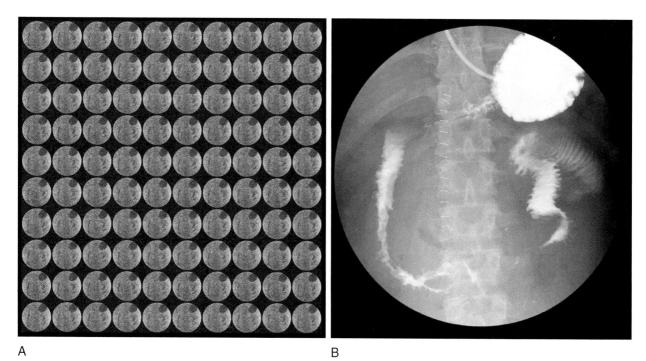

A   B

Fig. 3.21 The amount of radiation used in 100 fluoroscopy frames (A) is the same as used to create one photospot image (B). Ten photospot images use the same radiation as 1 minute of 1 fluoroscopy, when the latter consists of approximately 1000 fluoroscopy frames (i.e., 15 frames per second × 60 seconds).

CASE 4.1

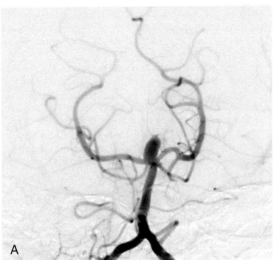
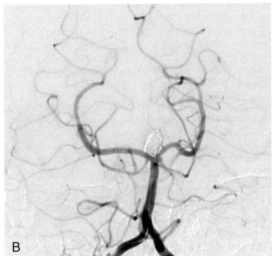

Fig. 4.1

1. How do the anode capacity (megajoules) and the corresponding x-ray tube filtration on an interventional radiology (IR) x-ray tube compare with those on a standard radiographic/fluoroscopic unit?
   A. Higher; higher
   B. Higher; lower
   C. Lower; higher
   D. Lower; lower

2. Which is the most likely type of detector used in flat panel detectors (FPDs) in IR suites?
   A. Scintillators (e.g., cesium iodide CsI)
   B. Photoconductors (e.g., selenium)
   C. Photostimulable phosphors (e.g., BaFBr)
   D. Ionization chambers (e.g., xenon)

3. What x-ray tube voltage (kV) will result in the highest attenuation by iodine in a blood vessel?
   A. 55
   B. 70
   C. 90
   D. 110

4. What is the primary benefit of the use of equalization filters (soft collimators)?
   A. Improved resolution
   B. Higher contrast
   C. Reduced data
   D. Lower doses

CASE 4.1

Flat Panel Detector Systems

Fig. 4.1 Digital subtraction angiograms pre (A) and post (B) stent-assisted coil embolization of a basilar tip aneurysm.

1. **A.** IR systems use fluoroscopy, digital photospot, and digital subtraction angiography (DSA) extensively, so that a lot of energy is deposited in the anode in a relatively short time. Accordingly, anode capacities on an IR system will be substantially higher than on a radiographic or fluoroscopic system. High filtration used in IR systems reduces the skin doses delivered in IR which are potentially very high.

2. **A.** Most FPDs used in IR are cesium iodide scintillators, which have excellent x-ray absorption properties at the energies used in IR. Photoconductors are made of selenium, which has poor x-ray absorption properties, and photostimulable phosphors are impractical for fluoroscopy and photospot/DSA imaging. Ionization chambers are rarely used in medical imaging.

3. **B.** An x-ray tube voltage of 70 kV results in an average energy that is just above the k-edge energy of iodine (33 keV). This will maximize the x-ray absorption by iodine, which will also result in the highest contrast of the blood vessel in comparison to the surrounding tissues. Use of higher and lower x-ray tube voltages would reduce the x-ray absorption by the iodine in vasculature and thereby reduce the contrast (visibility) of the blood vessels.

4. **D.** Equalization filters (also known as soft collimators or wedge filters) reduce the radiation incident on structures adjacent to the region of interest. The region of interest, which receives the full x-ray beam intensity, is visualized with the lowest amount of mottle. The surrounding tissues that are needed only to landmark position receive less radiation and therefore more mottle. There are generally no changes in the resolution, contrast, or the amount of data being generated when equalization filters are applied.

Comment

IR systems have high-quality x-ray generators and tubes mounted on agile gantries that can be maneuvered into any desired orientation. Modern systems use FPDs, rather than image intensifiers (IIs), which offer a smaller footprint and improved maneuverability. Furthermore, FPDs can accurately detect very small and very large amounts of radiation (i.e., wide dynamic range). Although an IR system is approximately $1,250,000, as compared with $350,000 for an II-based radiographic/fluoroscopic system, the clinical needs of IR procedures justify the additional costs.

FPDs used in IR are no different to those used in digital radiography. However, when FPDs are also used for fluoroscopy, it is essential for vendors to minimize electronic noise because the dose hitting the image receptor in IR will be approximately 100 times lower than in radiography. The size of an FPD used in IR is typically 40 cm × 40 cm with a CsI scintillator that is used to capture the primary x-ray pattern emerging from the patient (Fig. 4.2). FPD pixels are approximately 170 μm, offering good resolution in photospot imaging. A TV camera or charge-coupled device is not required because FPD data are read out electronically for real-time processing and/or display on a monitor.

The field of view (FOV) can be selected by the operator (e.g., 32 × 32 cm, 20 × 20 cm, 16 × 16 cm). Selecting a smaller FOV in this manner usually results in an increase in air kerma (K_{air}) at the image receptor which is designed by the vendor to reduce the level of perceived mottle. Increases in image receptor K_{air} with a reduced FOV are programmed in by the vendor. However, the image receptor K_{air} is not generally increased when collimation is used to reduce the field size. It is important that operators understand this important difference between FOV selection and x-ray beam collimation because these impact both patient doses and the perceived image quality (mottle).

IR equipment has more sophisticated hardware than does the radiographic/fluoroscopic unit used to perform gastrointestinal (GI)/genitourinary (GU) imaging. The gantry on an IR system can orient the imaging plane along any desired axis. IR systems use more powerful x-ray tubes and have anodes with greater heat capacities that are suitable for performing

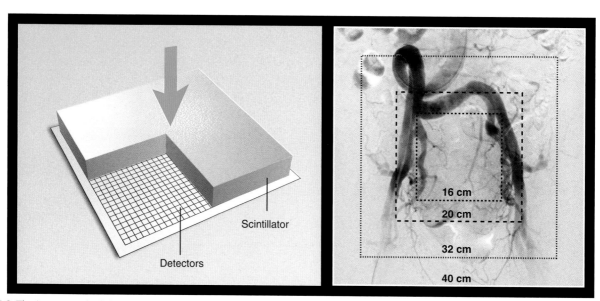

Fig. 4.2 The image on the left shows a schematic of the x-ray detector, which normally consists of a cesium iodide scintillator that absorbs the incident x-rays. The resultant light is detected by a two-dimensional array of light detectors. The image on the right shows four available fields that may be selected by the operator. As the field of view is reduced, the vendor has preset values that increase the amount incident radiation to ensure satisfactory image quality.

extended interventional procedures. All IR systems offer the ability to add copper (Cu) filtration, typically 0.1 mm or more, that increases beam quality (penetration) to reduce skin and effective doses. In addition, IR systems also provide equalization filters that are mixtures of lead and plastic. These filters reduce transmission of x-rays to peripheral regions and thus limit the doses to organs and tissues adjacent to the structures of interest.

References

Bushberg JT, Seibert JA, Leidholdt Jr EM, Boone JM. Fluoroscopy. In: *The Essential Physics of Medical Imaging*. 3rd ed. Philadelphia: Wolters Kluwer; 2012:290-291.
Huda W. Interventional radiology. In: *Review of Radiological Physics*. 4th ed. Philadelphia: Wolters Kluwer; 2016:133-134.

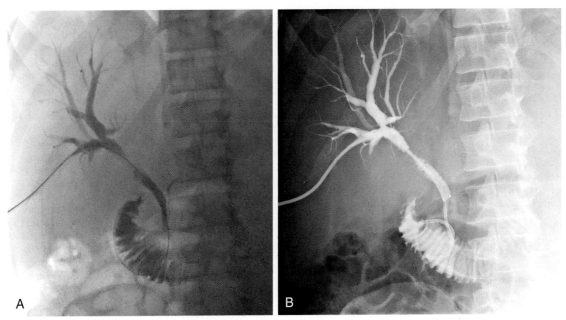

A   B

Fig. 4.3

1. How does the electronic signals from an FPD in fluoroscopy compare with those obtained during radiography/photospot imaging?
   A. 100× bigger
   B. 10× bigger
   C. 10× smaller
   D. 100× smaller

2. What image property increases when detector elements are binned during image acquisition with an FPD?
   A. Acquired data
   B. Perceived mottle
   C. Spatial resolution
   D. None (all reduced)

3. What is most likely the same for a fluoroscopy frame and photospot image obtained using an FPD?
   A. Lesion contrast
   B. Image mottle
   C. Patient dose
   D. A, B, and C

4. What is the most likely exposure index (EI) for a photospot image obtained using an FPD?
   A. 1
   B. 10
   C. 100
   D. 1000

CASE 4.2

Fluoroscopy With Flat Panel Detectors

Fig. 4.3 Flat panel detector fluoroscopic image (A) and photospot image (B) during a cholangiogram demonstrating narrowing of the common bile duct and nonvisualization of the left hepatic ductal system.

1. **D.** The amount of radiation used to create a fluoroscopy frame (~ 0.01 µGy) is approximately 100 times less than the radiation used to create a diagnostic quality image (~1 µGy) in photospot imaging. This is achieved by using fluoroscopy tube currents of a few milliamperes, which is a hundred or so times lower than used in photospot imaging.

2. **D.** Binning is often used when FPDs are used to perform fluoroscopy with a large FOV. Four adjacent pixels are combined into one larger pixel because fluoroscopy at a large FOV does not require high-resolution performance. Binning halves spatial resolution and quarters amount of data. Perceived mottle is also halved because each pixel now has four times as many photons hitting it.

3. **A.** Differences between a fluoroscopy frame and a photospot image relate to the amount of radiation used to create an image. The photospot image will have less mottle, but this requires the use of much more radiation. Lesion contrast will be the same, although lesion visibility will generally be worse in the fluoroscopy frame because of higher mottle.

4. **C.** The radiation incident on an FPD to create one photospot image is approximately 1 µGy, which can be translated into an EI of 100. This is a little less than the dose used to create a chest x-ray which normally would have an EI of 300.

Comment

Assuming a 40-cm FOV and normal mode setting, fluoroscopy performed with FPD and IIs would likely have similar K_{air} rates at the image receptor (and patient). FPDs do not use electronic magnification and do not use automatic brightness control (ABC) that IIs use. Accordingly, when the FPD FOV is reduced, there is no technical reason why image receptor K_{air} needs to be increased. When the FPD FOV is reduced, K_{air} is increased by the vendor to reduce mottle perceived by the operator (a design choice). Halving the FOV in an II-based fluoroscopy system using ABC would likely quadruple all K_{air} values but only double K_{air} values for an FPD.

As the FOV increases, it becomes difficult for the human eye to perceive tiny (high-resolution) structures due to their decreasing size. FPD systems take advantage of this by binning adjacent detector elements together (typically 4 pixels are binned into a single new pixel) (Fig. 4.4). This reduces the resolution, which is acceptable because very high resolutions are not visible to the human eye at a large FOV. Binning reduces mottle, as well as the amount of data. However, binning would not be used when FPDs are used to generate photospot images because mottle is already reduced in such images due to increased dose and because full-resolution images are desirable in such cases.

In DSA, a mask image without contrast is subtracted from the corresponding image with contrast to remove patient anatomy. A DSA image will thus simply depict the opacified structure of interest. In this way, DSA imaging improves vasculature visibility because the clutter of "anatomical background" is of often greater importance than quantum mottle itself. IR systems provide a road-mapping capability that permits an image to be captured and displayed on a monitor, while at the same time a second monitor shows live images. Road mapping is also used to capture images with contrast material to overlay onto a live fluoroscopy image.

Tube voltages of 70 kV are generally used in angiography and DSA imaging because these result in an average photon energy that matches the energy of k-edge of iodine (i.e., 33 keV) and thus gives the highest contrast visibility. High tube currents are used (e.g., 400 mA) with relatively short exposures times (e.g., 50 ms). The total technique is thus approximately 20 mAs, or similar to that used in an abdominal radiograph. The x-ray beam intensity at the image receptor in angiography and DSA is similar to that of radiographic imaging, which is between 1 and 3 µGy per image (i.e., EI 100 to 300). Frame rates currently used in angiography and DSA imaging can range up to 4 frames per second, with a typical run of up to 10 seconds, thereby resulting in the acquisition of up to 40 images.

References

Huda W. Interventional radiology. In: *Review of Radiological Physics*. 4th ed. Philadelphia: Wolters Kluwer; 2016:134–135.

Nickoloff EL. AAPM/RSNA physics tutorial for residents: physics of flat-panel fluoroscopy systems survey of modern fluoroscopy imaging: flat-panel detectors versus image intensifiers and more. *Radiographics* 2011;31:591–602.

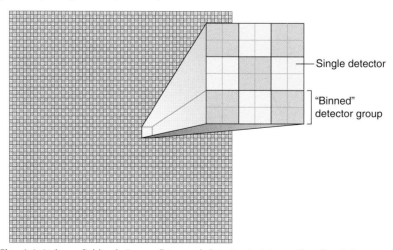

Fig. 4.4 At large fields of view, a flat panel detector in interventional radiology uses binning during fluoroscopy procedures. Note that individual pixels are not easily visualized and thus that level of resolution is not necessary. The detector adds (i.e., bins) 4 pixels together, which reduces resolution but reduces perceived mottle and decreases the amount of image data.

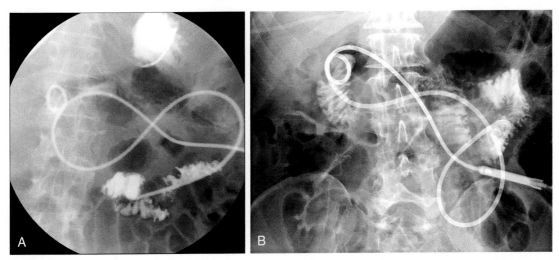

Fig. 4.5

1. How does the limiting spatial resolution of an FPD in an IR suite compare with that of a radiographic/fluoroscopic system that uses an II/TV combination operated in normal mode?
   A. Superior
   B. Comparable
   C. Inferior

2. How does the contrast-to-noise ratio (CNR) of an FPD in an IR suite compare with that of an II-based radiographic/fluoroscopic system?
   A. Superior
   B. Comparable
   C. Inferior

3. Which artifacts are most likely observed on an FPD in an IR suite?
   A. Vignetting
   B. Pincushion
   C. S-distortion
   D. None of the above

4. How does quartering the selected image area affect the incident K_{air} on an II-based system and on an FPD, respectively?
   A. Quadruples; quadruples
   B. Quadruples; doubles
   C. Doubles; quadruples
   D. Doubles; doubles

CASE 4.3

Flat Panel Detectors Versus Image Intensifiers

Fig. 4.5 Percutaneous gastrojejunostomy catheter placement preformed using an image intensifier (A) and a flat panel display (B).

1. **A.** The limiting resolution of an FPD is typically 3 line pairs per mm (lp/mm). Although an II can resolve approximately 5 lp/mm, images are viewed through a standard 500-line TV system that reduces the liming resolution to approximately 1 lp/mm during fluoroscopy.

2. **B.** The same detector (cesium iodide) is used in both IIs and FPDs, which will absorb the same number of photons and have the same noise. The contrast of a lesion is generally independent of detector type. Accordingly, the CNR in fluoroscopy performed with an II and FPD should be essentially the same.

3. **D.** IIs display a range of artifacts due to their unique design using electrons traversing an evacuated space and the use of a curved input surface. Vignetting, pincushion, and S-distortion are all unique to II-based systems and cannot occur in FPD systems.

4. **B.** In Mag 2 mode the input FOV is halved and the irradiated area is quartered. The ABC must quadruple the K_{air} to maintain a constant brightness at the II output. When an FPD area is a quarter, manufacturers will generally double the image receptor K_{air} because this will produce a level of mottle that has been empirically found to be acceptable to radiologists.

Comment

Contrast is primarily determined by the choice of x-ray tube voltage (kV) but may also be reduced when scatter is present. Accordingly, contrast will be *identical* for both IIs and FPD because voltages are similar and the amount of scatter will not be markedly different. Noise (mottle) is determined by the number of photons captured by an image detector and used to create any image. The detectors used in IIs in both FPD and II-based systems are essentially the same material (cesium iodide) and a similar thickness (~0.4 mm), so the number of photons absorbed (i.e., mottle) is identical. However, FPDs have a carbon cover which transmits slightly more x-rays than the 1.5 mm aluminum that is currently used in IIs. As a result, an FPD detector will detect slightly more photons than an II-based detector. Photon detection is estimated to be approximately 90% for an FPD versus 75% for an II at 70 kV.

The limiting resolution of an FPD is determined by the pixel size. This is typically 170 μm (0.17 mm), which corresponds to a sampling rate of 6 pixels per mm. The limiting resolution, called the Nyquist frequency, is half the sampling rate or 3 lp/mm. This resolution is fixed, unlike the electronic magnification modes of an II-based system where a reduction in the FOV can improve the achievable limiting resolution. Although image intensifiers have a high intrinsic limiting resolution (5 lp/mm), resolution is reduced when the images are viewed using a TV system. Standard fluoroscopy resolution is approximately 1 lp/mm using a 500-line TV, whereas photospot imaging using a 1000-line TV system typically achieves 2 lp/mm when operated in the normal mode.

Large differences are found in the artifacts that occur in IIs and FPDs. An II-based system may exhibit artifacts such as pincushion distortion and vignetting (Fig. 4.6). In the presence of external magnetic fields, S-distortion can occur when electrons are deflected from their normal paths of travel. At higher intensities, glare and saturation effects manifest (Fig. 4.7). The fidelity of images obtained using FPDs is excellent, without any of the problems that occur daily in images created with an II. For example, saturation does not typically occur with FPDs because these have a linear response and large dynamic range that is markedly superior to the sigmoidal response of an II. The response of an II is analogous to that of film (i.e., sigmoidal) and has the same limitations as film at both low and high exposures.

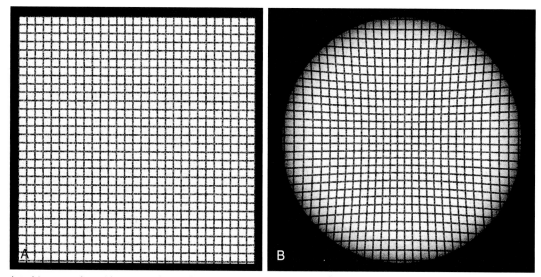

Fig. 4.6 Simulated images of a grid pattern obtained with a flat panel detector (FPD) (A) and an image intensifier (B) illustrate the excellent fidelity of FPD images.

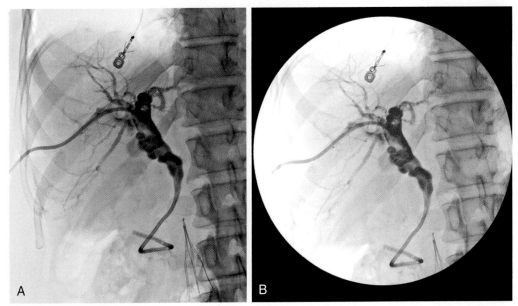

Fig. 4.7 Simulated image quality with an image intensifier (B) is reduced compared with a flat panel display (A). Veiling glare artifact and the limited sigmoidal response of the image intensifier create a clearly inferior image.

References

Bushberg JT, Seibert JA, Leidholdt Jr EM, Boone JM. Fluoroscopy. In: *The Essential Physics of Medical Imaging*. 3rd ed. Philadelphia: Wolters Kluwer; 2012:282-311.

Huda W. Interventional radiology. In: *Review of Radiological Physics*. 4th ed. Philadelphia: Wolters Kluwer; 2016:133-134.

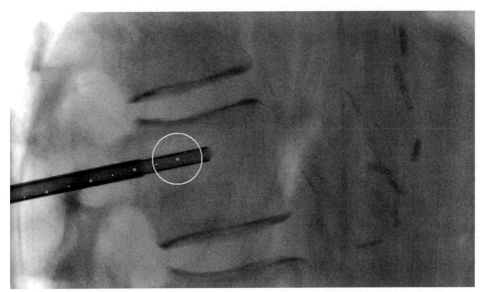

Fig. 4.8

1. What is the typical source-to-image receptor distance (SID) (cm) for IR procedures performed using an FPD?
   A. 75
   B. 100
   C. 125
   D. 150

2. What is the benefit of using geometric magnification in an IR suite?
   A. Sharper images
   B. Lower patient skin doses
   C. Reduced operator doses
   D. None

3. How much will patient skin dose rate increase (%) when switching from a posteroanterior to a lateral projection in a 75-kg adult abdomen assuming automatic exposure control (AEC) is active?
   A. 10
   B. 30
   C. 100
   D. 300

4. By how much will entrance K_{air} rate be reduced if a grid were removed (%) during infant examination performed with AEC?
   A. <10
   B. 10
   C. 50
   D. 90

CASE 4.4

Irradiating Patients

Fig. 4.8 Lateral fluoroscopic view of the lumbar spine during vertebroplasty using laser-guided needle placement.

1. **B.** Most of radiographic and fluoroscopic imaging is performed at a SID of 100 cm. If the SID is reduced, there is an increase in distortion due to geometric magnification and increase of patient skin dose due to the inverse square law. If the SID is increased, exposure times and corresponding motion blur will increase.

2. **D.** Geometric magnification occurs when the patient is moved away from the image receptor, as in magnification mammography. With geometric magnification, resolution is decreased due to focal spot blur and skin dose will increase according to the inverse square law. An air gap results in scatter radiation "missing" the image receptor and thereby irradiating any adjacent operator.

3. **C.** A patient has an oval cross-section that has the larger diameter in the lateral direction. A good rule of thumb is that when the projection in an abdominal radiograph is changed from the shorter anteroposterior thickness (23 cm) to the longer lateral thickness (30 cm), the amount of radiation used is generally doubled. This is based on empirical evidence because the tube voltage is likely increased to improve patient penetration when attenuation increases.

4. **C.** When a grid is used with AEC active, contrast and patient dose increase. For an infant, approximately half the radiation is attenuated by the grid (Bucky factor of 2). This means that removing the grid will halve the required radiation to keep the receptor dose constant.

Comment

The optimal SID in IR is approximately 100 cm (Fig. 4.9). For an average-sized adult patient with an abdominal AP dimension (thickness) of 23 cm, the resultant source-to-skin distance (SSD) will be close to 75 cm. If the SID is reduced, this will increase skin doses due to the inverse square law. Image distortion will also occur due to a variable magnification factor between the anterior and posterior sides of the patient. If the SID is increased, this will result in longer image exposure times to keep receptor dose constant, and this could result in increased motion blur.

Geometric magnification is highly undesirable in any type of fluoroscopic or radiographic imaging. A gap between the patient and the image receptor introduces focal spot blurring that reduces resolution and increases skin doses because the patient is closer to the x-ray tube focal spot. In fluoroscopy, an air gap will also permit scatter exiting the patient to irradiate nearby operators who will be present for these types of examinations. Reducing the gap between the patient and the image receptor will minimize geometric magnification in IR imaging.

During fluoroscopy, the entrance K_{air} for a 23-cm thick adult is approximately 10 mGy/min. In angiography and DSA imaging, the patient entrance K_{air} will be approximately 1 mGy per acquired image in a normal-sized patient, so that 10 acquired images are likely to deliver the same amount of radiation as 1 minute of fluoroscopy. Entrance rates can be up to an order of magnitude higher in very large patients and can be an order of magnitude lower in small patients, as illustrated in Table 4.1. X-ray tube voltages must be raised to higher than 100 kV in large patients to achieve adequate penetration, and this is generally performed automatically depending on patient size. These nominal values may be markedly affected by the changes in the choice of FOV, pulse fluoroscopy frame rate, and selected dose mode (low to high).

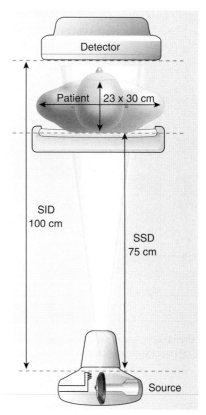

Fig. 4.9 This figure shows that the ideal irradiation geometry in interventional radiology would use a source-to-image receptor distance *(SID)* of 100 cm and a minimal air gap. When the patient is moved closer to the x-ray tube, an air gap is introduced, which increases both focal blur and patient skin doses.

Table 4.1 HOW NOMINAL DOSES LIKELY CHANGE WITH PATIENT SIZE FOR FLUOROSCOPY AND PHOTOSPOT IMAGING IN INTERVENTIONAL RADIOLOGY

SMALL PATIENTS		LARGE PATIENTS	
Size	**Relative Dose (%)**	**Size**	**Relative Dose (%)**
15 cm × 20 cm Pediatric	40	30 cm × 40 cm Large Adult	200
10 cm × 10 cm Neonate	20	40 × 50 cm Obese	400
10 × 10 cm No grid	10	50 cm × 50 cm Extreme obese	800

Entrance K_{air} rates in lateral projections are normally twice those of AP projections due to increased patient thickness. It is important that operators use a lateral projection when clinically appropriate even though the radiation delivered to the patient is increased. However, it is also important that the operator reverts to lower dose projections when clinically indicated. In infants and thin body parts (e.g., knees), where the attenuating thickness is less than 12 cm, the use of grids in IR procedures is optional. When a grid is removed in an examination performed using AEC, radiation intensities are likely halved for an infant examination (i.e., Bucky factor is 2).

References

Huda W. Interventional radiology. In: *Review of Radiological Physics*. 4th ed. Philadelphia: Wolters Kluwer; 2016, pp 135–136.

The Image Gently Alliance. Steps for radiation safety in pediatric interventional radiology; 2014. http://www.imagegently.org/LinkClick.aspx?fileticket=dKcIRoL3eJE%3d&tabid=765&portalid=6&mid=1989. Accessed March 1, 2018.

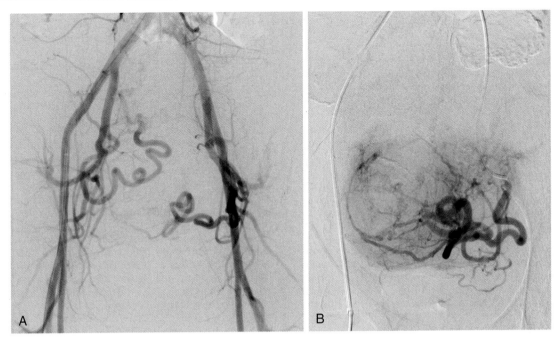

Fig. 4.10

1. What is the minimum distance (cm) that a patient can be from the focal spot of a fixed IR imaging system?
   A. <30
   B. 30
   C. 38
   D. >38

2. How much scatter radiation (%) is most likely transmitted through a 0.5-mm lead shield in an IR suite?
   A. <5
   B. 5 to 15
   C. 15 to 30
   D. >30

3. How many fluoroscopy images will be generated by the time the first 5-minute warning alarm is sounded on the control panel of an IR system?
   A. 50
   B. 500
   C. 5000
   D. 50,000

4. Which of the following are regulatory requirements for fluoroscopy imaging systems used in US IR suites?
   A. Last image hold available
   B. Table <2 mm aluminum equivalent
   C. A and B
   D. Neither A nor B

CASE 4.5

Equipment Safety

Fig. 4.10 (A and B) Digital subtraction angiograms obtained during uterine artery embolization. The procedure is performed using ipsilateral and contralateral tube angulation (in addition to collimation) during treatment of each respective uterine artery in order to reduce peak skin dose.

1. **C.** To ensure that the skin cannot get too close to the x-ray tubes, fluoroscopy units use a physical spacer that acts as a barrier to limit the distance of closest approach. For stationary systems, this distance is 38 cm, and for mobile C-arms, it is 30 cm. The US Food and Drug Administration (FDA) requires all fluoroscopy units sold in the United States to have such spacers installed.

2. **A.** Very little x-ray radiation will be transmitted through a lead apron or shield that has a 0.5-mm lead equivalence thickness. It is safe to say that less than 10% will be transmitted, and in practice, the measured levels are generally between 1% and 5%, depending on the scattered photon energies.

3. **C.** Modern equipment makes use of pulsed fluoroscopy, which "freezes" motion and improves spatial resolution performance. For most clinical applications, an image acquisition rate of 15 pulses per second has been found to be satisfactory, which corresponds to nearly 1000 per minute (i.e., 900). The number of acquired images in 5 minutes is thus nearly 5000 (i.e., 5 × 900).

4. **C.** The FDA requires that when the exposure is stopped the last acquired image is available for view without need to further expose the patient (last image hold). FDA also mandates a maximum table 2 mm or less of aluminum equivalence, which will attenuate approximately one-third of the incident radiation (half-value layer [HVL] is typically 3 mm Al). For AP projections, this radiation would be removed after it has hit the patient but before it hits the receptor.

Comment

During IR procedures, operators are exposed to radiation scattered from patients. At 1 m from the patient, scattered radiation intensities are approximately 0.1% of the radiation intensity that is incident on the patient. One meter from an average patient scatter is 10 µGy/min for fluoroscopy and 1 µGy per photospot image. Operator effective doses generally change in direct proportion to those of patient effective doses. Large patients require higher intensities and larger fields that will also increase operator dose, and vice versa.

Where the x-ray beams enter into a patient is the principal source of scatter in all of x-ray imaging, including IR. Under the table x-ray tubes will result in lower operator doses, but an over-table x-ray tubes will result in higher doses. Specifically, eye lens doses will be high for this type of exposure geometry. For lateral projections, operators should stand on the receptor side and avoid the x-ray tube location where radiation intensities will be substantially increased. Regardless of what projection or positioning is used, a spacer cone fixed to the x-ray tube prevents the patient from being too close to the x-ray source (Fig. 4.11).

Lead drapes can be attached to patient tables, which will help to reduce scatter from an under-table x-ray tube, as well as from the patient. Leaded plastic made of transparent material drapes positioned between the patient and the operator will allow the patient to be readily observed and substantially reduce operator doses. Moveable lead barriers can be placed to protect ancillary staff that may need to remain in the room. Most lead protective shielding will use 0.5-mm lead, which should attenuate most (i.e., >90%) of the incident radiation. The precise attenuation is energy dependent, with more shielded at low energies and vice versa.

Shielding should ensure that workers and members of the public do not exceed current regulatory dose limits. Medical physicists, not radiologists, are trained to perform shielding calculations which take into account factors such as IR facility workload and the occupancy factors at selected locations. Approximately 2 mm of lead is usually added to the walls of most IR rooms, up to a height of 2 m. Exposures to staff and members of the public outside any IR room will be extremely low and of no practical concern.

References

Huda W. Interventional radiology. In: *Review of Radiological Physics.* 4th ed. Philadelphia: Wolters Kluwer; 2016:135–137, 141–142.

The Image Gently Alliance. Step lightly checklist; 2014. http://www.imagegently.org/Portals/6/Procedures/ImGen_StpLight_Chcklst.pdf. Accessed March 1, 2018.

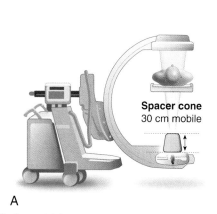
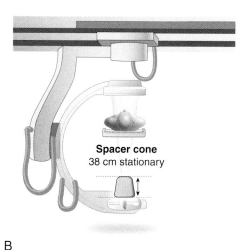

Fig. 4.11 The mobile C-arm (A) has a spacer to ensure that the patient cannot be closer than 30 cm to the focal spot. The interventional radiology imaging chain (B) has a spacer to ensure that the closest the patient can get to the focal spot is 38 cm.

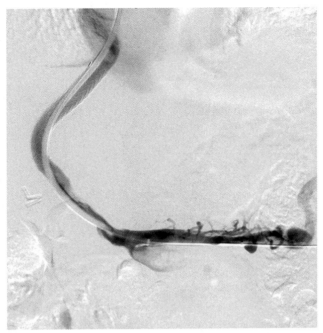

Fig. 4.12

1. What is the minimum lead equivalence (mm) for aprons worn by operators working in an IR suite?
   A. 0.1
   B. 0.25
   C. 0.5
   D. 1

2. Which radiation risk is reduced the most by wearing thyroid shields?
   A. Carcinogenic
   B. Eye cataract
   C. Skin burn
   D. Genetic

3. In practice, what is the most likely dose reduction (%) to the eye lens achieved by wearing leaded glasses during IR procedures?
   A. 30
   B. 65
   C. 90
   D. >90

4. What is the most likely dose reduction (%) to the hands achieved by wearing lead-impregnated surgical gloves during IR procedures?
   A. 25
   B. 50
   C. 75
   D. 90

CASE 4.6

Protective Clothing

Fig. 4.12 Digital subtraction angiography image obtained during transsplenic transjugular intrahepatic portosystemic shunt (TIPS).

1. **B.** The required limit for lead aprons in IR is 0.25 mm, although in practice most lead aprons are 0.5-mm lead equivalent. There are lightweight lead aprons that claim to be equivalent to lead, which decrease the muscle strain and injury that can occur due to the weight of lead aprons.

2. **A.** Thyroid shields reduce the thyroid dose, an organ that is very sensitive to carcinogenesis, especially in young females. The wearing of thyroid shields thus reduces the risk of cancer induction in the operator.

3. **B.** When such protective glasses are worn, the eye lens also receives scatter radiation from x-rays that are incident on the operator around the eye. As a result, the dose reduction is not the 90% predicted by x-ray transmission data but is instead found to be approximately 65%.

4. **B.** Lead impregnated surgical gloves are available and would likely reduce extremity doses by approximately 50%. They are not recommended because the use of such gloves can make operators "overconfident," encouraging poor technique that actually increases extremity dose.

Comment

In IR, wrap-around aprons provide 360-degree protection and are therefore preferred (Fig. 4.13). The weight of lead aprons can be spread between the hips and shoulders by using a skirt and vest combination. Lighter, non-lead style aprons are unlikely to be as effective at reducing operator exposures compared with lead aprons but are nonetheless likely to provide a sufficient level of protection to IR operators. Apron-like enclosures suspended from the ceiling are available and offer good protection without heavy weight being borne by the operator, but such systems are very expensive. In general, any dosimeter that is worn under lead aprons in IR is likely to record very low values (often indicating dose is less than the minimum detectable).

To reduce the dose to the radiosensitive thyroid, IR staff should wear a thyroid protective collar, which may be a regulatory requirement when operators are double badged. Heavy leaded gloves are generally not worn because they are bulky and also difficult to sterilize. Lead-impregnated sterile gloves may attenuate up to 50% of the incident radiation, but these may also provide a false sense of security. This can encourage poor practice and result in increased operator dose. IR operators must ensure their hands are never placed directly into the x-ray beam incident on the patient, which is the best protective practice that is currently available.

Not all leaded glasses are the same. Protective glasses should include lateral shields or wrap-around geometry because operators are normally viewing the monitors rather than patients. Although a direct beam on the glasses would be reduced by 90% or more, dose reduction to the eye lens when using leaded glasses is only 65% in practice. This is because x-rays

Lead glasses

Thyroid sheild

Lead apron

Fig. 4.13 The essential protective equipment that should be worn routinely by any operator working in interventional radiology. Note that the two-piece lead apron provides 360-degree protection.

are backscattered into the eye lens so that eye lens dose reductions are much lower than that predicted by the transmission of x-rays. Eye lens doses can further be reduced by placing a "hang down" transparent leaded-glass barrier between the operator and the patient. Such barriers should be placed as close to the source of the scatter as possible (usually the patient) to ensure maximum efficacy.

References

Huda W. Radiation protection. In: *Review of Radiological Physics*. 4th ed. Philadelphia: Wolters Kluwer; 2016:82-83.

Huda W. Interventional radiology. In: *Review of Radiological Physics*. 4th ed. Philadelphia: Wolters Kluwer; 2016:141–142.

Radiological Society of North America. RSNA/AAPM physics modules. Fluoroscopy: 2.0 radiation dose and safety in interventional radiology. https://www.rsna.org/RSNA/AAPM_Online_Physics_Modules_.aspx. Accessed March 1, 2018.

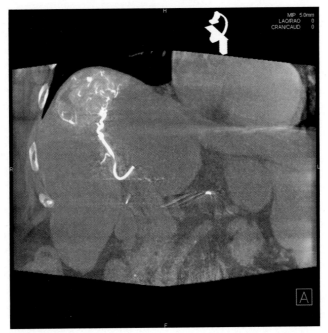

Fig. 4.14

1. By how much will halving the number of x-ray images, and the corresponding fluoroscopy time, reduce operator doses (%) during an IR procedure?
   A. <50
   B. 50
   C. >50

2. What is the most likely increase in dose when the distance to an x-ray source is reduced from 10 to 1 cm?
   A. 3×
   B. 10×
   C. 30×
   D. 100×

3. What is the most effective way of reducing doses to operators who perform IR procedures?
   A. Reduce fluoroscopy time
   B. Increase distance to patient
   C. Use lead shielding
   D. A, B, and C comparable

4. What are the effective doses of the most highly exposed IR operators in comparison with current regulatory dose limits (%)?
   A. 10
   B. 20
   C. 50
   D. >50

CASE 4.7

Operator Doses

Fig. 4.14 Coronal reformat of a cone beam CT performed during transarterial chemoembolization of a right hepatic hepatocellular carcinoma. Patient dose is comparable to or less than conventional CT. Image quality is generally much worse (artifacts and mottle), although it may be sufficient for certain tasks. Cone beam CT is performed with the staff in the control room due to high scatter.

1. **B.** The operator radiation doses will be directly proportional to the patient dose. When the amount of radiation incident on the patient is halved, the operator dose will also be halved.

2. **D.** The variation in intensity of any x-ray beam with the distance from the x-ray source (focal spot) obeys the inverse square law. When the distance from a radiation source is reduced by a factor of 10, the intensity will increase 100-fold (10 squared).

3. **C.** Operator doses are directly proportional to the time of exposure (i.e., 50%). Doubling the distance from a source will reduce the operator exposure to a quarter (i.e., 25%). Placing a lead shield with a lead equivalence of 0.5 mm will reduce dose to less than one-tenth of the initial radiation intensity (i.e., <10%).

4. **A.** The current US regulatory dose limit for a radiologist is 50 mSv/year. In practice, IR fellows receive the highest occupational radiation doses. Two badges are worn (above and below a lead apron) to obtain a reliable estimate of such doses, which are generally approximately 5 mSv/year (i.e., 10% of the dose limit).

Comment

At relatively high operator doses, as occurs to IR fellows, operators are generally required to wear two dosimeters (i.e., double badge). One dosimeter is worn under the lead apron and one dosimeter above the lead apron. Clearly the use of two dosimeters can provide an improved estimate of the operator effective dose. Pregnant workers can continue to work in an IR environment but will need to wear a dosimeter under their lead apron to monitor the fetal exposure. Current US occupational worker dose limits are 50 mSv/year, and a worker fetal dose limits is 0.5 mSv/month.

Operators reduce their exposures by limiting the amount of radiation used to perform IR procedures. Collimating x-ray beams to the anatomy of interest reduces patient and operator doses, reduces scatter, and increases image contrast. Increasing the distance from the patient reduces operator doses due to the inverse square law. A doubling of the distance reduces the operator dose to a quarter. For the same reason, the operator should also stand at the image receptor side of the imaging chain when performing lateral views (Fig. 4.15). Shielding is generally the most effective way to reduce all operator doses because less than 10% will penetrate 0.5 mm of lead (Fig. 4.16).

Eye doses are of concern in IR; several studies have demonstrated increased precataract eye lens deformations in interventional radiologists. The most accurate way to monitor eye lens doses is placing a dosimeter close to the eye on the frame of the operator's glasses. A dosimeter that is worn above a lead apron at the level of the collar will most likely overestimate eye lens doses. Current IR operator eye lens doses are kept well below the US regulatory eye lens dose limit (150 mSv/year). However, keeping eye lens dose limits below the dose limit recommended by the International Commission on Radiological Protection (ICRP) in 2011 (i.e., 20 mSv/year) is more difficult to achieve.

Annual effective doses to the most highly exposed operators in IR are currently approximately 5 mSv. This dose is comparable to the most highly exposed operators at nuclear power plants. It is interesting to note that the additional radiation dose received by all air crew from flying at 30,000 feet each year (where cosmic background radiation is high) is also

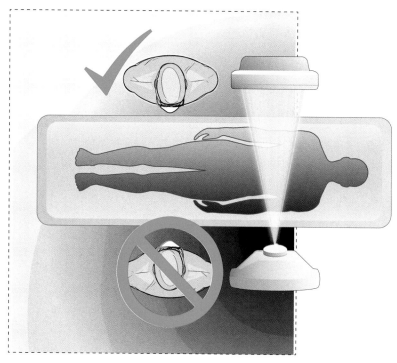

Fig. 4.15 Operators should stand at the image receptor end of the imaging chain rather than the x-ray tube end to minimize their radiation doses. Operator dose is primarily radiation that is scattered out of the patient, with very little due to x-ray tube leakage.

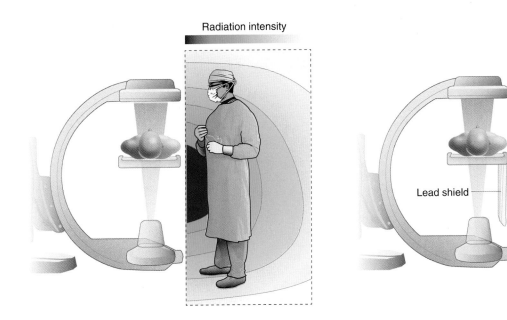

Radiation intensity

Radiation intensity

Lead shield

Fig. 4.16 Adding a lead shield can help to reduce the operator doses.

approximately 5 mSv, but air crew are generally not classified as "occupationally exposed workers." Operator doses vary depending on procedure complexity. Technologists generally receive badge readings of between 1 and 2 mSv, with lower doses to ancillary staff.

References

Huda W. Radiation protection. In: *Review of Radiological Physics*. 4th ed. Philadelphia: Wolters Kluwer; 2016:81-83.

Huda W. Interventional radiology. In: *Review of Radiological Physics*. 4th ed. Philadelphia: Wolters Kluwer; 2016:135–137.

Miller DL, Balter S, Schueler BA, et al. Clinical radiation management for fluoroscopically guided interventional procedures. *Radiology*. 2010;257(2):321–332.

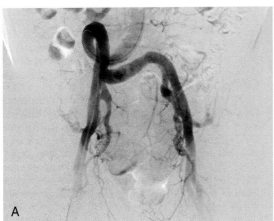
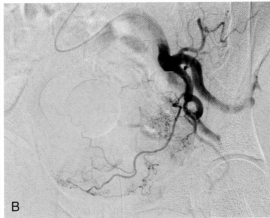

A

B

Fig. 4.17

1. Where is the interventional reference point (IRP) located for an IR system?
   A. 15 cm from the image receptor
   B. 15 cm from the isocenter toward receptor
   C. 0 cm from isocenter (i.e., at isocenter)
   D. 15 cm from the isocenter toward the source

2. What would be the most likely display value of IRP K_{air} (mGy) after 1 minute of fluoroscopy during a procedure in an IR suite?
   A. 1
   B. 10
   C. 100
   D. 1000

3. What would be the most likely display value of IRP K_{air} (mGy) after 10 photospot images have been acquired during a procedure in an IR suite?
   A. 1
   B. 10
   C. 100
   D. 1000

4. What is the most likely median IRP K_{air} value (mGy) for patients undergoing IR procedures?
   A. 10
   B. 100
   C. 1000
   D. 10,000

CASE 4.8

Air Kerma at the Interventional Reference Point

Fig. 4.17 (A and B) Digital subtraction angiograms obtained during embolization of a large acutely bleeding bladder mass.

1. **D.** The IRP is located 15 cm closer to the x-ray tube than the isocenter around which the gantry rotates. Doses displayed on the monitor to the operator are generally based on measurements predicted by the vendor free in air (with no patient backscatter) and independent of the presence of any patient table (no table attenuation taken into account).

2. **B.** For a normal-sized patient (23 cm), the most likely entrance K_{air} rate is 10 mGy/min, which is a value that is based on the system being operated in the large (normal) FOV with a pulse rate of 15 frames per second.

3. **B.** A typical photospot image of a normal-sized adult patient (23 cm) would likely use an entrance K_{air} of 1 mGy. Ten photospot images have the same incident K_{air} as 1 minute of fluoroscopy on a normal-sized patient (10 mGy).

4. **C.** Approximately half of IR procedures at US academic medical centers will be less than 1 Gy (1000 mGy), and the other half will be greater than 1 Gy. Values of IRP K_{air} of 100 mGy are more representative of GI and GU examinations than of IR, and radiographic examinations would rarely exceed 10 mGy. An IRP K_{air} of 10 Gy would virtually guarantee a deterministic skin burn.

Comment

IR gantries rotate around the gantry isocenter, which is generally located close to the center of the patient. An IRP has been defined as a point that is 15 cm closer to the focal spot than the system isocenter. For a 24-cm-thick patient, with the imaging chain isocenter selected to be the patient center,

the IRP would at a point 3 cm from the patient entrance as one moves toward the focal spot (see figure). Measurements based on this location would be conservative, because the patient entrance K_{air} would be slightly lower because of the inverse square law when the patient is further away from the x-ray tube than the IRP.

Manufacturers display values of the radiation intensity in Gray (Gy) at this IRP location (i.e., IRP K_{air}) and also provide a summary at the end of each procedure (Fig. 4.18). The value of IRP K_{air} generated by the vendor uses selected techniques (kV and mA) and the distance from the focal spot to the IRP. Values of IRP K_{air} exclude patient backscatter and attenuation of the table and do not account for any use of multiple fields. It has been estimated that half of patients undergoing IR procedures will have an IRP K_{air} that exceeds 1 Gy, and 5% will have values that exceed 5 Gy. It should be noted that patient doses at academic medical centers are likely higher than average because they perform relatively complex IR procedures and also because of the use of trainees to perform these procedures.

Radiation intensity for any given diagnostic task can be modified by the operator. In normal mode of operation, and for an average-sized patient, the entrance K_{air} rate during fluoroscopy would likely be 10 mGy/min and the entrance air for each photospot image would 1 mGy. Low-dose mode can normally be selected for easy diagnostic tasks such as detecting the location of a swallowed coin in a pediatric patient. In low-dose mode, the radiation intensity incident on a patient would likely be 50% lower than in normal mode. High-dose mode can normally be selected for more demanding diagnostic tasks such as detecting subtle low-contrast lesions. In high-dose mode, the radiation intensity incident on a patient would likely be 50% higher than in normal mode. Radiation output can also be influenced by changing the pulse rate in fluoroscopy, which can be reduced when temporal resolution is not important.

References

Huda W. Interventional radiology. In: *Review of Radiological Physics.* 4th ed. Philadelphia: Wolters Kluwer; 2016:135–137.

Miller DL, Balter S, Noonan PT, Georgia JD. Minimizing radiation-induced skin injury in interventional radiology procedures. *Radiology.* 2002;225(2):329–336.

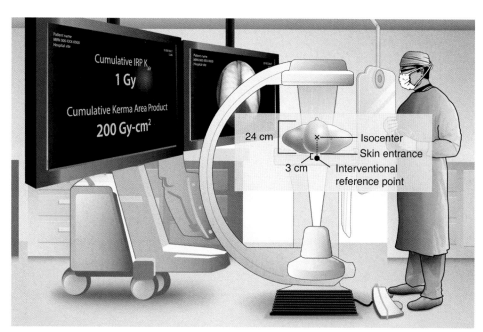

Fig. 4.18 Operators are provided with information about the interventional reference point (IRP) radiation intensity (Gy) and kerma area product (Gy-cm²). In this example, the IRP is 3 cm below the patient entrance and will therefore (slightly) overestimate the patient entrance air kerma.

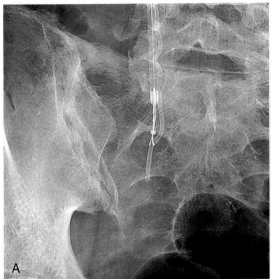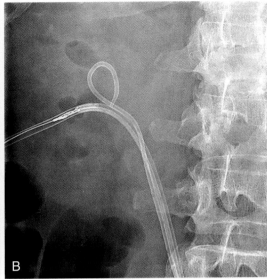

Fig. 4.19

1. How does moving the K_{air} measurement location closer to the focal spot affect the kerma area product (KAP)?
   A. Increases
   B. Does not affect
   C. Reduces

2. What is the KAP (Gy-cm²) for an IR photospot image if the incident K_{air} is 1 mGy and the corresponding x-ray beam area is 1000 cm²?
   A. 0.01
   B. 0.1
   C. 1
   D. 10

3. What is the most likely median value of KAP (Gy-cm²) for patients undergoing IR procedures?
   A. 1
   B. 10
   C. 100
   D. 1000

4. What is the most likely fraction of radiation delivered to patients from fluoroscopy in IR procedures?
   A. <⅓
   B. ⅓
   C. ⅔
   D. >⅔

CASE 4.9

Kerma Area Product in Interventional Radiology

Fig. 4.19 Photospot images obtained during double J ureteral stent retrieval.

1. **B.** If the distance from the source is doubled, the K_{air} value will be reduced to one-quarter because of the inverse square law. However, the x-ray beam area will be quadrupled. As a result, the KAP that is incident on the patient is independent of the measurement location.

2. **C.** The K_{air} can be thought of as the number of photons per square millimeter (i.e., not the total number of photons). To obtain a quantity that is directly related to the total number of photons (i.e., total energy), the intensity has to be multiplied by the corresponding area. If the intensity is 1 mGy and the beam area is 1000 cm^2, the incident KAP is 1 $Gy\text{-}cm^2$.

3. **C.** In IR procedures, the total amount of radiation used would likely be 100 times more than in a simple radiographic examination (1 $Gy\text{-}cm^2$) and 10 times more than a GI/GU procedure (10 $Gy\text{-}cm^2$). The median IR procedure is 200 $Gy\text{-}cm^2$, which is close to 100 $Gy\text{-}cm^2$. Very few IR procedures will have KAP values as low as 10 $Gy\text{-}cm^2$ or as high as 1000 $Gy\text{-}cm^2$.

4. **B.** When no other information is available, most medical physicists would assume that fluoroscopy is responsible for one-third of the incident radiation, but this is simply an "informed guess."

Comment

When the distance from the source is reduced, the K_{air} value increases according to the inverse square law. However, the x-ray beam area will be reduced by exactly the same amount as the increase in the K_{air}, which is a requirement of the principle of the conservation of energy. This is because the KAP is essentially a measure of the total energy in the beam, which has to be the same irrespective of where a measurement is made. When KAP is measured using an ionization chamber, this is located in the x-ray tube housing (collimator) where the area will be small but the K_{ai} value will be high. Because K_{air} is measured in Gy, and beam areas are measured in square centimeter, the unit of KAP is the $Gy\text{-}cm^2$.

KAP values can be computed from the IRP K_{air} value (see earlier) and corresponding x-ray beam areas in both fluoroscopy and photospot imaging ($Gy\text{-}cm^2$). Cumulative IRP K_{air} values will always include contributions from both fluoroscopy and photospot imaging, and the latter is generally greater than the former. Historically, the fluoroscopy time has always been recorded in virtually all IR procedures. It is only relatively recently that measures such as the cumulative IRP K_{air} and total KAP have become available. In the absence of any specific information, it is customary to assume that approximately one-third of the total radiation delivered to patients is from the fluoroscopy component of any IR procedure.

The median KAP in an academic medical center is estimated to be approximately 200 $Gy\text{-}cm^2$. Approximately 50% of KAP values for IR procedures at an academic medical center would be between 100 and 300 $Gy\text{-}cm^2$. A simple IR procedure such as a nephrostomy obstruction would be likely to have a KAP of approximately 25 $Gy\text{-}cm^2$. By contrast, typical values for complex procedures such as spine arteriovenous malformation embolization can likely have a KAP greater than 500 $Gy\text{-}cm^2$. IR has KAP values that are approximately two orders of magnitude higher than radiography and an order of magnitude higher than in fluoroscopy-guided procedures performed in GI/GU examinations.

The KAP can be reduced by using increased collimation or by selecting a smaller FOV (Fig. 4.20). These two modes of limiting the x-ray beam area do not affect the KAP in the same way (see next figure). When a smaller FOV is selected by the operator, the radiation intensity is increased automatically by the vendor to obtain the desired level of image quality (mottle). However, when collimation is increased by the operator, the radiation intensity is unchanged. For the same changes in x-ray beam area, the KAP is reduced more by collimation than by selecting a smaller FOV.

References

Huda W. Interventional radiology. In: *Review of Radiological Physics*. 4th ed. Philadelphia: Wolters Kluwer; 2016:137–141.

Huda W. Kerma-area product in diagnostic radiology. *AJR Am J Roentgenol.* 2014;20(3):W565–W569.

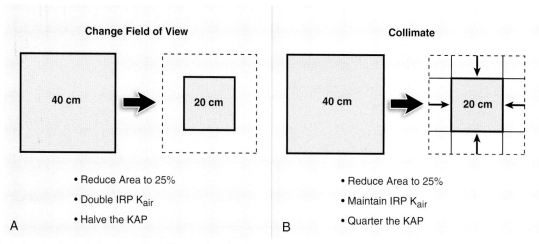

Fig. 4.20 Reducing the field of view from 40 to 20 cm (A) and collimating from 40 to 20 cm (B) both reduce the area by 25%. However, reducing the field of view from 40 to 20 cm reduces the kerma area product by 50%, whereas collimating from 40 to 20 cm reduces the KAP by 75%.

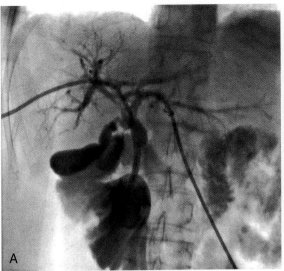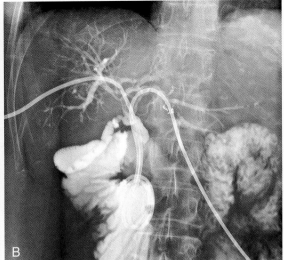

Fig. 4.21

1. Which values should always be recorded in the medical chart of any patient who undergoes an IR procedure?
   A. IRP K_{air} (Gy)
   B. KAP (Gy-cm^2)
   C. A and B
   D. Neither A nor B

2. Which patient risk is best predicted by IRP K_{air}?
   A. Carcinogenesis
   B. Hereditary
   C. Skin burn
   D. A and B

3. Which patient risk is best predicted by the KAP?
   A. Skin burn
   B. Cataracts
   C. Carcinogenesis
   D. A and B

4. Which is subject to a regulatory limit in the United States?
   A. Fluoroscopy K_{air} rate (mGy/min)
   B. Fluoroscopy time (minutes)
   C. IRP K_{air} (Gy)
   D. KAP (Gy-cm^2)

CASE 4.10

Interpreting Interventional Reference Point Air Kerma and Kerma Area Product

Fig. 4.21 Percutaneous transhepatic cholangiogram in a patient with primary sclerosing cholangitis showing a fluoroscopic image (A) and photospot (B). Subtle findings are much easier to see on the photospot image because the radiation dose in a photospot image is 100 times higher than in a fluoroscopy frame.

1. **C.** Medical physicists are trained to convert IRP K_{air} and KAP into patient doses and their corresponding radiation risks. It is therefore of paramount importance that IRP K_{air} and KAP data are always available in the patient chart for any future dose and/or radiation risk assessment purposes.

2. **C.** IRP K_{air} is directly related to the chance of a radiation burn. KAP is related to the total energy deposited in the patient and thus directly related to the total stochastic radiation risks (genetic effects + carcinogenesis).

3. **C.** The incident KAP is directly related to organ doses, and the latter are used to estimate the carcinogenic radiation risk. Cataract risks require an estimate of the eye lens dose, and radiation burns require an estimate of the peak skin dose (PSD), both of which are directly related to IRP K_{air}.

4. **A.** In fluoroscopy imaging, which is generally not diagnostic, there is currently a limit of approximately 100 mGy/min in normal mode. There are no regulatory limits on fluoroscopy time, IRP K_{air}, or total KAP.

Comment

The radiologist is generally responsible for the amount and quality of the radiation incident on the patient. Medical physicists are able to convert these incident radiation quantities into patient doses, so it is essential that IRP K_{air} and KAP data are always available in the patient chart (Fig. 4.22). Medical physicists can use such IRP K_{air} and KAP data, together with information on the procedure being performed, to compute *any* dose and/or radiation risk.

IRP K_{air} is the radiation intensity incident on the patient (photons/mm^2) and is directly related to the chance of a radiation burn. Converting IRP K_{air} into an accurate assessment of

PSD is very difficult and can be performed only by a trained medical physicist. In the absence of any additional information, a radiologist should always assume that the IRP K_{air} value is the nominal PSD. For values of IRP K_{air} less than 2 Gy, skin burns are unlikely. However, at greater than 2 Gy, radiation burns will be possible, and operators should be advised to take steps to limit additional exposures. For example, the angulation may be changed to limit the dose to the initial site of the entrance beam.

The incident KAP is directly related to the total energy incident on the patient, two-thirds of which will be deposited in the patient. For most patients, two-thirds of the incident energy will be absorbed by the patient. For a given patient and type of examination, the KAP will be directly proportional to the stochastic radiation risk, primarily carcinogenesis. Radiation risks will increase when more sensitive regions are irradiated (e.g., chest) and decrease when less sensitive regions are irradiated (e.g., head and extremities). AP projections will result in higher risks than will PA and lateral projections because most radiosensitive organs are located frontally (e.g., breast). For a given KAP, patient risks will increase as the patient size is reduced, and vice versa.

In fluoroscopy imaging there is a limit of approximately 100 mGy/min in normal mode. Current regulations make use of non-SI units, and the entrance exposure limit (10 roentgens/min) translates into 87.6 mGy, which can be safely approximate to 100 mGy/min. This entrance K_{air} rate can be increased to approximately 200 mGy/min when high level control (HLC) is activated in large patients. With HLC active, there is a requirement for audible and/or visual alarms to operate because dose adds up quickly in this mode. Fluoroscopy images have a lot of mottle and are therefore rarely diagnostic, so this limit is designed to minimize the likelihood of radiation burns from excessive fluoroscopy exposures. When diagnostic images are being acquired, there are no patient dose limits, and it is the radiologist who is responsible for deciding "how much radiation" should be used.

References

Huda W. Interventional radiology. In: *Review of Radiological Physics*. 4th ed. Philadelphia: Wolters Kluwer; 2016:135–141.

Miller DL, Balter S, Cole PE, et al. Radiation doses in interventional radiology procedures: the RAD-IR study part I: overall measures of dose. *J Vasc Interv Radiol*. 2003;14(6):711–727.

Miller DL, Balter S, Cole PE, et al. Radiation doses in interventional radiology procedures: the RAD-IR study part II: skin dose. *J Vasc Interv Radiol*. 2003;14(8):977–990.

IR Dose Summary Report

1. Photospot Acquisitions

Run (#)	IRP K_{air} (Gy)	KAP (Gy-cm^2)
1	0.09	18
2	0.07	14
...	...	...
12	0.08	16

Number of Digital Acquisitions	12
Total Photospot Radiation IRP K_{air} (Gy)	1
Total Photospot KAP (Gy-cm^2)	200

2. Fluoroscopy

Fluoroscopy IRP K_{air} (Gy)	0.5
Fluorscopy Photospot KAP (Gy-cm^2)	100

3. Photospot Acquisitions and Fluoroscopy

Total Examination IRP K_{air} (Gy)	1.5
Total Examination KAP (Gy-cm^2)	300

Fig. 4.22 Total interventional reference point K_{air} is directly related to the peak skin dose, which is used to estimate the likelihood of a radiation-induced skin burn. Total examination kerma area product *(KAP)* is related to the effective dose and can be used to estimate the stochastic (cancer) risk when patient demographics are taken into account.

CASE 5.1

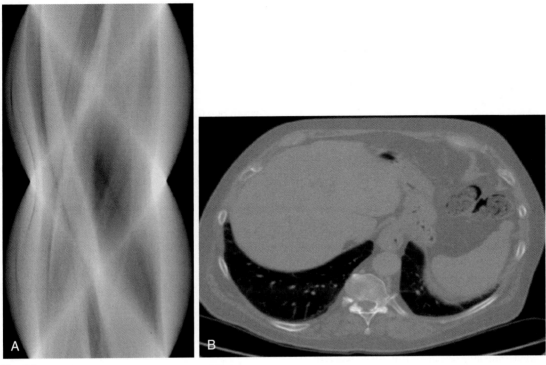

Fig. 5.1

1. Which generation CT scanner is typically used clinically?
   A. 3rd
   B. 4th
   C. 5th
   D. All (3rd, 4th, and 5th)

2. Which will affect the cone angle on a multidetector CT scanner?
   A. Number of detector arrays
   B. Width of each detector array
   C. Depth of each detector element
   D. A and B

3. What is the typical geometrical magnification of a lesion at scanner isocenter for a single projection (angle) of the CT?
   A. 1.1
   B. 1.5
   C. 2
   D. 3

4. What do CT scanners actually measure?
   A. Linear attenuation (μ)
   B. Physical density (ϱ)
   C. Mass attenuation (μ/ϱ)
   D. B and C

CASE 5.1

Image Data Acquisition

Fig. 5.1 Sinogram showing projections horizontally and x-ray tube angles vertically (A) with corresponding axial CT image of the upper abdomen (B).

1. **A.** The acquisition geometry of the projection data on CT scanners is referred to as the "generation." Virtually all current clinical CT scanners are third generation, where the x-ray tube and detector array rotate about the CT scanner isocenter. Higher generations exist but are not clinically used for reasons related to performance, cost, and durability.

2. **D.** The cone beam angle is determined by the total width of the detector array. A modern 64-slice CT scanner has 64 detector arrays, each of which typically has a width of 0.625 mm. The total beam width (also called collimation or coverage) is thus 40 mm (64 × 0.625). If the number of detector arrays and/or the width of each individual detector array increases, so does the cone beam angle, and vice versa.

3. **C.** The distance from the x-ray tube to the CT scanner isocenter is approximately half the distance between the x-ray tube focal spot and the detector array. Consequently, the geometrical magnification at the CT isocenter is approximately a factor of two for each projection.

4. **A.** When an x-ray beam passes through a patient, x-rays are lost from the beam because of absorption (photoelectric effect) and scatter (Compton and coherent). Consequently, the physical quantity that is being measured is the total linear attenuation in each pixel. CT images are two-dimensional arrays of linear attenuation coefficients that have been scaled for easy storage and interpretation (Hounsfield units [HU]).

Comment

CT scanner "generation" defines the acquisition geometry. The first-generation scanner used by Godfrey Hounsfield in 1973 acquired individual rays using a single movable detector along a projection and then rotated the x-ray tube gantry before acquiring the next projection. In a third-generation scanner there is a rotating x-ray tube aligned to a rotating detector that is used by virtually all current clinical CT systems. The gantry (i.e., image receptor, electronics, and x-ray tube) spins in a circle about a fixed point halfway between the tube and image receptor (i.e.,

isocenter), with the fastest rotation times being approximately 0.3 seconds. As the gantry spins, x-ray images (called projections) are taken at each angle. On a clinical CT, the distance from the x-ray tube focus to isocenter is approximately 60 cm, resulting in a magnification close to a factor of two. The presence of a significant air gap helps to minimize the amount of scatter that reaches the detector array.

An x-ray fan beam angle of 50 degrees provides a field of view (FOV) of 50 cm diameter. Up to 1000 detectors are arranged in an arc 120 cm from the focal spot. Central rays of projections acquired at 0 and 180 degrees are offset by half a detector width to improve data sampling and spatial resolution. Typical multidetector CT scanners are capable of generating 64 axial slices in a single rotation and are known as "64 slice" systems, although systems exist with 256 or more slices. With a detector width of 0.625 mm, the total beam width will be 40 mm in the craniocaudal direction. A 64-slice CT scanner covers a 40-cm long abdomen/pelvic scan in 10 x-ray tube rotations. In general, the higher the slice number, the faster the scan is obtained. The width of the detector array defines the cone beam angle. At large cone beam angles (cone beam CT), anatomical coverage is reduced, scatter increases, and the image reconstruction artifacts become of increasing importance.

At each location of the x-ray tube, an individual detector will measure the x-ray transmission through the patient, which is known as a ray (Fig. 5.2). A single ray measures the total attenuation in the patient along a line passing through the patient from the focus to the x-ray detector and generates one data point. Consequently, all CT images show the two-dimensional pattern of x-ray attenuation coefficients within the selected slice through a patient. All the rays obtained from the detector array at a given tube angular position is known as a projection. For clinical CT scanners, each projection will have 1000 individual data points for an array of 1000 individual detectors (one for each ray). To produce a clinical CT image, approximately 1000 projections will be acquired in one rotation of the x-ray tube. A plot of projection as a function of x-ray tube angle (0 to 360 degrees) is known as a sinogram, which is the input into a reconstruction algorithm to create the final CT image.

References

Dalrymple NC, Prasad SR, Freckleton MW, et al. Introduction to the language of three-dimensional imaging with multidetector CT. *Radiographics.* 2005;25(5):1409–1428.

Huda W. CT I. In: *Review of Radiological Physics.* 4th ed. Philadelphia: Wolters Kluwer; 2016:147–150.

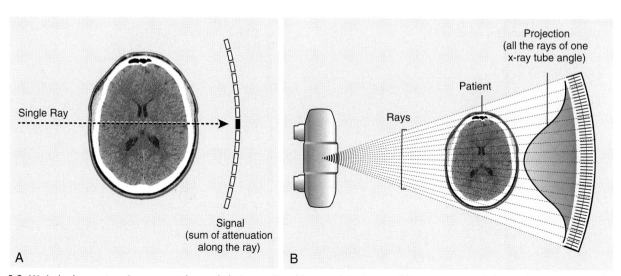

Fig. 5.2 (A) A single ray at a given x-ray tube angle being captured by one detector providing a measure of the total attenuation within the patient. The sum of all of the rays at one x-ray tube angle is termed a projection (B).

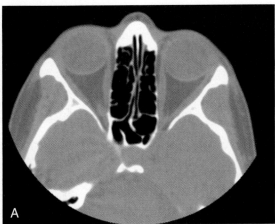 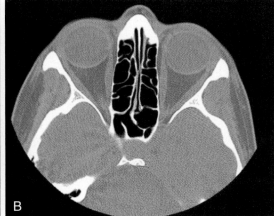

A

B

Fig. 5.3

1. How many units of attenuation are allocated to each pixel when the total attenuation of 500 units is back-projected over 500 pixels?
   A. <1
   B. 1
   C. >1
   D. Indeterminate

2. Which CT reconstruction kernel (mathematical reconstruction filter) would likely be best to obtain CT images of liver metastases?
   A. Detail
   B. Lung
   C. Soft tissue
   D. Bone

3. What image quality metrics are "traded" when choosing a reconstruction kernel in filtered back projection (FBP) CT imaging?
   A. Contrast and noise
   B. Contrast and resolution
   C. Noise and resolution
   D. Artifact and contrast

4. Which is most likely to remain unchanged when FBP is replaced by an iterative reconstruction (IR) CT algorithm?
   A. Lesion contrast
   B. Image noise
   C. Noise texture
   D. Image artifacts

CASE 5.2

Image Reconstruction

Fig. 5.3 CT images at the level of the orbits produced using a soft tissue filter (A) compared with a bone reconstruction kernel (B).

1. **B.** If 500 units are allocated to 500 pixels, then using back projection would allocate exactly 1 unit of attenuation to each pixel. When images are generated in this manner, the reconstruction is not accurate and results in blurry images. In practice, projections are first filtered and then are subsequently back projected.

2. **C.** Detail, lung, and bone reconstruction kernels would produce images that have high resolution (sharp) but also include a lot of random image noise. A soft tissue (AKA standard) kernel would produce an image that was smoother but lacking in fine detail. Because liver metastases are relatively low-contrast lesions with HU values similar to liver tissues, a soft tissue kernel is appropriate, which will increase the lesion contrast-to-noise ratio.

3. **C.** In FBP reconstruction, a range of reconstruction kernels are offered on all commercial CT scanners that trade "resolution performance" with "image noise." Detection of small high-contrast lesions (e.g., lung nodule) would use a kernel with excellent resolution properties because noise is irrelevant when imaging a high-contrast lesion. For large low-contrast lesions, resolution performance is not important, but the amount of random noise should be minimized by using a soft tissue kernel.

4. **A.** The choice of image reconstruction algorithm should not affect the lesion contrast in CT imaging. A fat nodule (HU = −100) in a fluid-like background (HU = 0) will always have a contrast of 100 HU between these two materials. Use of IR algorithms would likely reduce the amount of image noise and the image artifacts (e.g., streak artifacts resulting from photon starvation) but not contrast.

Comment

Modern scanners use FBP algorithms to reconstruct CT images. Acquired projections are first multiplied by a reconstruction kernel to generate filtered projections. Filtered projections are then back-projected so that the total attenuation value in the projection is allocated equally to every pixel along ray traced through the patient (Fig. 5.4). This can be visualized as taking the attenuation and "smearing" it back along the ray. FBP is a very fast process capable of reconstructing millions of data points in less than 1 second using array processors.

CT scanners typically offer six or seven reconstruction kernels. Vendors may use descriptive names such as "bone" or "soft tissue" or more technical names such as "H30," which refers to a head kernel. Each kernel defines a resolution and noise performance value, but when resolution improves, noise increases and vice versa. Detection of large low-contrast liver lesions would use a kernel that reduces noise at the price of reduced resolution performance. Alternatively, imaging a bone/air interface would use kernels that offer the best resolution but also produce noisy images. This is of little importance

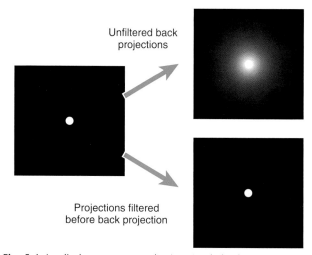

Unfiltered back projections

Projections filtered before back projection

Fig. 5.4 A cylinder reconstructed using simple back projection (*upper right*) would show a smeared out version of the scanned object. When the projections are multiplied by a mathematical filter and subsequently back projected (i.e., filtered back projection), it is possible to obtain a perfect reconstruction of the cylinder (*lower right*).

because the intrinsic air/bone contrast is very high. It is always possible to reconstruct the original x-ray transmission data (i.e., acquired sonogram) using any number of kernels without extra dose to the patient, but this would require a radiologist viewing multiple series of images.

IR algorithms commence with an assumed image which permits multiple projections to be obtained. These projections can be compared with the acquired projections from the patient and (any) differences noted. Improvement factors, obtained from the differences between a calculated and acquired (patient) projection, are then applied to improve the reconstructed image. This process is repeated (iterated) until the agreement between the projections derived from the reconstructed image and the corresponding acquired projections is acceptable. IR can help to reduce random noise and streak artifacts. Noise texture differences make any direct comparison between IR and FBP reconstruction modes problematic, and the measurement of IR image quality is generally difficult.

IR algorithms are computationally and time intensive but can offer improved image quality. IR algorithms can also be used to reduce the patient's radiation dose because a smaller amount of radiation is needed to generate an image with the same level of diagnostic performance. The initial IR algorithms used clinically were termed statistical techniques, and an IR images could be combined with FBP (e.g., 50% IR + 50% FBP). Recently developed model-based IR algorithms improve the fidelity of CT images by incorporating specifics of the CT scanner geometry. This is very computationally intensive but can markedly reduce random noise and improve overall image fidelity. As computers become more powerful and calculation time decreases, IR will likely become the default reconstruction method for CT.

References

Bushberg JT, Seibert JA, Leidholt Jr EM, Boone JM. Computed tomography. In: *The Essential Physics of Medical Imaging*. 3rd ed. Philadelphia: Wolters Kluwer; 2012:350–358.

Huda W. CT I. In: *Review of Radiological Physics*. 4th ed. Philadelphia: Wolters Kluwer; 2016:151–152.

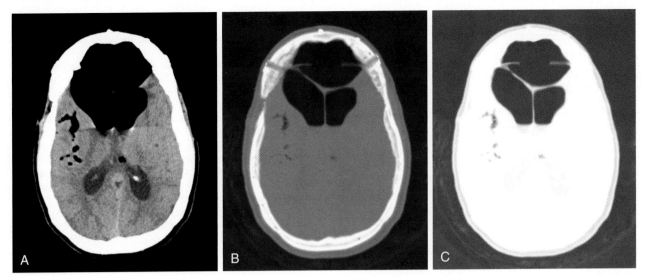

Fig. 5.5

1. What is the fraction of x-rays transmitted through 1 cm of a material with an attenuation coefficient of 0.1 cm^{-1}?
   A. 0.01
   B. 0.09
   C. 0.1
   D 0.9

2. What is the x-ray attenuation of gray matter compared to white matter in HU?
   A. −20
   B. −5
   C. +5
   D. +20

3. Which x-ray tube voltage will result in the highest HU value for dilute iodine in the aorta?
   A. 80
   B. 120
   C. 140
   D. HU is kilovoltage independent

4. Which most likely increases when x-ray photon energy in CT imaging is increased?
   A. Patient transmission
   B. Linear attenuation
   C. Tissue HU
   D. Lesion contrast

CASE 5.3

Hounsfield Units

Fig. 5.5 CT of the head in a patient with pneumocephalus secondary to cerebrospinal fluid leak after meningioma resection. Using different window and level settings, radiologists can evaluate tissues and structures of varying Hounsfield units such as brain (A), bone (B), and air (C).

1. **D.** An attenuation coefficient of 0.1 cm^{-1} means that 10% of the incident photons will be lost from the x-ray beam when traveling a distance of 1 cm so that the fraction that is transmitted is 0.9. In the next centimeter, 10% of the incident number of photons (now 0.9) will be attenuated (i.e., 0.09), so the total fraction emerging after traveling through 2 cm will be 0.81, and so on.

2. **C.** The attenuation of gray matter (~45 HU) is slightly higher than that of white matter (~40 HU), which is attributed to a slighter higher density of gray matter. A difference in HU of 5 HU means that the attenuation of gray matter is approximately 0.5% higher than that of white matter.

3. **A.** As the average photon energy increases, the difference in x-ray attenuation between a high Z material and that of water is reduced, and vice versa. Consequently, using an x-ray tube voltage of 80 kV will result in the highest HU value for dilute iodine, and using a voltage of 140 kV will result in the lowest HU value for dilute iodine.

4. **A.** Increasing x-ray photon energy will generally reduce linear attenuation coefficients, tissue HU, and lesion contrast. However, an increase in energy will increase the percentage of x-ray photons transmitted through a patient. Transmission of photons through an adult patient in any CT scan is always low (<1%), with most of the incident energy at any voltage being attenuated by the patient.

Comment

CT images are two-dimensional maps of the tissue attenuation coefficients. In clinical CT, attenuation coefficients (μ) are expressed as CT numbers or HU. They are named after Sir Godfrey Hounsfield, who was a clinical CT pioneer in the early 1970s. The HU of a material is the attenuation of the material when compared with that of water at the same photon energy.

A positive HU shows attenuation that is greater than water, whereas a negative HU shows that attenuation is less than that of water.

By definition, the HU value for water is 0, and that of air is −1000 (Fig. 5.6). The HU for fat is approximately −100, for soft tissue is approximately 50, and for bone is very high and generally greater than 1000. A lesion with a HU value of +10 will attenuate 1% more than water, and a value of +20 will attenuate 2% more than water. Differences between gray and white matter are approximately 5 HU, showing that these two tissues differ in x-ray attenuation by only approximately 0.5%. This illustrates the power of CT, which can distinguish tissues that differ in x-ray attenuation properties of less than 1%! Because of this ability, CT can be used to characterize tissue. For example, a lesion with HU of −100 indicates that the lesion consists of fat.

Because attenuation coefficients depend on photon energy, HU values also depend on the photon energy. Principle factors that affect photon energy in CT are the x-ray tube voltage and the corresponding total x-ray beam filtration. Positive HU values increase as photon energy decreases, whereas negative HU values become more negative. Soft tissue at 40 keV has an HU value of 60, but this is reduced to 50 HU when the photon energy is doubled to 80 keV. Fat has an HU of −150 at 40 keV but is only −70 HU when the photon energy is doubled to 80 keV.

HU changes with photon energy are strongly dependent on the material atomic number (Z). HU changes with energy will be very large for iodine ($Z = 53$), which has an atomic number very different to that of water ($Z = 7.5$). For example, diluted iodine with an HU of 400 at 80 kV would likely have a markedly reduced HU (e.g., 100) at 140 kV. It is important that all angiography is performed at the lowest possible voltage to achieve the highest image contrast of iodine. In angiography with larger patients, the tube voltage will likely need to be increased to achieve a satisfactory penetration (ensuring enough radiation reaches the detector), even though this increase in tube voltage will reduce the amount of vasculature contrast.

References

Huda W. CT I. In: *Review of Radiological Physics*. 4th ed. Philadelphia: Wolters Kluwer; 2016:152–153.

Radiological Society of North America (RSNA). RSNA/AAPM physics modules. Computed tomography: 1. CT systems. https://www.rsna.org/RSNA/AAPM_Online_Physics_Modules_.aspx. Accessed March 1, 2018.

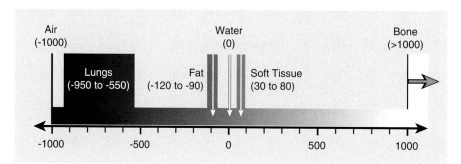

Fig. 5.6 Hounsfield units depict the attenuation of any material relative to that of water so that water is always zero and air is always −1000. Materials that attenuate less than water are negative (e.g., fat), and those that attenuate more than water are positive (e.g., soft tissues and bone).

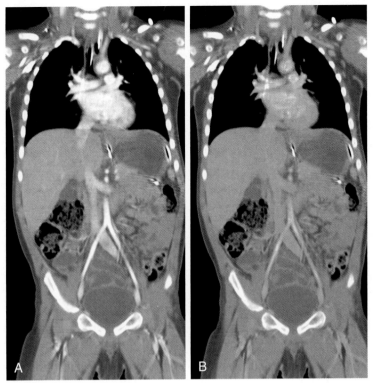

Fig. 5.7

1. What are focal spot sizes on CT x-ray tubes relative to those on conventional radiographic x-ray tubes?
   A. Bigger
   B. Comparable
   C. Smaller

2. How does the anode capacity of a CT x-ray tube compare with that of a normal R/F x-ray tube?
   A. Smaller
   B. Comparable
   C. Bigger

3. How does the CT x-ray beam half-value layer compare with an x-ray tube used for abdominal radiography (80 kV)?
   A. Smaller
   B. Comparable
   C. Higher

4. What is most likely reduced by the addition of a beam shaping (bow tie) filter in CT imaging?
   A. Patient radiation dose
   B. Image dynamic range
   C. A and B
   D. Neither A nor B

CASE 5.4

X-Ray Tubes

Fig. 5.7 CT angiography of a 2-year-old male performed at 80 kV (A) compared with a simulated image at 140 kV (B). Note the loss of contrast of the administered iodine in B.

1. **B.** Focal spots in CT are the same as those that are used on conventional radiographic and fluoroscopy units. The large focal spot is 1.2 mm, and the small focal spot is 0.6 mm. Most CT imaging would make use of the large focal spot to minimize the image acquisition time, and a small focal spot might be used to improve spatial resolution (e.g., inner ear).

2. **C.** The anode heat capacity of a CT x-ray tube is approximately 5 MJ, which is an order of magnitude higher than for a normal radiography and/or fluoroscopy x-ray tube (0.5 MJ). CT scanners use the same power as normal radiography systems (i.e., 80 or 100 kW), but the x-ray tubes are "on" for much longer periods of time, which requires much larger anode heat capacities.

3. **C.** X-ray beam qualities in CT are 5 to 10 mm aluminum, which are much higher than those in conventional abdominal radiography (e.g., 3 mm Al). CT x-ray tubes have much higher filtration than in conventional radiography and fluoroscopy. The use of high-quality x-ray beams in CT scanning minimizes beam-hardening artifacts.

4. **C.** With no beam shaping filters, the signal at the center of the head is low (high attenuation), whereas the signal at the periphery (e.g., through the ear lobe) is very high (low attenuation). Adding a beam-shaping filter (Fig. 5.8) will result in signals through all body parts to be approximately similar. Use of beam-shaping filters also reduces doses because less radiation will be incident on the patient at the patient periphery.

No Filter

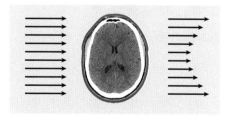

Higher dose, wide dynamic range

Beam-Shaping Filter (Bowtie)

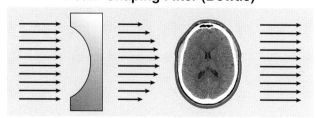

Lower dose, narrow dynamic range

Fig. 5.8 The use of a beam-shaping filter will reduce the radiation incident on the patient and thereby lower the patient dose. The transmitted radiation intensities are also more equal, which markedly reduces the detector dynamic range (i.e., maximum to minimum detected signals), which improves signal detection accuracy.

Power is supplied to the CT x-ray tube using a slip ring that allows the gantry to stay powered while rotating. High voltages range from 80 to 140 kV and are delivered through contact rings (slip rings) in the gantry. Lower voltages (70 kV) have recently been introduced for pediatric imaging, and higher voltages (150 kV) are also now available that improve penetration in obese patient imaging. Tube currents can range up to 1000 mA. X-ray output is the product of the tube current (mA) and the x-ray tube rotation time (s) and is known as the milliampere-second. The milliampere-second is a relative indicator of radiation output (higher milliampere-second means higher output), but the exact value of radiation output per milliampere-second depends on tube design as well as tube voltage.

Clinical CT scanners have a large focal spot (1.2 mm) and a small focal spot (0.6 mm). X-ray tubes are oriented so that the anode-cathode axis is perpendicular to the imaging plane (i.e., in the craniocaudal direction), which will minimize the consequences of the heel effect. Filtration on CT x-ray tubes is relatively high and can use high-Z materials, including copper and tin, to remove low-energy photons from the beam before they interact with the patient. CT requires heavy filtration to reduce artifacts that are caused by x-ray beam hardening (especially in the head).

A patient is shaped like an oval; thus for a given projection angle, much more radiation is blocked from the detector in the center of the projection (where the patient is thickest) than at the edge (where the patient is thin). The detector on a CT scanner functions best if there is relatively little variation in the radiation hitting it. A bow tie–shaped filter is used in CT to compensate for the shape of the human body (see Fig. 5.8). This filter blocks little of the beam in the center of the projection but a lot of the beam at the edge. The result is a more uniform exposure and better image quality. Furthermore, because the bow tie filter reduces the incident radiation beam intensity on thinner body parts, this will reduce the patient dose.

CT scanners normally have several bow tie filters of varying sizes, and the appropriate one must be selected for each patient and type of examination. Bow tie filters are generally available for scanning adult heads, adult bodies, and pediatric patients, as well as for cardiac imaging. To be effective, each bow tie filter must be correctly positioned relative to the patient undergoing a CT examination. This will require that the patient is accurately centered in the CT gantry isocenter. Poor image quality and unnecessary patient radiation dose are the most likely consequences of incorrect patient centering.

References

Bushberg JT, Seibert JA, Leidholdt Jr EM, Boone JM. Computed tomography. In: *The Essential Physics of Medical Imaging*. 3rd ed. Philadelphia: Wolters Kluwer; 2012:312–334.

Huda W. CT I. In: *Review of Radiological Physics*. 4th ed. Philadelphia: Wolters Kluwer; 2016:147–149.

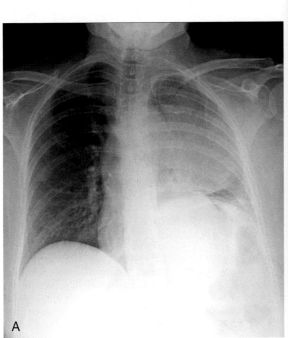

A

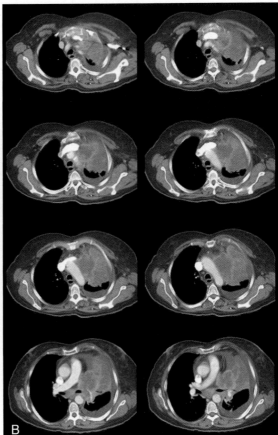

B

Fig. 5.9

1. Which is the most likely x-ray tube voltage used to scan patients on current commercial CT scanners?
   A. 80
   B. 100
   C. 120
   D. 140

2. How do the tube currents in CT x-ray tubes compare with those currently used in radiographic imaging?
   A. Smaller
   B. Comparable
   C. Larger

3. How many radiologists' homes, each requiring 3 kW, could be supplied by a typical generator used with current commercial CT scanners?
   A. 1
   B. 3
   C. 10
   D. 30

4. What is the average heat dissipation rate on a CT scanner with an x-ray tube powered by a 100-kW generator (kW)?
   A. <100
   B. 100
   C. >100

CASE 5.5

Generators and Energy Deposition

Fig. 5.9 Frontal radiograph of the chest (A) showing a large opacity in the left chest and CT of the chest (B) showing a left upper lobe mass invading the mediastinum. The total energy deposited in the anode in the CT scan (>500,000 J) is much greater than the chest x-ray exam (<500 J).

1. **C.** Most scans in current CT practice are performed at 120 kV. Lower voltages are used in pediatric CT to reduce dose and angiography to improve vascular contrast. Higher voltages (e.g., 140 kV) are used in large patients to achieve sufficient patient penetration and reduce beam-hardening artifacts.

2. **B.** Both radiography and CT imaging make use of x-ray tube currents that are several hundred milliamperes. The power limit on x-ray tubes is currently approximately 100 kW so that the maximum x-ray tube current at a nominal tube voltage of 100 kV would be 1000 mA. The major difference between radiography and CT is that exposure times in radiography are short (i.e., milliseconds), whereas those in CT are much longer (i.e., seconds).

3. **D.** A typical home in North America uses approximately 3 kW, which is approximately 30 times less than that used in most radiography and CT systems. Although CT uses high power levels, these are only on for relatively short periods of time, a few seconds or so for each patient scan, so that energy costs will be relatively modest.

4. **A.** The rate at which heat is dissipated on a conventional CT x-ray tube is up to approximately 10 kW, which is 10 times less than the rate at which energy flows into the x-ray tube anode, where 99% of the electrical energy in x-ray tubes is converted to heat. What this means is that sooner or later the energy (heat) storage in the anode reaches its maximum limit and no more exposures can be made until the x-ray tube anode cools down.

Comment

The maximum power that is applied to a CT x-ray tube is approximately 100 kW, which is four times higher than the 25 kW in the 1980s and double the 50 kW in the 1990s. When operated at their full power rating, x-ray tubes will have 100 kJ deposited into the anode for *every second of operation*. X-ray tube anode heat capacities are generally 5 MJ, which is approximately 10 times higher than for normal radiographic and fluoroscopy x-ray tubes. When focal spot blur is minimized by using a small focal spot of 0.6 mm (e.g., inner ear), the power used in the CT scan must be reduced. Because the smaller focal spot has only 25% the area of a large focal spot, the applied power needs to be reduced to 25 kW.

Without any heat losses, an anode will become saturated in less than 1 minute of operation (100 kJ deposited each second and 5000 kJ capacity). However, heat is lost from standard CT

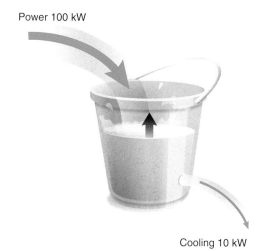

Power 100 kW

Cooling 10 kW

Fig. 5.10 The anode of a CT x-ray tube can be compared with a bucket with a small hole in it. Energy is deposited in the anode (bucket) at the rate of 100,000 J every second (100 kW) but dissipated through the "hole" at a much lower rate of 10,000 J every second (10 kW). If the tube is left on for a long period, the anode becomes saturated analogous to a full bucket.

x-ray tubes at the rate of approximately 10 kJ every second (i.e., 10 kW), 10 times less than the normal rate of heat deposition (Fig. 5.10). Because CT x-ray tubes are designed to tolerate high power for extended times and efficiently dissipate heat, these are very sophisticated pieces of engineering that can cost well over $200,000. When x-ray tubes are operated for long times, the anode will eventually become overheated. When this happens, it will generally require "several minutes" to dissipate this stored energy.

A novel x-ray tube design has a rotating envelope tube (RET) that offers high heat dissipation rates (>50 kW). This x-ray tube has substantially increased the achievable cooling rates by having the anode disk directly in contact to the oil cooling medium. The anode disk is part of the tube envelope, which results in the rotation of the entire tube with respect to the anode axis. This type of design is thus markedly different to rotating anode tubes, where the tube envelope is stationary. It should be noted that even when the x-ray tube heat dissipation rate is increased fivefold from 10 to 50 kW, tube overheating in clinical practice may still occasionally occur with heavy workloads because the anode heat loading progressively increases without having time to cool down.

References

Bushberg JT, Seibert JA, Leidholdt Jr EM, Boone JM. Computed tomography. In: *The Essential Physics of Medical Imaging*. 3rd ed. Philadelphia: Wolters Kluwer; 2012:312–324.

Huda W. CT I. In: *Review of Radiological Physics*. 4th ed. Philadelphia: Wolters Kluwer; 2016:147, 154–156.

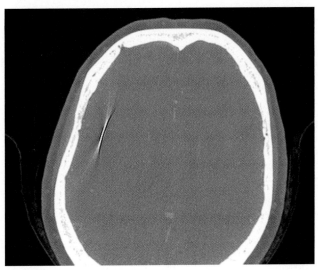

Fig. 5.11

1. What type of x-ray detectors are most likely used on current commercial CT scanners?
   A. Scintillators
   B. Photoconductors
   C. Photostimulable phosphors
   D. Gas ionization chambers

2. How many of the incident x-ray photons (%) are likely absorbed by the detector on a typical commercial CT scanner?
   A. <10
   B. 10 to 30
   C. 30 to 90
   D. >90

3. How will miscalibrated detectors appear in an image of a uniform phantom?
   A. Increased noise
   B. Blurred image
   C. Reduced contrast
   D. Ring artifacts

4. How will a long light decay time in a CT scintillator detector affect image quality in the resultant CT images?
   A. Blurry images
   B. Noisy images
   C. Reduced contrast
   D. Increased artifacts

CASE 5.6

Detectors

Fig. 5.11 CT of the head showing a detector outage. The artifact is a bright arc that traces a circle as one scrolls through the images.

1. **A.** Virtually all detectors currently used on commercial clinical CT scanners are scintillators. In the 1980s many CT scanners made use of high-pressure xenon detectors. Photoconductors are not used because these have low atomic numbers and absorb too few photons, and photostimulable phosphor technology would be impractical.

2. **D.** CT scintillator detectors generally absorb virtually all of the incident photons (>90%) in most of CT scanning. Imaging scientists describe CT imaging as being quantum mottle limited. This means that there is no technical way of reducing image mottle apart from using more photons, which will clearly increase the patient dose.

3. **D.** Miscalibrated detectors will result in numerous ring artifacts. Recalibration of the detector array is performed by the CT engineer by scanning phantoms, which is performed at all x-ray tube voltages and x-ray tube current values. Recalibration can take up to 1 to 2 hours but will invariably solve the phantom ring artifact problem.

4. **A.** If the light from a scintillator persists for a long time (relative to the rotation speed), this will result in blurry-looking images. Instantaneous emission of scintillator light would result in the sharpest possible images. On modern scanners, where rotation speeds continue to increase, the decay of light after absorption of x-rays is one of the most important technical specifications of interest to the imaging community.

Comment

CT scanners generally have antiscatter collimation in the form of thin lamellae (e.g., 0.1-mm tantalum or lead sheets) located between the detector elements. When sheets are located between detector columns (long patient axis) but not between rows of detectors, they are known as a "1D" antiscatter collimators. In multislice scanners, scatter removal is improved when shielding is placed between both columns and rows of detectors. In this case, both directions need to be focused to the x-ray source, and these are known as 2D antiscatter collimators.

In the early days of CT, many commercial systems used high-pressure xenon gas detectors. Currently, CT image receptors contain detector elements using scintillators that produce light when x-ray photons are absorbed (Fig. 5.12), which are coupled to light detectors. Most commercial detectors are proprietary and are designed to have low afterglow characteristics. Long light lag introduces blur when the x-ray tube rotation speeds increase and reduce spatial resolution performance.

Current scintillator detectors will absorb most of the incident x-ray photon energy and are therefore deemed by physicists to have a high quantum detection efficiency (fraction of incident photons absorbed). Scintillators in commercial CT scanners generally have detection efficiencies greater than 90%. Current commercial CT scanners are quantum limited, which means that quantum mottle is the only source of random noise.

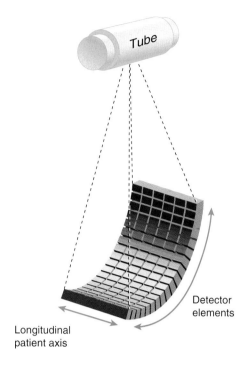

Fig. 5.12 The detector elements are arranged in an arc to collect projection data, with a number of such arcs lined up along the long patient axis to create multiple slices for rotation of the detector array. A 64-slice CT scanner with 800 detectors in each row would have approximately 50,000 individual detectors.

Given that CT detectors generally absorb all the incident photons, there are no detector improvements that are possible to reduce dose. However, image processing and advanced reconstruction techniques can be used to improve image quality, which may permit the use of less radiation without affecting diagnostic performance.

In CT image receptors, an electric signal is produced that is directly proportional to the incident radiation intensity, which is digitized and sent to the computer. It is important that the detectors are properly calibrated to give the correct brightness per unit detector dose. If detectors are out of calibration (or broken), then this will create arcs and rings in the images (see Fig. 5.11). Recalibrating or replacing the detector elements fixes this. When only a very small number of photons reach the detector, the signal will be extremely small and can even cause artifacts to appear. Whenever an obese patient has a body CT scan, the resultant image will generally be very noisy. Along the thickest projection angles, streaks of noise may be seen because so little radiation was transmitted to the CT detectors at those angles. This is referred to as "photon starvation." When x-ray signals are low, electronic noise can become significant, and x-ray detection is no longer quantum mottle limited.

References

Huda W. CT I. In: *Review of Radiological Physics*. 4th ed. Philadelphia: Wolters Kluwer; 2016:149–150.

Huda W. CT radiation exposure: an overview. *Curr Radiol Rep.* 2015;3:80.

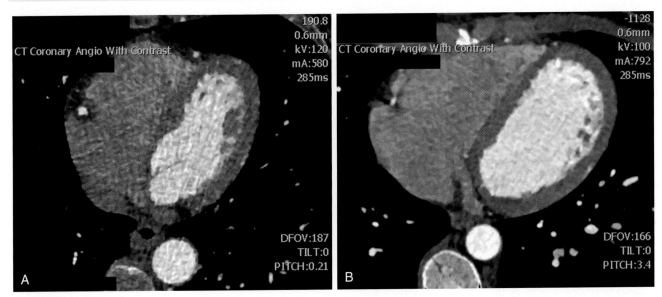

Fig. 5.13

1. When the x-ray table moves 60 mm in one 360-degree rotation of the x-ray tube, what is the pitch on a CT scanner with a 40-mm wide detector array?
   A. 0.25
   B. 0.5
   C. 1
   D. 1.5

2. What is the major benefit of helical CT compared with axial scanning?
   A. Reduced blur
   B. Faster scans
   C. Improved contrast
   D. Fewer artifacts

3. Which gives the lowest patient dose?
   A. 100 mAs and 0.5 pitch
   B. 200 mAs and 1 pitch
   C. 400 mAs and 2 pitch
   D. A, B, and C equal

4. Which likely has the lowest image noise?
   A. 200 mAs and 0.5 pitch
   B. 300 mAs and 1 pitch
   C. 400 mAs and 2 pitch
   D. A, B, and C equal

CASE 5.7

Helical Computed Tomography

Fig. 5.13 Retrospective gated cardiac CT with low pitch (A) compared with a dual source cardiac CT with high pitch (B).

1. **D.** Pitch is defined as the distance the table moves during a single 360-degree rotation of the x-ray tube divided by the corresponding x-ray beam width. In this example, the table moves 60 mm and the beam width is 40 mm, so that the CT pitch is 1.5.

2. **B.** The major benefit of helical CT scanning is that the total examination time will be reduced compared with axial CT. In axial scanning, after each 360-degree rotation of the x-ray tube the table has to be incremented before acquiring the next axial slice. This stop-and-start motion takes time.

3. **D.** Patient dose is directly proportional to the effective milliampere-second, which is defined as the actual milliampere-second divided by the pitch. The effective milliampere-second is exactly 200 in all three examples, so that the patient doses will be the same.

4. **A.** The higher the effective milliampere-second, the lower the image noise, and vice versa. The highest effective milliampere-second is option 1 (400), which will have the lowest level of noise. Option 3 has the lowest effective milliampere-second (200), which will have the highest level of noise. Option 2 has the median effective milliampere-second (300) and will also have the median level of noise.

Comment

In helical scanning, the patient continuously moves through the CT gantry as the x-ray tube rotates. The table increment distance during a 360-degree rotation of the x-ray tube divided by the nominal x-ray beam width is the key parameter in helical CT and is called CT pitch (Fig. 5.14). For example, when a table moves 40 mm per 360-degree rotation of the x-ray tube on a 64–multidetector CT (MDCT) scanner where each detector element is 0.625 mm wide (i.e., beam width 40 mm), the pitch is 1. A table movement of 60 mm corresponds to a pitch of 1.5, and table movement of 20 mm would correspond to a pitch of 0.5. The major benefit achieved by the use of helical scanning is to reduce the total CT scan time, which is important to minimize blur that results from voluntary and involuntary motion, especially in young children.

At any given craniocaudal (z-axis) position, data acquired during a helical acquisition will provide only one of the 1000 projections that are required to create a CT image. A total of 999 lines will be "missing," and these are obtained by the use of a suitable interpolation algorithm. At a given x-ray tube angle, use is made of projections that are available "upstream" and "downstream" of the z-axis location of interest. After interpolation has been performed, a complete sinogram can be generated at each z-axis location where image reconstruction is to be performed. One important advantage is that the location of a reconstructed slice is arbitrary, which permits reconstructing slices at intervals that are smaller than the nominal slice thickness (although not smaller than the individual detector element size). Interpolation in MDCT is performed automatically by the vendor and is not taken into account in developing imaging protocols.

To understand how mottle and patient dose are affected by the pitch in helical CT imaging, it is helpful to understand the concept of "effective milliampere-second." In helical scanning,

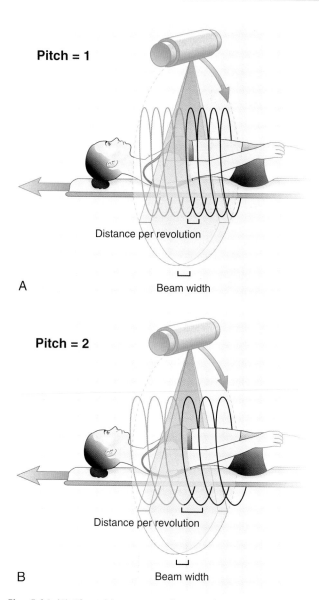

Pitch = 1

Distance per revolution

Beam width

A

Pitch = 2

Distance per revolution

Beam width

B

Fig. 5.14 (A) The table moves a distance that is exactly equal to the total CT x-ray beam width in one x-ray tube rotation so that the pitch is 1. (B) The table moves a distance that is double the total CT x-ray beam width in one x-ray tube rotation, so that the pitch is 2.

effective milliampere-second is the actual milliampere-second divided by the CT pitch. If the milliampere-second is 300 and the CT pitch is 1.5, then the effective milliampere-second is 200 (i.e., 300/1.5). The value of effective milliampere-second predicts how dose and image quality change in helical CT when technique factors are modified. Patient dose is directly proportional to the effective milliampere-second, whereas CT image noise is halved when effective milliampere-second quadruples. Thus, when the effective milliampere-second increases from 100 to 400 mAs, the patient dose will be quadrupled and the CT image noise will be halved.

References

Huda W. CT I. In: *Review of Radiological Physics*. 4th ed. Philadelphia: Wolters Kluwer; 2016:150–151.

Mahesh M. Search for isotropic resolution in CT from conventional through multiple-row detector. *Radiographics*. 2002;22(4):949–962.

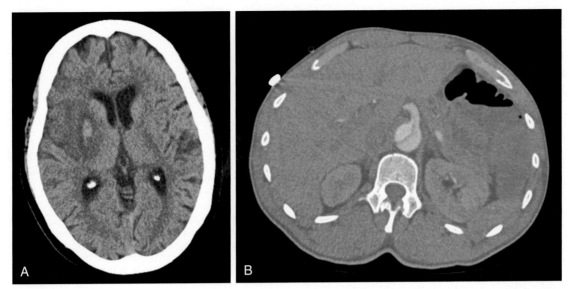

Fig. 5.15

1. What is the most likely pixel size (mm) in a CT image of an adult head?
   A. 0.1
   B. 0.2
   C. 0.5
   D. 1.0

2. How many shades of gray can be displayed using 1 byte (8 bits) to code for each pixel in a CT image?
   A. <256
   B. 256
   C. >256

3. What is the thinnest achievable slice thickness (mm) on current commercial CT scanners?
   A. 0.1
   B. 0.2
   C. 0.5
   D. 1.0

4. How much memory is required (MB) to store 20 reconstructed slices of an abdominal CT examination?
   A. 0.1
   B. 1
   C. 10
   D. 100

CASE 5.8

Computed Tomography Images

Fig. 5.15 CT head (A) with a 25-cm field of view (FOV) (0.5 mm pixel) in a patient with a hemorrhagic stroke. CT abdomen (B) with a 50-cm FOV (1.0 mm pixel) in a patient with aortic dissection.

1. **C.** The pixel size is the FOV divided by the matrix size. A typical head CT will use an FOV of 250 mm, and the matrix size in most current clinical CT is 512 × 512. Accordingly, in a head CT, each pixel will be 0.5 mm.

2. **B.** When each pixel can contain 1 bit, the bit can take on values of either 0 or 1, so that this could display two shades of gray (mathematically 2^1). Eight shades of gray can thus take on 256 shades of gray, which can be expressed mathematically as 2^8.

3. **C.** A 320-slice CT scanner has detectors that are generally 0.5 mm wide, and a 64-slice CT scanner likely has detectors with a width of 0.625. Accordingly, the correct answer is 0.5 mm, with slice widths of 0.1 and 0.2 mm being far too small and 1 mm being far too wide.

4. **C.** The matrix size of clinical CT images is 512 × 512, so that there are approximately 250,000 individual pixels. Each pixel is coded for using 2 bytes (16 bits). Accordingly, each reconstructed CT slice requires 0.5 MB, and 20 slices requires 10 MB, which is also the amount of data storage space required for a single chest x-ray.

Comment

The FOV is the diameter of the circular area that is being imaged and through which x-rays pass. The FOV in adult head CT scans is 250 mm, so the resultant pixel size for a 512 × 512 matrix is 0.5 mm (i.e., 250/512). The FOV in an adult body CT scan is likely 400 mm, so the corresponding pixel size is 0.8 mm (i.e., 400/512 mm). The physical factors that limit CT resolution are focal spot size and x-ray detector. Increasing the matrix size beyond 512 × 512 will not typically improve resolution because the inherent limits have been reached. However, reducing the matrix size to less than 512 × 512 would reduce the spatial resolution (although this is never actually done in clinical practice).

Current multislice scanners have an acquired slice thickness of approximately 0.6 mm, which is determined by the length of the detector along the craniocaudal patient axis (z-axis). Accordingly, a 360-mm long chest CT scan, acquired with a 0.6-mm slice thickness, could generate 600 contiguous images, which would take far too long to review. Use of such thin slices would also result in extremely noisy images because of the low number of photons that make up any given slice. Consequently, current clinical CT practice uses images with a nominal slice thickness of 2.5 to 5 mm, generated by combining several

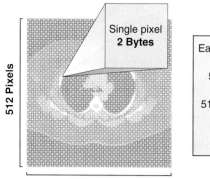
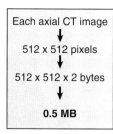

Fig. 5.16 The matrix size of CT images is generally 512 × 512, which means that there are 250,000 pixels. The Hounsfield unit in each pixel is coded using 2 bytes (1 byte = 8 bits), so the data content of a CT image is 0.5 MB (250,000 pixels × 2 bytes per pixel). By contrast, a chest x-ray requires of 10 MB of memory for storage.

"thinner" acquired slices. Very thin slices are used only in specialized scanning such as search for pulmonary embolism.

Each 512 × 512 pixel CT image generally contains a quarter of a million pixels (Fig. 5.16). Each individual pixel is coded using 12 bits, which requires 2 bytes of computer memory, even though only 12 of the available 16 bits (2 bytes) are used. When 12 bits are used to code for a pixel, this will produce up to 4096 shades of gray (i.e., 2^{12}). This is sufficient to display each HU value from air (−1000) to dense bone (>1000 HU). Each CT image thus requires 0.5 MB (quarter of a million × 2 bytes per pixel). An abdomen/pelvis CT examination with 50 images requires 25 MB of storage space.

Acquired (raw) data have a much higher storage requirement than processed images. A sinogram with up to 1000 projections and 1000 rays in each projection would use up 2 MB of data storage for each slice. In addition, there may also be 10 times more acquired slices that are "thin" than reconstructed images that are "thick." For this reason, acquired data are not sent to a picture archiving and communication system (PACS) because of extra storage. However, it is important to understand that (raw) acquired data are essential when images require additional processing to modify the FOVs (zoom), change slice thickness, or make use of an alternative reconstruction kernels.

References

American College of Radiology. ACR–AAPM–SIIM technical standard for electronic practice of medical imaging. Revised 2017 (Resolution 41). https://www.acr.org/-/media/ACR/Files/Practice-Parameters/elec-practice-medimag.pdf?la=en. Accessed March 1, 2018.

Huda W. CT I. In: *Review of Radiological Physics*. 4th ed. Philadelphia: Wolters Kluwer; 2016:152–154.

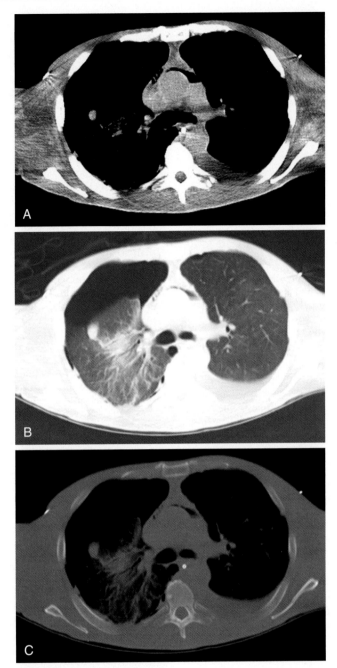

Fig. 5.17

1. Which of the following tissues will appear the darkest on any conventional CT image?
   A. Lung
   B. Fat
   C. Tissue
   D. Bone

2. Which display setting most likely has the lowest level setting?
   A. Chest (lung)
   B. Chest (mediastinum)
   C. Abdomen (liver)
   D. Head (brain)

3. Which display setting most likely has the widest window width setting?
   A. Chest (lung)
   B. Chest (mediastinum)
   C. Abdomen (liver)
   D. Head (brain)

4. Which can be used to characterize the functioning of a diagnostic workstation used for CT displays?
   A. Digital Imaging and Communications in Medicine (DICOM)
   B. Society of Motion Pictures and TV Engineers (SMPTE)
   C. Gray scale display function (GSDF)
   D. A, B, and C

CASE 5.9

Display

Fig. 5.17 CT of the chest windowed in mediastinal (A), lung (B), and bone (C) windows. Images demonstrate a right pneumothorax, pneumomediastinum, and a right upper lobe pulmonary nodule.

1. **A.** The more attenuating a pixel value is, the brighter (whiter) it will appear on a CT image, and vice versa. Because lung has by far the lowest attenuation (−700 HU) and much lower than that of fat (−100 HU) and tissue (40 HU), it will always appear darkest on CT images (see Fig. 5.17).

2. **A.** The level setting for a chest/lung display will be approximately −700 HU, which is the average lung HU, and would appear gray on CT images using this level setting. Mediastinum, liver, and brain all have a level that will be close to 40 HU, which corresponds to the CT number of soft tissue.

3. **A.** The window width setting in a lung display will be very wide, which permits all of the lung tissues, ranging from air (−1000 HU) to soft tissue (40 HU), to be visualized. The window width settings for a chest/mediastinum, abdomen (liver), and head (brain) will all be relatively narrow to permit discrimination of soft tissues with similar HU values.

4. **D.** DICOM is a standard for transferring and storing images. It is widely used in radiology. An SMPTE pattern is used to visually assess the display of the diagnostic workstation. GSDF is the optimal way to convert a range of pixel values (0% to 100%) into monitor brightness (black to white) based on psychophysics studies of human perception.

Comment

CT examinations can be satisfactorily viewed on 2 megapixel (MP) monitors, which are capable of displaying up to eight 512 × 512 images at full resolution. The monitor brightness is directly related to the pixel HU value. In CT images, higher HU values will always appear brighter, which reflects increased x-ray attenuation. However, the display brightness of any pixel can always be modified by adjusting the display level settings on the reader's workstation. Similarly, the difference in the relative appearance of two pixels can be modified by adjusting the display window width setting. However, HU values of any pixel in a CT image never change no matter how the operator adjusts the display settings.

Choice of level (AKA center) affects the brightness of a given pixel, and as the level increases, tissue brightness is reduced, and vice versa. When the level is set to 50 HU, soft tissues appear gray, and tissues with HU greater than 50 appear progressively whiter, and tissues with HU less than 50 appear progressively darker. In a chest CT image, setting the level at −600 HU would ensure that the average lung appears gray. Choice of window width affects the contrast of a pixel relative to its neighbors (Fig. 5.18). A window width of 100 HU at window level of 0 HU has all HU less than −50 appearing black and all HU greater than 50 appearing white. If the window width were increased from 100 to 200, HU less than 100 would appear black and greater than 100 would appear white. Examples of commonly used window width and level presets can be seen in Table 5.1.

It would be impossible to identify a small ground-glass pulmonary nodule when viewing a chest CT using a mediastinal window because each lung would look uniformly black. All of the contrast and image detail of the low attenuation lungs is lost because the lowest HU value using a mediastinal window is set to −150 HU. Every value less than −150 HU will be assigned zero brightness and appear black when displayed on a radiology workstation. Lowering the level and expanding the window will assign different brightness values for pixels in the lung window range, accentuating the contrast between tissue and air and therefore allowing the radiologist to detect the nodule.

References

Bell DJ, Murphy A, et al. Windowing (CT). Radiopaedia. https://radiopaedia.org/articles/windowing-ct. Accessed March 1, 2018.

Huda W. CT I. In: *Review of Radiological Physics.* 4th ed. Philadelphia: Wolters Kluwer; 2016:153–154.

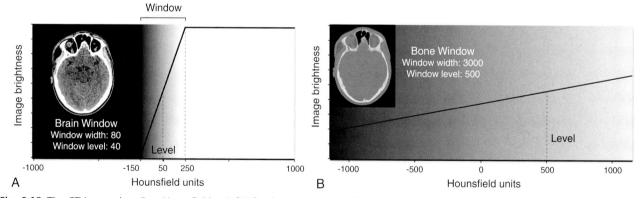

Fig. 5.18 The CT image data (i.e., Hounsfield unit [HU] values) are converted into monitor brightness using a line that is characterized by a window level and a window width. The level affects the image brightness and the window width the image contrast. The same HU values in a head CT can be displayed with a narrow window width (A) to optimize display of soft tissues, or a wide window width (B) to depict the bone structures.

Table 5.1   TYPICAL WINDOW AND LEVEL PRESENTS USED IN CLINICAL CT

Setting	Window Width[a]	Level[a]
Brain	80	40
Subdural	150	75
Mediastinum	350	50
Lung	1500	−600
Abdomen	450	20
Liver	150	60

[a]The level serves as middle gray, and the window width dictates the range of values from black to white.

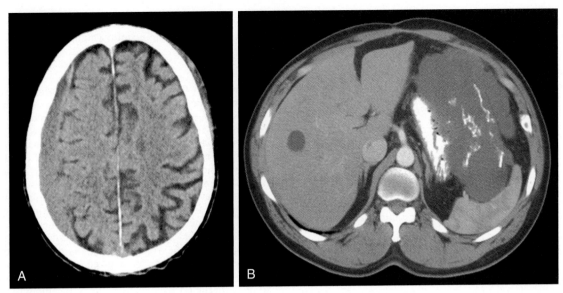

Fig. 5.19

1. Which metric best describes the radiation output of a CT scanner?
   A. mAs
   B. kV
   C. Volume CT dose index (CTDI$_{vol}$)
   D. A and B

2. What most likely improves when the CTDI$_{vol}$ used in a CT scan is increased?
   A. Resolution
   B. Mottle
   C. Contrast
   D. Patient dose

3. Which CT dose metric is most analogous to the kerma area product (KAP; dose area product) used in radiography and fluoroscopy?
   A. kV and mAs
   B. CTDI$_{vol}$
   C. Dose length product (DLP)
   D. Size-specific dose estimate (SSDE)

4. Which tube parameter can be used to change the x-ray beam quality (half-value layer in millimeters of aluminum) on a CT scanner?
   A. Voltage (kV)
   B. Current (mA)
   C. Rotation time (s)
   D. A, B, and C

CASE 5.10

Quantity and Quality

Fig. 5.19 CT head with a volume CT dose index (CTDI$_{vol}$) 60 mGy (A) showing an isodense right cerebral convexity subdural hematoma and CT abdomen with a CTDI$_{vol}$ 15 mGy (B) demonstrating marked gastric wall thickening in a patient with a gastrointestinal stromal tumor.

1. **C.** The voltage determines the average photon energy, and milliampere-second is a relative (not absolute) indicator of x-ray tube output. CTDI$_{vol}$ is a universal measure of CT output and is used by accreditation bodies such as the American College of Radiology (ACR) to specify "how much" radiation should be used to perform a selected type of examination (e.g., CTDI$_{vol}$ <80 mGy for routine head CT scan).

2. **B.** The choice of CTDI$_{vol}$ will influence the amount of mottle in the resultant CT image. When the selected value of CTDI$_{vol}$ is increased by a factor of four, random image mottle in the CT image will be halved. However, increasing CTDI$_{vol}$ by a factor of four will also quadruple the patient dose.

3. **C.** In projection x-ray imaging, the intensity of the x-ray beam (air kerma in milligray) is analogous to CTDI$_{vol}$. The total radiation incident on a patient undergoing an x-ray examination is the KAP. In CT, the DLP quantifies the total amount of radiation incident on the patient. The SSDE factors patient size into CTDI$_{vol}$ but does not account for scan length.

4. **A.** The beam quality on a CT scanner can be increased by increasing the x-ray tube voltage. Increasing the current (mA) and/or the rotation time (s) increases only the quantity but not the quality (average energy). When the tube voltage is increased, the beam quantity (number of photons) also increases.

Comment

Vendors specify the output of CT scanners using measurements made in acrylic cylinders that have diameters of either 16 cm (small [S]) or 32 cm (large [L]). These measurements explicitly account for CT pitch and are expressed as a CTDI$_{vol}$ in milligray. CTDI$_{vol}$ better reflects CT output than milliampere-second because CTDI$_{vol}$ takes into account x-ray tube characteristics, as well as the selected x-ray tube voltage. Using identical techniques (kilovolt and milliampere-second), a small phantom records twice the phantom dose as a large phantom so that CTDI$_{vol}$(S) of 10 mGy and CTDI$_{vol}$(L) of 5 mGy indicate approximately the same CT output. This change of measured dose with size is explicitly factored into a quantity called the size-SSDE.

At constant tube voltage, CTDI$_{vol}$ is directly proportional to milliampere-second. For a given patient examination, an increase in CTDI$_{vol}$ indicates that more radiation is being used to perform the CT examination. When the amount of radiation increases, this will reduce the amount of mottle in the resultant image, and vice versa. Increasing CTDI$_{vol}$ has no effect on image contrast or spatial resolution but will increase patient dose. CTDI$_{vol}$ is thus analogous to air kerma used in radiographic examinations, which is adjusted solely to modify the amount of mottle in the resultant projection image.

DLP is the product of CTDI$_{vol}$ and scan length expressed in milligray-centimeter (Fig. 5.20). DLP measures the total amount of radiation used to generate a CT scan. DLP is analogous to the KAP in projection imaging, which is also directly associated with the patient stochastic radiation risk. Both DLP and KAP can be converted into patient effective doses by use of

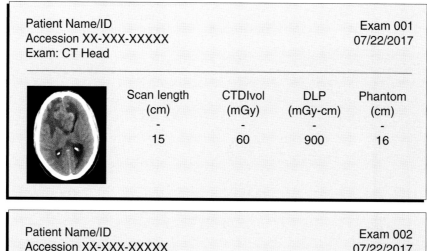

Fig. 5.20 All CT scanners provide a dose summary sheet that contains information on the scan length (centimeter), volume CT dose index (CTDI$_{vol}$), and the phantom in which it is measured. The product of the scan length and CTDI$_{vol}$ is the dose length product (DLP). CTDI$_{vol}$ is related to the mottle in the images, whereas the DLP is directly related to the patient effective dose and corresponding stochastic radiation risk.

appropriate conversion factors. In CT, effective dose conversion factors have to take into account the body region being irradiated, as well as the age and size of the patient undergoing the CT examination.

CT x-ray beam quality is adjusted varying by x-ray tube voltage because the filtration is fixed and cannot be changed by the operator. CT beam qualities are high because these beams use heavy filtration. At 80 kV, for example, the typical CT beam quality is 4.5 mm aluminum, which is higher than the 3 mm normally obtained in abdominal radiography at the same tube voltage. Use of higher x-ray beam qualities minimizes beam-hardening artifacts and also increases patient penetration. In CT, it is important to note that increasing CT tube voltage will reduce patient doses only when tube output (milliampere-second) is reduced to maintain the radiation intensity at the CT detectors.

References

Huda W. CT I. In: *Review of Radiological Physics*. 4th ed. Philadelphia: Wolters Kluwer; 2016:154-157.

McNitt-Gray MF. AAPM/RSNA physics tutorial for residents: topics in CT. Radiation dose in CT. *Radiographics*. 2002;22(6):1541-1553.

Sensakovic WF, Warden DR. What is the CT dose report sheet and why is it useful? *AJR Am J Roentgenol*. 2016;207(5):929-930.

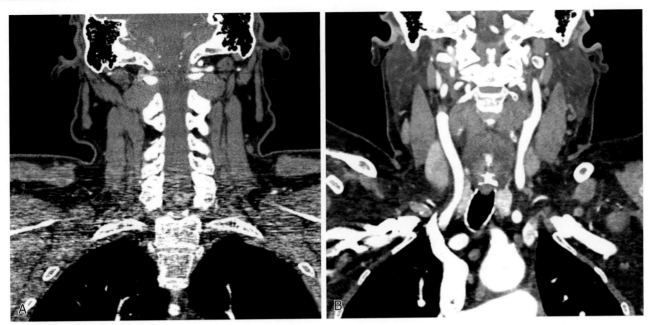

Fig. 5.21

1. Which imaging system is least likely to use a radiation detector to assess the amount of x-rays transmitted to the image receptor?
   A. Radiographic
   B. Mammographic
   C. Fluoroscopic
   D. CT

2. Who selects the milliampere-second used for an average-sized patient undergoing a given type of CT examination?
   A. Automatic exposure control (AEC) (vendor)
   B. Technologist
   C. Physicist
   D. Radiologist

3. Which is adjusted by an AEC system on current commercial CT scanners?
   A. mA
   B. kV
   C. s
   D. A, B, and C

4. What is the major benefit of using an AEC system on a commercial CT scanner?
   A. Eliminates unnecessary radiation
   B. Reduces scanning time
   C. Reduces manufacturing costs
   D. B and C

CASE 5.11

Automatic Exposure Control

Fig. 5.21 Coronal reconstructions of two CT angiograms of the neck. A was performed without tube modulation, while tube modulation was turned on in B. Notice the streak artifact in the lower part of A caused by the shoulders.

1. **D.** Radiography, mammography, and fluoroscopy generally have an AEC system that is built around a radiation detector (e.g., ionization chamber) at the image receptor. In this way the amount of radiation used in any examination can be "fixed." By contrast, there is no such chamber in any current commercial CT scanner. Instead, tube current modulation is based on a prescan localizer radiograph.

2. **B.** It is the responsibility of the radiologist to select technique factors for an average-size patient, to ensure the right amount of radiation is used in clinical CT. This differs in radiography, mammography, and fluoroscopy, where the radiation intensity at the image receptor is measured (and controlled) using a radiation detector.

3. **A.** The amount of radiation used to create a CT image for a normal-sized patient undergoing a specified type of CT examination must be selected by the radiologist (kilovolt and milliampere-second). The CT AEC will then increase the milliampere used when the patient is more attenuating (e.g., larger) and reduce the milliampere when the patient is less penetrating (e.g., smaller). To increase the milliampere-second in a CT scan, it is possible to change only the milliampere because the rotation time is fixed.

4. **A.** The CT AEC system primarily eliminates unnecessary radiation by increasing the milliampere for more attenuating body parts (e.g., pelvis) and reducing the milliampere for less attenuating body parts (e.g., chest). The scan time will not be dependent on the use of AEC but is instead dependent on the pitch and detector size.

Comment

CT does not have an AEC system of the type used in radiography, mammography, and fluoroscopy. In CT, operators must manually set either a tube current (milliampere) or noise value (i.e., HU standard deviation) that generates acceptable image quality for a given diagnostic task. The role of the CT automatic exposure control (AEC_{CT}) system is to increase the milliampere when patient attenuation increases and to reduce the milliampere when patient attenuation is reduced. In this way the key role of AEC in CT imaging is to eliminate unnecessary radiation by ensuring that the right amount is used to achieve the desired level of image mottle.

Tube currents are modulated as the x-ray tube rotates around the patient, which is known as angular modulation. Tube current for a lateral projection through the shoulders will thus be (much) greater than an anteroposterior (AP) projection because the lateral projection attenuates more than an AP projection. Tube currents are modulated as the patient moves through the CT gantry, which is known as longitudinal modulation (Fig. 5.22). The applied tube current in the chest will be much lower than in the abdomen because the attenuation by the lungs is much lower than the pelvis.

AEC_{CT} can also modify the radiation used when patient size changes. When infants are scanned (i.e., less attenuation) CT tube currents are generally reduced, and they are increased when above average sized patients are scanned. However, it is often necessary to manually adjust the tube voltage when patient size changes. For example, scanning obese patients may require an increased tube voltage to increase penetration and get enough radiation to the detector. If 0% of the incident beam is transmitted through an extremely large patient using a tube current of 500 mA, increasing the tube current has no effect on the transmitted radiation intensity.

Manufacturers use a variety of methods for tube modulation with different user control options (Table 5.2). These AEC_{CT} systems provided by the major vendors of CT differ markedly in the way they operate, and it is essential that the radiologist take ownership for how these are used in clinical scanning. This can be achieved by paying close attention to the CTDI-vol and DLP data that are shown in a dose summary sheet that accompanies each patient scan that is sent to a PACS system and is readily available for review during the reporting process.

References

Huda W. CT I. In: *Review of Radiological Physics*. 4th ed. Philadelphia: Wolters Kluwer; 2016:155–156.

Raman SP, Johnson PT, Deshmukh S, et al. CT dose reduction applications: available tools on the latest generation of CT scanners. *J Am Coll Radiol.* 2013;10(1):37–41.

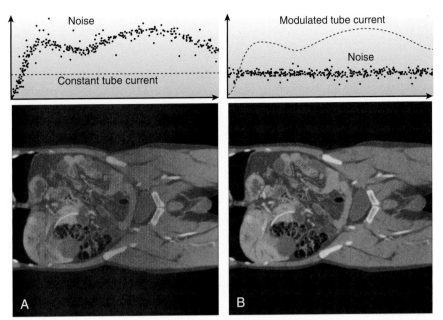

Fig. 5.22 When the tube current is kept constant (A), the amount of radiation transmitted through the patient will vary and result in low levels of mottle in the "thin" chest and high levels of mottle in the "dense" pelvis. With a modulated tube current that keeps the radiation at the CT detectors constant, the level of mottle is constant in each image slice (B).

Table 5.2 TUBE-CURRENT MODULATION SYSTEMS CURRENTLY AVAILABLE ON COMMERCIAL CT SCANNERS

	Siemens	**Philips**	**Toshiba**	**GE**
Name	CareDose 4D	DoseRight	Sure Exposure 3D	Auto mA, Smart mA
Modulation	Depends on body region	Longitudinal and transverse	Longitudinal and transverse	Auto mA: longitudinal Smart mA: longitudinal and transverse
User control	Reference mAs; adaptation curve	DoseRight index; reference size	Global quality setting, standard deviation, max mA, min mA	Noise index, max mA, min mA

CASE 6.1

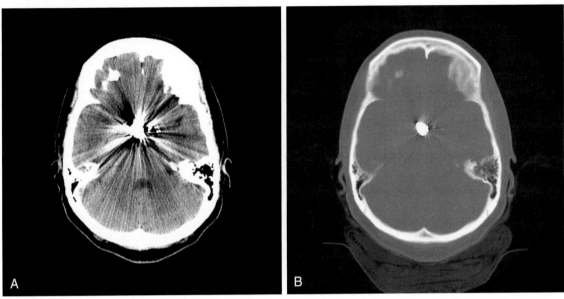

Fig. 6.1

1. Which voltage is most likely to reduce beam-hardening arti-
   facts in CT imaging of the posterior fossa (kV)?
   A. <120
   B. 120
   C. >120

2. What would be the most likely volume CT dose index
   (CTDI$_{vol}$) (mGy) for a routine adult head CT examination
   for images reconstructed using a filtered back projection
   algorithm?
   A. 15
   B. 30
   C. 60
   D. 120

3. For a routine adult head CT examination, what is the
   most likely value of the total dose length product (DLP)
   (mGy-cm)?
   A. 250
   B. 500
   C. 1000
   D. 2000

4. After a routine head CT examination performed at 120 kV,
   how should the x-ray tube voltage be adjusted for a CT head
   angiogram?
   A. Increased
   B. Kept the same
   C. Decreased

CASE 6.1

Adult Head

Fig. 6.1 CT scan of the head of a patient with a history of intracranial aneurysm coiling. Note the streak artifact produced by the coil mass.

1. C. In CT imaging, Hounsfield unit (HU) values are generally lower behind a strong attenuator such as bone (beam-hardening artifacts). Beam-hardening artifacts are generally reduced by increasing the x-ray tube voltage and/or increasing the amount of x-ray beam filtration. Scanning the posterior fossa at a high voltage (e.g., 140 kV) likely reduces beam-hardening artifacts.

2. C. Routine head CT examinations typically use a $CTDI_{vol}$ of 60 mGy, which is lower than the American College of Radiology (ACR) dose limit of 80 mGy. Iterative reconstruction algorithms generally have lower levels of image noise and may be able to reduce the amount of radiation used without adversely affecting imaging performance.

3. C. A typical routine head CT scan would likely use a $CTDI_{vol}$ of 60 mGy and have a scan length of 16 to 17 cm. The DLP is obtained by multiplying $CTDI_{vol}$ by the corresponding scan length and is thus approximately 1000 mGy-cm.

4. C. Iodine has a k-edge of 33 keV, so x-ray absorption is a maximum at this energy, which corresponds to a tube voltage of approximately 80 kV. Accordingly, all angiography should be performed at lower voltages (80 or 100 kV) compared with the routine scan performed at 120 kV.

Comment

CT examinations begin with a projection radiograph, also known as scout views, topograms, scanograms, or localizers. Projection radiographs are obtained using a low milliampere at the planned scan tube voltages (kV) while the patient is advanced through the gantry with the tube and detector array in a fixed position. Projection radiographs can ensure the patient is correctly centered, and patient attenuation data are used by the CT automatic exposure control (AEC_{CT}) to modulate x-ray beams during subsequent CT scans. Radiation delivered to patients during projection radiographs is modest compared with most CT examinations and will increase patient effective doses by no more than a few precentage points for adult scanning.

In head CT scans, x-ray tube voltages are usually 120 kV. In the posterior fossa, tube voltages (140 kV) may be increased to help minimize beam-hardening artifacts. Reducing the x-ray tube voltage is customary in angiographic imaging because this improves the visibility of iodinated contrast media. At a voltage of 80 kV, the average photon energy will be close to the iodine k-edge energy, which will maximize x-ray absorption by iodine and thereby increase the vasculature contrast. At 120 kV, the average photon energy (60 keV) is well above the k-edge energy, which reduces x-ray absorption and will markedly reduce vasculature contrast.

CT intensities for adult head scans are generally measured in a small 16-cm diameter phantom, whereas adult body scans have the CT output measured in a large 32-cm diameter phantom. High x-ray beam intensities, typically $CTDI_{vol}$ 60 mGy, are used in head CT scans, which are required to reduce mottle and visualize subtle soft tissue structures. The ACR accreditation $CTDI_{vol}$ limit for adult head CT examinations is currently 80 mGy, and facilities that submit values exceeding this value will automatically fail. Iterative reconstruction has been reported to allow facilities to the reduce intensities used in head CT scans ($CTDI_{vol}$). For specific imaging tasks such as imaging high-contrast structures (e.g., airways and bony structures), $CTDI_{vol}$ can easily be reduced because increased noise will be of negligible importance.

A typical head CT scan will likely have a $CTDI_{vol}$ of 60 mGy (Fig. 6.2) and a scan length of 16 cm. Accordingly, the DLP for a routine head CT examination is approximately 1000 mGy-cm (60 mGy × 16 cm scan length). Values of $CTDI_{vol}$ and DLP are displayed on the dose summary sheet and available on every patient examination at the picture archiving and communication system (PACS) workstation. The total DLP will include all the exposure the patient has received and will be directly related to the patient effective dose and corresponding stochastic (carcinogenic) radiation risk.

References

American Association of Physicists in Medicine. Adult routine head CT protocols. Version 2.0. 2016. http://www.aapm.org/pubs/CTProtocols/documents/AdultRoutineHeadCT.pdf. Accessed March 1, 2018.

Huda W. CT II. In: *Review of Radiological Physics*. 4th ed. Philadelphia: Wolters Kluwer; 2016:161.

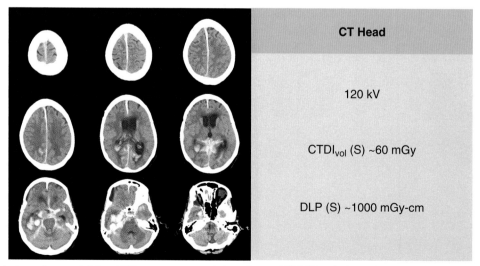

Fig. 6.2 Typical techniques for an adult head CT scan are an x-ray tube voltage of 120 kV and a beam intensity of a $CTDI_{vol}$ of 60 mGy measured in a small (S) phantom. The total amount of radiation used to perform this examination is a dose length product of 1000 mGy-cm.

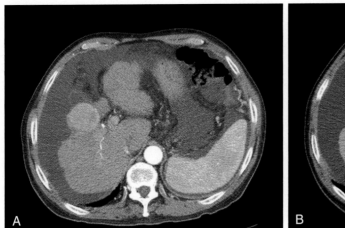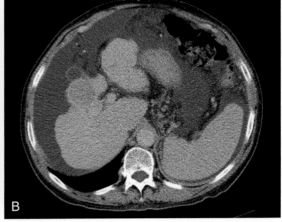

Fig. 6.3

1. For a routine adult abdominal CT examination in a normal-sized adult, what would be the most likely x-ray tube voltage (kV)?
   A. 80
   B. 100
   C. 120
   D. 140

2. For a routine adult abdominal CT examination in a normal-sized adult, what would be the most likely x-ray tube output CTDI$_{vol}$ (mGy)?
   A. 4
   B. 8
   C. 16
   D. 24

3. For a routine adult chest/abdomen/pelvis CT examination in a normal-sized adult (length of 50 cm), what would be the most likely total DLP (mGy-cm)?
   A. 150
   B. 300
   C. 750
   D. 1500

4. During a chest/abdomen/pelvis examination, when the scan moves from the abdomen to the chest region, how would an optimized CTDI$_{vol}$ value be adjusted?
   A. Increased
   B. Left unchanged
   C. Reduced

CASE 6.2

Adult Body

Fig. 6.3 CT scan of an abdomen demonstrating an arterially enhancing mass in the liver (A) that washes out on delayed images (B). In a cirrhotic patient, these findings are diagnostic for hepatocellular carcinoma.

1. **C.** For an adult of average size, routine abdominal CT examinations use 120 kV. When imaging soft tissue–type lesions without contrast, there is no advantage in using low energies. This is because when the photon energy is reduced, more radiation is required to achieve a given level of patient penetration and the corresponding increase in soft tissue contrast is very modest.

2. **C.** The x-ray tube output for a routine abdominal CT examination would likely be approximately 16 mGy, where the output is measured in a large-diameter (32 cm) phantom. The ACR CT accreditation program has a $CTDI_{vol}$ limit of 30 mGy for an abdominal CT examination in normal-sized adults.

3. **C.** A typical routine body CT scan would likely use a $CTDI_{vol}$ of 15 mGy (large phantom) and have a scan length of 50 cm. The DLP is obtained by multiplying $CTDI_{vol}$ by the corresponding scan length and is thus approximately 750 mGy-cm for an abdominal scan.

4. **C.** When moving from the abdomen region to the chest region, the total attenuation in the patient is reduced. Accordingly, in the chest region, less radiation in needed so x-ray tube output (i.e., $CTDI_{vol}$) is reduced by lowering the mAs.

Comment

Tube voltages in abdominal and pelvic CT scanning are typically 120 kV (Fig. 6.4). Whenever possible, the tube voltage should be reduced if iodinated contrast media have been administered. In chest CT scans, for example, tube voltages are likely reduced to 100 kV to increase the visibility (contrast-to-noise ratio [CNR]) of the iodinated material. The problem with lowering tube voltage is that this will reduce penetration in adult patients. For larger patients, increased voltages are essential to ensure that there is sufficient penetration. At a constant x-ray tube voltage of 120 kV, an additional 4 cm of tissue will attenuate half of the intensity of a CT x-ray beam.

An average-sized adult undergoing a routine abdominal and/or pelvic CT scan would use a $CTDI_{vol}$ of 15 mGy. For the specific task of detecting soft tissue lesions (e.g., liver scan), the $CTDI_{vol}$ might be increased to 20 mGy. The ACR CT accreditation $CTDI_{vol}$ limit for adult abdominal CT examinations is currently 30 mGy; higher values will result in an automatic failure. Because the chest is less attenuating than the abdomen, the $CTDI_{vol}$ of a chest CT scan is reduced (e.g., 9 mGy). When detecting high-contrast lesions in the lung, the amount of radiation used to perform a chest CT scan can be further reduced because an increase in image mottle will not adversely affect the detection of a high-contrast lesion.

Total DLP for an abdominal scan is 300 mGy-cm, with a similar value for a pelvic examination. A combined abdominal/pelvic CT scan would likely have a DLP of 500 mGy-cm. DLP values for a single-phase chest CT examination would likely be 300 mGy-cm (i.e., scan length of 30 cm). For multiphase exams, values of $CTDI_{vol}$ are unlikely to change, whereas the total patient DLP will be progressively increased as the number of phases used increases. The DLP is a measure of the total amount of radiation used to perform any CT examination. It can be readily converted into organ doses, and the corresponding effective dose when account is taken of the body region, patient age, and patient size.

The CT system's AEC will adjust the x-ray tube output when patient size changes. The average thickness of an abdomen is approximately 28 cm of tissue. Because the size of an average chest measuring 24 cm is approximately 1 tissue half-value layer (HVL) (4 cm) less than the abdomen, the AEC_{CT} will likely use half the radiation of that used in the abdomen. $CTDI_{vol}$ and DLP are always increased when larger patients are scanned using an AEC_{CT} system, and vice versa. Accordingly, when $CTDI_{vol}$ parameters on the dose summary sheet are being reviewed, "patient size" should be taken into account. Values of $CTDI_{vol}$ should be higher than average in larger patients and lower in smaller patients.

References

American Association of Physicists in Medicine. Routine adult abdomen/pelvis CT protocols. Version 1.1. 2015. https://www.aapm.org/pubs/CTProtocols/documents/AdultAbdomenPelvisCT.pdf. Accessed March 1, 2018.

Huda W. CT II. In: *Review of Radiological Physics*. 4th ed. Philadelphia: Wolters Kluwer; 2016:161.

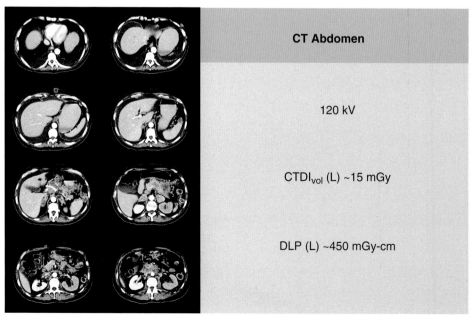

CT Abdomen
120 kV
$CTDI_{vol}$ (L) ~15 mGy
DLP (L) ~450 mGy-cm

Fig. 6.4 Typical techniques for an adult abdomen CT scan are an x-ray tube voltage of 120 kV and a $CTDI_{vol}$ of 15 mGy measured in a large (L) phantom. The total amount of radiation used to perform this examination is a dose length product *(DLP)* of 450 mGy-cm.

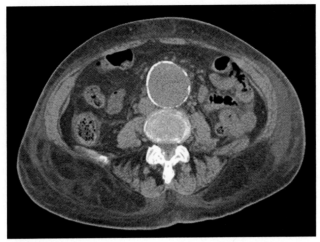

Fig. 6.5

1. How much tissue (cm) is needed to attenuate half an x-ray beam generated at 120 kV?
   A. 1
   B. 2
   C. 4
   D. 8

2. How many tissue HVLs are the anteroposterior (AP) and lateral projections of an average-sized adult abdomen (cm)?
   A. 6 and 6
   B. 6 and 8
   C 8 and 6
   D. 8 and 8

3. How much radiation is most likely transmitted through a 40-cm patient (%)?
   A. 0.01
   B. 0.1
   C. 1
   D. 10

4. Which parameter MUST be increased when scanning obese patients?
   A. mA
   B. s
   C. kV
   D. A and B

CASE 6.3

Patient Size and X-Ray Transmission

Fig. 6.5 Partially calcified abdominal aortic aneurysm.

1. **C.** The tissue HVL at the energies used in CT imaging is approximately 4 cm. When the tissue thickness increases by 4 cm, the radiation incident on the patient must be doubled to achieve the same radiation intensity at the CT detectors.

2. **B.** The AP dimension of an average-sized adult is approximately 24 cm, which translates into six tissue HVLs. The lateral dimension is greater (e.g., 32 cm), which translates into eight tissue HVLs.

3. **B.** For 40 cm of tissue, which corresponds to 10 tissue HVLs, the transmission will be approximately 0.1% (i.e., $1/2^{10}$). This is a very low transmission rate and explains why CT images of obese patients tend to be very noisy (mottled).

4. **C.** When very few x-rays are transmitted through a patient, it is the penetrating power of the x-ray beam that needs to be increased. X-ray beam penetrating power can be increased by increasing the x-ray tube voltage. Increasing the x-ray beam intensity (mAs) cannot improve "patient penetration."

Comment

Patient transmission is affected by the CT beam quality and physical characteristics of the patient. These latter include the attenuator atomic number (Z), physical density (ρ), and total thickness. Patients are generally modeled as soft tissue that has an atomic number of 7.5 and 1 g/cm³ density. For soft tissues, and the high beam qualities used in most CT, the tissue HVL is approximately 4 cm (Fig. 6.6), which is higher than in mammography (1 cm) and abdominal radiography (3 cm). This increase in the soft tissue HVL is partly because of the use of

higher x-ray tube voltages (e.g., 120 kV) and partly because of the use of high filtration to reduce beam-hardening artifacts.

The adult head has a mass equivalence of a water cylinder with a diameter of 18 cm. The size of an adult head is approximately independent of age, sex, and patient size. Average-sized adult abdomens have a mass equivalence of water cylinders with diameters of 28 cm and chests with diameters of 24 cm. When patient size changes in body CT, radiation techniques must be adjusted to maintain a constant intensity at CT detectors. At 120 kV, reducing patient size by 4 cm requires the milliampere-seconds to be halved, whereas increasing patient size by 4 cm requires the milliampere-seconds to be doubled. An obese adult (50 cm) will transmit 1000 times less radiation than a neonate (10 cm) when the beam quality is kept the same (Fig. 6.7). This is the reason that CT images of large patients are invariably unsatisfactory because of their very high levels of noise, which is the direct result when very few x-ray photons are detected.

X-ray tube voltage (penetration) may be reduced when scanning infants and young children. For larger patients, it is essential to increase penetration by increasing the tube voltage. If a low-voltage beam has zero penetration through a "large patient," any increase in x-ray beam intensity (mAs) will still result in zero penetration through the patient. Reducing the tube voltage from 120 to 80 kV for a 26-cm patient would require a fivefold increase in milliampere-seconds to maintain the same radiation intensity at the image receptor. In this same patient, increasing tube voltage from 120 to 140 kV would require milliampere-seconds to be halved. Because maintaining a constant amount of radiation at the image receptor in obese patients is very difficult, the resultant images will be very mottled and reduce the visibility of even large low-contrast lesions.

References

Fursevich DM, LiMarzi GM, O'Dell MC, et al. Bariatric CT imaging: challenges and solutions. *Radiographics*. 2016;36(4):1076–1086.

Huda W. CT II. In: *Review of Radiological Physics*. 4th ed. Philadelphia: Wolters Kluwer; 2016:162–163.

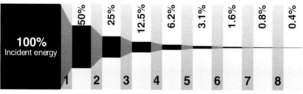

Tissue Half-Value Layers

Fig. 6.6 The incident radiation intensity is attenuated through increasing soft tissue half-value layers and is reduced to 0.4% by a patient 32 cm thick (tissue equivalent).

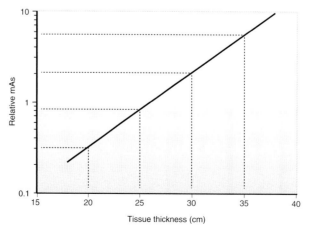

Fig. 6.7 The milliampere-seconds needs to be increased with increasing patient thickness to maintain a constant air kerma at the CT detector, assuming a CT x-ray spectrum generated at 120 kV.

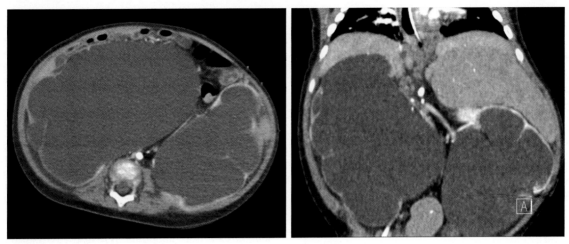

Fig. 6.8

1. For an infant noncontrast CT examination, what is the most likely optimum x-ray tube voltage (kV)?
   A. 80
   B. 100
   C. 120
   D. 140

2. When specifying the output of an x-ray tube for a pediatric body CT examination, which phantom size is typically used by vendors (16 or 32 cm)?
   A. 16
   B. 32
   C. A or B

3. For a 5-year-old (20-kg) pediatric patient undergoing a routine abdominal/pelvic CT examination, what is the $CTDI_{vol}$ limit in a small 16-cm phantom currently used for ACR accreditation (mGy)?
   A. 10
   B. 15
   C. 20
   D. 25

4. When switching from an adult to a pediatric CT examination, which parameter is most likely to be increased?
   A. Voltage (kV)
   B. Current (mA)
   C. Rotation time (s)
   D. CT helical pitch

CASE 6.4

Pediatric Computed Tomography

Fig. 6.8 CT scan of an abdomen demonstrating severe bilateral hydronephrosis in a patient with megacystis microcolon intestinal hypoperistalsis syndrome.

1. **A.** Empirical evidence has shown that the use of 80 kV is appropriate for children who weigh less than approximately 10 kg. For children weighing 10 to 20 kg, the voltage should be increased to 100 kV, and for all noncontrast studies in children greater than 20 kg, an x-ray tube voltage of 120 kV is optimal.

2. **C.** The radiation output of a CT scanner can be measured either in a small (16-cm diameter) phantom or in a large (32-cm diameter) phantom. Most operators specify the CT output for pediatric body exams in a small phantom (e.g., ACR accreditation program), although some specify the CT output for pediatric body exams in a large phantom (e.g., Siemens).

3. **C.** The current ACR CT accreditation $CTDI_{vol}$ limit for a 5-year-old (20-kg) child undergoing an abdominal CT exam is 20 mGy in a small phantom, which corresponds to 10 mGy when the CT output is specified in a large phantom.

4. **D.** X-ray tube voltage (kV) and tube current (mA) would likely be reduced in pediatric CT for dose reasons, and short examination times minimize motion artifacts so rotation time will not be increased. When CT helical pitch is increased, this will reduce the total examination time thereby minimizing motion blur.

Comment

Infants undergoing noncontrast abdominal studies should use the lowest voltages available, which is typically 80 kV. This decreases patient radiation dose. X-ray tube voltage should be increased with increasing patient size to maintain sufficient penetration. Infants between 10 and 20 kg would likely use 100 kV, and voltages comparable to adults (i.e., 120 kV) would likely be used on larger children. However, angiographic examinations typically use the lowest available tube voltage (i.e., 80 kV) because it improves the signal (i.e., brightness) of the contrast media. When patient penetration at 80 kV is inadequate, 100 kV would be used. The use of high tube voltages (120 or 140 kV) should always be avoided in pediatric CT angiography (CTA) of young and average-sized patients.

Tube current and rotation time, which are used to control the $CTDI_{vol}$, should be adjusted to ensure that image mottle is satisfactory (Fig. 6.9). Empirical studies have shown that for most diagnostic imaging tasks, the acceptable level of mottle in pediatric CT is lower than that in adult CT. This has been reported as being due to smaller structures and reduced fat planes delineating structures of interest. It is important to ensure that when scanning infants, short x-ray tube rotation times are used to minimize motion blur. Limiting the scan length is also very important in pediatric CT because it eliminates unnecessary radiation exposure to the child. One of the best dose reduction methods for typical clinical practice is to eliminate unnecessary phases in contrast-enhanced CT.

The current ACR CT limit on $CTDI_{vol}$ level for a 5-year-old abdomen CT scan (20 kg) is 20 mGy. However, the ACR expects values to be less than 15 mGy and would advise facilities that use between 15 and 20 mGy to investigate their protocols to ensure they are using best practices. Since an optimized protocol would likely expose a child of this age to a $CTDI_{vol}$ (small phantom), that is less 10 mGy. Some vendors specify the CT output for pediatric body scans using a 32-cm (large) phantom. When the large phantom replaces the small phantom to measure CT output, the recorded values are generally halved. Thus the ACR accreditation limit of 20 mGy in a small phantom corresponds to 10 mGy in the large phantom. An optimized pediatric body CT scan in a 5-year-old should therefore be using approximately 5 mGy $CTDI_{vol}(L)$ when a large phantom is used to specify the CT output.

References

Huda W. CT II. In: *Review of Radiological Physics*. 4th ed. Philadelphia: Wolters Kluwer; 2016:162–163.

Huda W, Vance A: Patient radiation doses from adult and pediatric CT. *AJR Am J Roentgenol*. 2007;188(2):540–546.

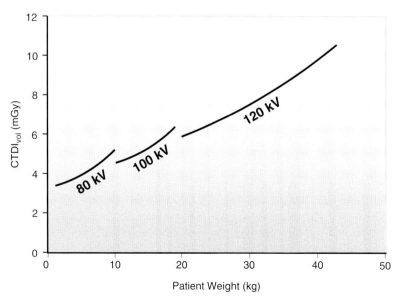

Fig. 6.9 Representative schematic as to how the x-ray beam quality (kV) and quantity ($CTDI_{vol}$) (16-cm phantom) should be adjusted with patient size for optimized noncontrast body CT imaging. For $CTDI_{vol}$ values specified in a large 32-cm phantom, the values on the y-axis would need to be halved.

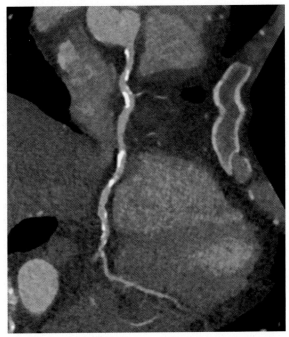

Fig. 6.10

1. Which type of scanner can be used to perform cardiac CTA?
   A. 64 slice
   B. 320 slice
   C. Dual source
   D. A, B, and C

2. What is the best x-ray tube voltage for performing coronary CTA on a patient weighing less than 90 kg?
   A. 100
   B. 120
   C. 140

3. When retrospective gating cardiac CT imaging is performed using helical acquisition, what is the most likely pitch?
   A. <1
   B. >1
   C. 1

4. Which scanner can capture the entire left ventricle in a single heartbeat?
   A. 64 slice
   B. 320 slice
   C. Dual source
   D. Both B and C

CASE 6.5

Cardiac Computed Tomography

Fig. 6.10 Three-dimensional reconstruction of a coronary CT angiography demonstrating high-grade stenosis of the right coronary artery.

1. **D.** Cardiac coronary CTA generally requires at least a 64-slice CT scanner. A 320-slice CT scanner can acquire projections covering the whole of the left ventricle in one single rotation. Dual-source CT scanners when operated at high pitch (e.g., >3) can acquire the projections required to create cardiac angiograms in a single heartbeat.

2. **A.** The Society of Cardiovascular Computed Tomography Guidelines recommend using a tube voltage of 100 kV when the patient is less than 90 kg and less than 30 kg/m² body mass index (BMI). The reason for this is that the average energy of an x-ray beam generated at lower kV is close to the iodine k-edge (33 keV), which maximizes absorption of x-rays by the iodine and results in the best vasculature contrast.

3. **A.** Retrospectively gated cardiac CTA uses a very low pitch, typically 0.2 to 0.3. This is required to ensure that the necessary projections for image reconstruction are acquired throughout the cardiac cycle.

4. **D.** A 64-slice CT scanner cannot capture images of the entire left ventricle during a single cardiac cycle because it will only cover 40 mm per rotation. A 320-slice CT scanner is able to accomplish this due to the higher beam width. A dual-source scanner has greater coverage due to increased pitch.

Comment

Cardiovascular CT imaging is applied for various applications, from coronary calcium scoring of the arteries to imaging for myocardial and pericardial disease. The heart is constantly moving, so it is important that scans have high temporal resolution to "freeze" motion and prevent blur. This is typically accomplished by electrocardiographic (ECG) gating, restricting imaging to cardiac phases that have the least motion, and using fast tube rotation times. For patients with a low resting heartbeat (<65 beats/min), the heart is moving the least during mid-diastole (~70% R-R interval). However, as heart rate increases, diastole shortens, and therefore patients with higher heart rates are often best imaged during end-systole (~40% R-R interval).

Use of 64-slice (or greater) multidetector CT (MDCT) is essential to ensure adequate coverage and reduce scan times. A 64-slice scanner will have 40 mm of coverage for each rotation and require several rotations to scan the entire heart. Several vendors now offer scanners with greater than 200 rows, which can scan the entire heart in 1 or 2 rotations. Good spatial resolution and contrast are also necessary, especially when looking at smaller objects such as the coronary arteries. Most vendors implement specific cardiac settings that use a smaller field-of-view and special cardiac bow tie filters to improve image quality. Cardiac scans are best performed at lower x-ray tube voltages (e.g., 100 kV) to improve contrast of calcium and contrast media. When low voltages cannot provide sufficient penetration in very large patients, the tube voltage must be increased.

Cardiac acquisitions are typically divided into two types—retrospective and prospective (Fig. 6.11). Retrospective gating uses low pitch (~0.2) to acquire the necessary projections throughout the heart at all phases of the cardiac cycle. This results in a relatively high dose scan because the pitch is low and the beam is on during all cardiac phases. One benefit of retrospective gating is that, because the x-ray beam is kept on throughout the cardiac cycle, images may be reviewed during any cardiac phase. This CT imaging mode therefore can provide functional information (e.g., left ventricle ejection fraction). It is also suitable for patients with high or irregular heartbeats because the user can retrospectively choose the best cardiac phase for interpretation.

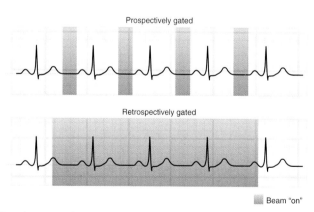

Fig. 6.11 In prospectively gated studies, the x-ray tube is switched on only for a selected cardiac phase. Retrospective gating acquires the necessary projections of the heart throughout the cardiac cycle.

Prospective gating offers an attractive alternative to retrospective gating because it substantially reduces patient dose. In prospective gating studies, the x-ray tube is switched on only for a selected cardiac phase and each part of the heart is scanned only once because the "effective pitch" is 1. CT scanners with markedly more than 64 slices (250 or 320 slice) have beam widths of 12 to 16 cm that can capture the heart in a single rotation of the x-ray tube. Dual-source CT scanners operated at high pitch values (>3.0) can acquire the complete cardiac data in a single heartbeat. Prospective gating is generally not suitable for patients with very fast or irregular heart rates because the ECG triggering may fail to capture the correct cardiac phase.

References

Halliburton SS, Abbara S, Chen MY, et al. SCCT guidelines on radiation dose and dose-optimization strategies in cardiovascular CT. *J Cardiovasc Comput Tomogr.* 2011;5(4):198–224.

Huda W. CT II. In: *Review of Radiological Physics.* 4th ed. Philadelphia: Wolters Kluwer; 2016:162.

Mahesh M, Cody DD. Physics of cardiac imaging with multiple-row detector CT. *Radiographics.* 2007;27(5):1495–1509.

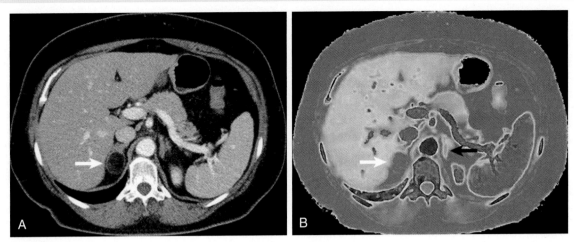

Fig. 6.12

1. Which is increased the most when using a dual-source CT scanner (vs. single-source CT)?
   A. Gantry rotation speed
   B. X-ray tube voltage
   C. Data acquisition rate
   D. Image matrix size

2. Which body region is likely to benefit the most when dual-source CT is introduced?
   A. Head
   B. Cardiac
   C. Abdomen
   D. Extremities

3. Which material will likely show the largest increase in x-ray attenuation when the tube voltage is reduced from 120 to 80 kV?
   A. Blood
   B. Tissue
   C. Liver
   D. Calcium

4. If the HU of fat at 120 kV is −100, what is the corresponding HU at 80 kV?
   A. −150
   B. −100
   C. −50

CASE 6.6

Dual Source and Dual Energy

Fig. 6.12 Right adrenal myelolipoma. (A) A right adrenal mass with macroscopic fat *(white arrow)*. (B) Z-effective map of the same level, depicting the lipid-rich lesion in the right adrenal *(white arrow)* compared with the normal left adrenal *(black arrow)*. (Courtesy Rivka Kessner, MD, University Hospitals Cleveland Medical Center/Case Western Reserve University.)

1. **C.** With two x-ray tubes that are offset by approximately 90 degrees, the rate at which projections are acquired is essentially doubled so that the time required to obtain the projections needed to reconstruct a CT image is essentially halved. Gantry rotation speed, x-ray tube voltage, and image matrix size are essentially the same in both single-source and dual-source CT scanning.

2. **B.** Fast scanning offered by a dual-source CT is particularly important in cardiac imaging, where a temporal resolution of approximately 75 ms can be achieved for a 300-ms x-ray tube 360-degree rotation time. For some heartbeat rates, it is now possible to image the whole heart in a single heart beam when very high pitch values (>3) are used.

3. **D.** For materials such as calcium with atomic number much higher than soft tissue, x-ray attenuation is high at low energies, where the photoelectric effect dominates. As the energy increases and Compton scatter becomes more important, differences in attenuation are much less pronounced. Consequently, the HU for high-Z materials such as calcium fall markedly as the CT x-ray tube voltage is increased.

4. **A.** When the photon energy is reduced, fat attenuation decreases more than that of water. Accordingly, the fat HU will diverge from that of water (0) and at 80 kV will be approximately −150. The HU of any material is its contrast relative to a water background, illustrating that the contrast of a fat lesion increases when voltages are reduced.

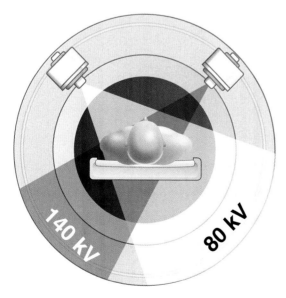

Fig. 6.13 Schematic of a dual-source CT scanner where x-ray tubes are displaced 90 degrees from each other. In dual-energy scanning, each x-ray tube can acquire projections at a different voltage, thereby enabling it to differentiate materials based on the material's atomic number (Z).

Comment

Dual-source CT has two x-ray tubes, as well as two detector arrays (Fig. 6.13). These two acquisition systems are arranged in a rotating gantry and offset from each other by approximately 90 degrees. The field of view one detector array is markedly smaller because of gantry space limitations. For this reason, the image quality (mottle) where the two fields of view overlap is markedly superior to that of the peripheral region. Compared with a single-source CT, the key characteristic of dual-source CT is the acquisition of twice as many CT projections in a given gantry rotation. This allows for high pitch (~3) and fast tables (approaching 50 cm/s) that substantially reduce scan time.

Dual-source CT offers substantial benefits in cardiac imaging due to the short scan times, and this is the principal reason why such a system would be acquired by a radiology department. The faster scanning may allow prospective scanning of patients at higher heart rates than can be scanned in a typical single-source scanner. Fast scanning is also important for patients who cannot hold their breath and infants who are unable to remain still. The faster scanning available using dual-source CT is an attractive alternative to the use of sedation to reduce motion artifacts.

Dual-energy CT requires the acquisition of projection data at two x-ray tube voltages (typically 80 and 140 kV). The attenuation data from each energy are then combined to differentiate materials with different proton content (Z) even if they have the same attenuation. The farther apart the energies, the better materials can be differentiated. Vendor methods of implementing dual energy typically fall into one of two categories—switching or dual source. In the switching method, the x-ray tube voltage is rapidly switched during scanning, resulting in similar projections acquired at two energies. In the dual-source method, two x-ray tubes acquire similar projection data with each tube set to a different energy.

In scans for gout, uric acid and calcium deposits cannot typically be differentiated based on attenuation. Using dual-energy scanning, the two materials can be identified because a low-Z material (uric acid) will exhibit a different change in attenuation at the two tube voltages than a high-Z material (calcium). Another application of dual-energy scanning is in CTA. Although bone can be manually removed from 3D reconstructions with some difficulty and time, removal of the skeleton is automatic using a dual-energy scan because bone is easily differentiated from contrast media. Dual-energy CT also permits removal of iodinated contrast in liver scans, which will allow a virtual unenhanced image to be generated.

References

Coursey CA, Nelson RC, Boll DT, et al. Dual-energy multidetector CT: how does it work, what can it tell us, and when can we use it in abdominopelvic imaging? *Radiographics*. 2010;30:4, 1037–1055.

Huda W. CT II. In: *Review of Radiological Physics*. 4th ed. Philadelphia: Wolters Kluwer; 2016:164–165.

McCollough CH, Leng S, Yu L, Fletcher JG. Dual- and multi-energy CT: principles, technical approaches, and clinical applications. *Radiology*. 2015;276:3:637–653.

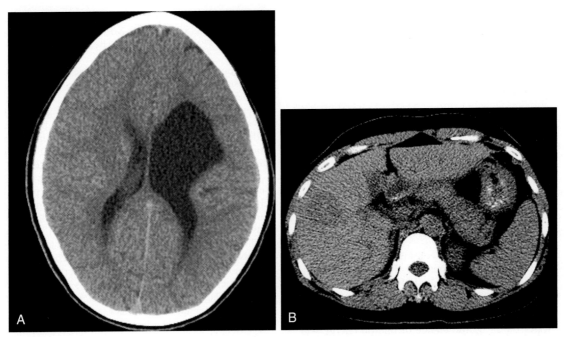

Fig. 6. 14

1. How much more attenuating than water is soft tissue with an HU value of 40 (%)?
   A. 0.04
   B. 0.4
   C. 4
   D. 40

2. How different are the linear attenuation coefficients of gray and white matter (%)?
   A. <0.1
   B. 0.1
   C. 0.5
   D. 2

3. Which lesion characteristic will likely affect its contrast in a soft tissue background?
   A. Atomic number
   B. Physical density
   C. Cross-sectional area
   D. A and B

4. As CT voltage is reduced, which HU likely increases the most?
   A. Water ($Z = 7.5$)
   B. Tissue ($Z = 7.5$)
   C. Bone ($Z = 12$)
   D. Iodine ($Z = 53$)

CASE 6.7

Lesion Contrast

Fig. 6.14 CT scan of a head demonstrating high-contrast frontoparietal porencephaly (A) and CT scan of an abdomen demonstrating a low-contrast metastatic liver mass (B).

1. **C.** Soft tissue, with an HU of 40, will attenuate an x-ray beam by 4% more than water. Every increase of 10 HU corresponds to an additional 1% x-ray attenuation so that an HU of −10 HU attenuates 1% less than water.

2. **C.** The CT number of gray matter is approximately 45, and that of white matter is approximately 40, so the difference between these two tissues is only 5 HU. White matter thus attenuates x-rays in CT approximately 0.5% less than gray matter and will thus appear slightly darker than gray matter.

3. **D.** The HU is, by definition, the contrast of a material when embedded in a water background (0 HU). The HU will increase with increasing density and atomic number but is independent of the cross-sectional area.

4. **D.** For materials with an atomic number that is similar to that of water (i.e., 7.5), the choice of photon energy will have only a modest impact on HU, and reducing the photon energy (kV) will result in only a modest increase in HU. However, for high-Z materials, the use of low energies (kV) will normally be expected to markedly increase the HU (attenuation).

Comment

The HU of any material is the difference in x-ray attenuation between this material and that of water (Table 6.1). When the HU of a soft tissue is 50, this means it attenuates 5% more than water, and this is the contrast when this type of lesion is imaged in a water background. Gray and white matter differ in x-ray attenuation by approximately 5 HU, which corresponds to a difference of only 0.5%. The wonder of CT imaging is the ability to clearly differentiate two types of tissue whose x-ray attenuating properties are so similar. For this reason, Godfrey Hounsfield was awarded the Nobel Prize in Medicine and Physiology in 1979, only 6 years after he introduced the first clinical system.

In general, CT contrast is simply the difference in HU value of a lesion relative to the background tissues. For example, a fat nodule with an HU of −100 in soft tissue that has an HU of 50 differs by approximately 150 HU. Accordingly, this shows that fat attenuates 15% less than soft tissues. A blood vessel with iodine that has an HU of 200 will have a contrast of 150 HU compared with soft tissues with an HU of 50. This shows that diluted iodine attenuates 15% more than soft tissues.

As in all radiographic imaging, contrast will change with selected photon energy (i.e., tube voltage). Increases in photon energy reduce contrast, whereas reductions in photon energy increase contrast (Fig. 6.15). However, the magnitude of such contrast changes critically depends on the lesion atomic number Z, as illustrated by the data in Table 6.1. When a lesion has an atomic number that is similar to that of water, the changes in contrast are very modest. Soft tissue lesions appear very similar at all current x-ray tube voltages because the changes in contrast are so small. However, large differences between iodinated contrast media atomic number and that of water (soft tissue) result in very large differences in contrast when photon energy is adjusted.

References

Huda W. CT II. In: *Review of Radiological Physics.* 4th ed. Philadelphia: Wolters Kluwer; 2016:166.

Yu L, Bruesewitz MR, Thomas KB, et al. Optimal tube potential for radiation dose reduction in pediatric CT: principles, clinical implementations, and pitfalls. *Radiographics.* 2011;31(3):835–848.

Table 6.1 CONTRAST (HOUNSFIELD UNIT) OF SPECIFIED MATERIALS RELATIVE TO THAT OF A WATER BACKGROUND AT 80 AND 120 KV

Material	80 kV	120 kV
Fat	−150	−100
Tissue	60	50
Dilute iodine	400	200

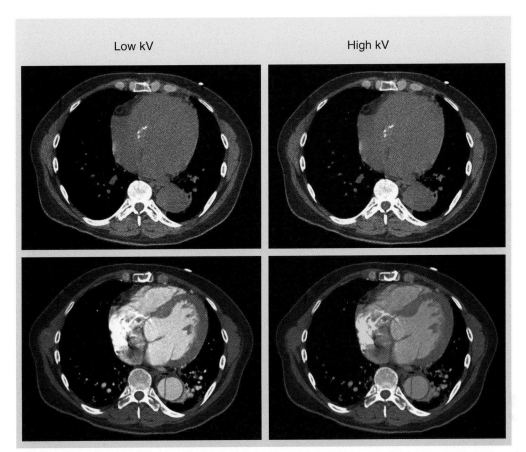

Fig. 6.15 Simulation demonstrating that when x-ray tube voltage on a CT scanner is increased *(right)*, contrast will always be reduced. However, the reduction in contrast for soft tissues *(top row)* is very modest and difficult to detect by eye. The reduction in iodine contrast *(bottom row)* is much more pronounced, explaining why angiography should not be performed at very high voltages (e.g., 140 kV).

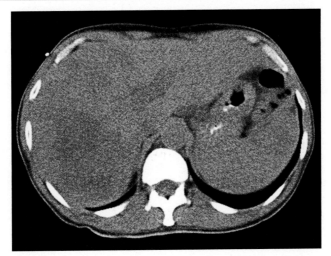

Fig. 6.16

1. When CT image noise increases, which lesion is most likely to become most difficult to visualize in the resultant image?
   A. Small
   B. Large
   C. High contrast
   D. Low contrast

2. What is most likely to happen to image mottle when the milliampere-second and CT pitch are both doubled?
   A. Increase
   B. Remain constant
   C. Decrease

3. When x-ray tube is reduced from 140 to 80 kV with no change in milliampere-second, which is most likely to increase the most?
   A. X-ray output
   B. Beam transmission
   C. Patient dose
   D. Image mottle

4. Which of the following affects image noise in CT imaging?
   A. Mathematical reconstruction filter
   B. Display window
   C. Display level
   D. B and C

CASE 6.8

CT Image Noise

Fig. 6.16 CT of the upper abdomen demonstrating a large right hepatic mass. Smaller left hepatic lesions are not well appreciated due to image noise.

1. **D.** Noise in CT is a random fluctuation of HU value around the mean value. In a CT image of a uniform water phantom with a noise of 3 HU, approximately 67% of the pixels would have HU values between −3 and + 3 HU. Noise inhibits the visibility of low-contrast lesions (e.g., liver metastases with HU that are very similar to those of liver).

2. **B.** When the milliampere-second and CT pitch are both doubled, the noise and effective mAs remain the same. Increasing milliampere-second decreases noise, whereas increasing pitch increases noise. In this example, the reduction in noise from doubling the milliampere-second is exactly offset by the increase in noise from doubling the pitch.

3. **D.** When the x-ray tube voltage is reduced, output and beam transmission are both reduced. Because fewer photons are incident on the patient and two-thirds of the incident energy is absorbed by the patient, patient doses are also reduced. Because there are few photons emitted and the percent transmission is reduced, the number of detected photons falls markedly, which will increase image mottle.

4. **A.** The choice of mathematical reconstruction filter (AKA reconstruction kernel) will affect the amount of noise (and resolution) in the reconstructed image. For example, a standard (soft tissue) filter will reduce image noise, whereas a detail (bone) filter will increase image noise. Choice of how images are displayed (i.e., window and level) will obviously affect the appearance of an image but not the intrinsic image properties such as noise.

Comment

The presence of noise in CT (AKA mottle) limits the radiologist's ability to detect low-contrast lesions. This will be true for both small and large lesions. When a lesion has high contrast, it can generally be detected irrespective of whether it is small or large, provided that the contrast is much higher than the level of random noise. Imaging scientists use the CNR as a relative indicator of the visibility of any lesion. When changing techniques (kV and mAs) increases the CNR, lesion visibility improves, and vice versa. The absolute value of any CNR metric has little meaning, whereas changes in CNR will indicate how the visibility of any lesion is expected to change (Fig. 6.17).

In CT, mottle represents random fluctuations in the attenuation coefficient expressed as HU. Consider an image of a water phantom with an average HU of zero. When the phantom is scanned, not every pixel has a value of zero and the measured noise for pixels in a region of interest would likely indicate a mean and standard deviation of 0 ± 3 HU. This means that 68% of the pixels have HU

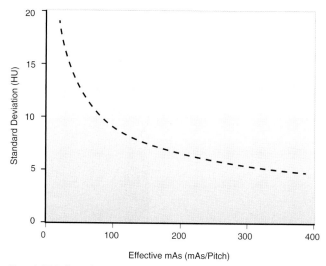

Fig. 6.17 When the effective milliampere-second (i.e., mAs/pitch) increases, the amount of measured noise in a CT image is reduced, and vice versa. By computing the effective milliampere-second, changes in noise can be readily obtained with explicit reference to CT pitch.

between −3 and +3, 95% between −6 and +6, and 99% between −9 and +9 HU. If the level of noise is ±3 HU, then a lesion that differed from the background by only 1 HU would be invisible, but one that differed by 10 HU would be visible when optimally displayed.

CT mottle is determined by the number of x-ray photons used to make the image, which is true in all x-ray imaging modalities. When four images are acquired of the same patient's anatomy and then averaged together, the signal is clearly quadrupled. Because noise is random in nature, the noise in the combined four images would only be doubled. Accordingly, when we have four times as many photons generating an image the CNR doubles. The number of photons used to create any CT image is directly proportional to the milliampere and the x-ray tube rotation time(s). It is inversely proportional to the pitch, so doubling the pitch also halves the number of photons creating the image. Quadrupling the slice thickness will also quadruple the number of x-ray photons in an image, which will halve the resultant image noise.

Mottle in reconstructed images is strongly influenced by choice of mathematical reconstruction filter. The use of filters with good resolution performance (e.g., lung and bone) will substantially increase CT mottle compared with a low-noise soft tissue (standard) filter. Iterative reconstruction is currently being introduced that reduces mottle in CT images, but these algorithms also changes noise texture and therefore how noise is measured.

References

Goldman LW. Principles of CT: radiation dose and image quality. *J Nucl Med Technol.* 2007;35(4):213–225.

Huda W. CT II. In: *Review of Radiological Physics.* 4th ed. Philadelphia: Wolters Kluwer; 2016:166.

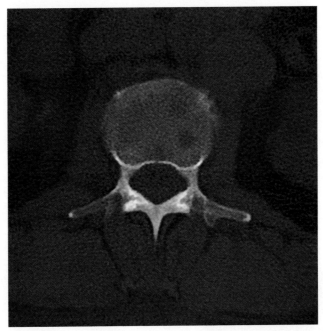

Fig. 6.18

1. Which of the following has the best spatial resolution performance?
   A. CT
   B. Magnetic resonance (MR)
   C. Ultrasound (US)
   D. Positron emission tomography (PET)

2. In CT imaging, increasing which factor could improve spatial resolution performance?
   A. Focal spot size
   B. Matrix size
   C. Rotation time
   D. Detector size

3. Changing which x-ray tube parameter would most likely affect the spatial resolution of a CT inner ear examination?
   A. Rotation time (s)
   B. Tube current (mA)
   C. X-ray voltage (kV)
   D. A, B, and C

4. Increasing which of the following parameters would likely degrade spatial resolution performance of a clinical CT scanner?
   A. Field of view
   B. Detector size
   C. Focal spot
   D. A, B, or C

CASE 6.9

Spatial Resolution

Fig. 6.18 Lytic lesion in a lumbar vertebral body.

1. **A.** Current clinical commercial CT scanners have a limiting resolution of approximately 0.7 line pairs (lp) per mm, which is better than that of MR (~0.3 lp/mm), axial resolution in adult abdominal US (~0.2 lp/mm), or PET (~0.1 lp/mm).

2. **B.** Increasing only the matrix size could improve the spatial resolution performance because this will result in smaller pixels for a given field of view. Increasing the focal spot and detector size both reduce the resolution performance, and an increase in rotation time will likely increase motion blur.

3. **A.** Motion blur will always be affected by the rotation time, where faster x-ray tube rotation speeds might improve image sharpness. Changing x-ray techniques of tube voltage (kV) and tube current (mA) will have no practical effect on spatial resolution performance in current clinical CT imaging.

4. **D.** A smaller field of view will improve the achievable resolution and is often used in clinical practice (zoom reconstruction using the original projections) as depicted in Fig. 6.19. Smaller detectors increase the detector sampling rate and will always improve spatial resolution. A small focal spot reduces focal blur, which is of particular importance given the relatively large geometric magnification (×2 at the isocenter) on commercial CT scanners.

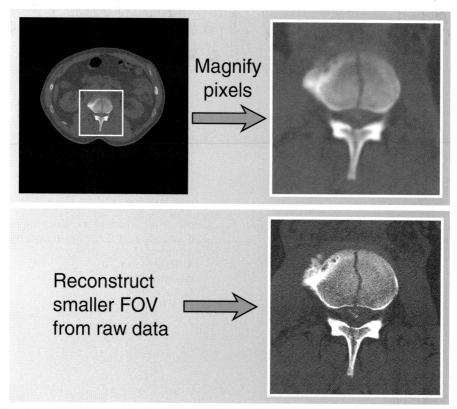

Magnify pixels

Reconstruct smaller FOV from raw data

Fig. 6.19 The effect of applying zoom reconstruction (i.e., reconstructing a smaller area of the image) versus magnifying on a workstation. Note the reduced blur and improved visualization of bone fragments in the spinal canal when the image is reconstructed using zoom as compared with magnifying on a workstation display. *FOV,* Field of view.

Comment

The fundamental limit on the achievable resolution on current CT scanners is determined by the sizes of the focal spot and detector element. In practice, using a 512 × 512 matrix yields the best achievable resolution, and larger matrix sizes would be of no practical benefit. CT spatial resolution improves in systems that use smaller focal spots and detectors with smaller pixels (e.g., dental cone beam CT). Current in-plane spatial resolution is approximately 0.7 lp/mm, which is markedly inferior to digital chest x-ray radiography (~3 lp/mm) and digital mammography (~7 lp/mm). The human eye, at a normal viewing distance of 25 cm, has a limiting resolution of approximately 5 lp/mm.

Detector length, typically 0.6 mm, determines resolution in the longitudinal direction that is oriented along the craniocaudal axis. Modern multidetector CT generally can achieve isotropic resolution (same in all dimensions), which is in marked contrast to CT scanners in the 20th century where the in-plane resolution was up to an order-of-magnitude better that in the longitudinal direction (slice thickness). Although modern CT scanners typically reconstruct images using thicker slices (3 to 5 mm), the underlying raw data have a more detailed intrinsic resolution. This was not the case in the days of single slice CT, where images were created from data that had a slice thickness between 5 and 10 mm.

Technical ways of improving spatial resolution relate to the choice of focal spot and the choice of x-ray tube rotation time. Use of the small focal spot improves spatial resolution for selected examinations such as the inner ear, where mottle (i.e., mAs) is of little concern. However, when the small focal spot is selected, the x-ray tube power loading will generally need to be reduced to 25%. Consequently, this may require a quadrupling of the scan time to maintain the same level of mottle. With longer scan times, there will always be an increased likelihood of more motion blur. This can be minimized by the use of patient immobilization devices. Reducing x-ray tube rotation times will help to reduce motion blur, which is particularly important in abdominal image and in pediatrics.

Spatial resolution can also be modified using "image processing" methods (Fig. 6.19). Detail-type reconstruction filters are used to achieve the best possible resolution, which is markedly better than soft tissue reconstruction filters. The price that is paid for using high-resolution filters is a corresponding increase in image noise. Use of zoom reconstruction takes the acquired projections and performs a filtered back projection algorithm using a reduced field of view. Use of zoom-reconstructed images in this way will improve resolution in the central image region. As shown in Fig. 6.19, magnifying images on the workstation will not improve the visibility of image detail.

References

Huda W. CT II. In: *Review of Radiological Physics.* 4th ed. Philadelphia: Wolters Kluwer; 2016:168.

Radiological Society of North America. RSNA/AAPM Physics Modules. Computed tomography: 2. CT image quality and protocols. https://www.rsna.org/RSNA/AAPM_Online_Physics_Modules_.aspx. Accessed March 1, 2018.

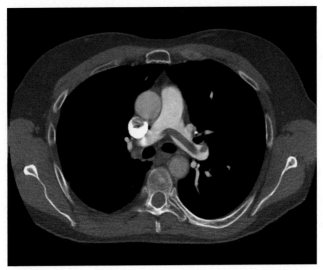

Fig. 6.20

1. For imaging the chest, which modality most likely has the best temporal resolution?
   A. Chest x-ray
   B. Fluoroscopy (frame)
   C. CT
   D. A, B, and C are equivalent

2. On a current single-source CT scanner, with an x-ray tube rotation speed of 4 rotations per second, what is the most likely best achievable temporal resolution (ms)?
   A. 500
   B. 250
   C. 125
   D. 60

3. When "fast" images are obtained using a nominal 180-degree x-ray tube rotation, what is likely degraded compared with "slower" images obtained using 360-degree x-ray tube rotation?
   A. Spatial resolution
   B. Image mottle
   C. Lesion contrast
   D. A and B

4. For dual-source CT scanner with an x-ray tube rotation time of 300 ms, what is the best achievable temporal resolution (ms)?
   A. 40
   B. 75
   C. 150
   D. 300

CASE 6.10

Temporal Resolution

Fig. 6.20 CT angiogram of the chest demonstrating a saddle pulmonary embolus.

1. **A.** The exposure time of a chest x-ray is very short (3 to 5 ms). A fluoroscopy frame takes approximately 33 ms to acquire. A routine CT image can normally be created in approximately 300 ms.

2. **C.** The best achievable temporal resolution on a single-source CT scanner is approximately half the gantry rotation time. If the x-ray tube rotates four times in 1 second, this corresponds to a gantry rotation time of 250 ms, so that the best possible temporal resolution would be 125 ms.

3. **D.** In normal CT imaging, projections at 0 degrees and 180 degrees are not identical but offset by a half detector distance, which improves data sampling and spatial resolution. Using only half the projections does reduce blurring associated with motion. When only half the projections are acquired, image mottle increases because "fewer x-rays" are being used to create a CT image. Lesion contrast will be identical for both 180- and 360-degree rotations.

4. **B.** For a dual-source CT scanner with the same gantry rotation time of 300 ms, the necessary projections to reconstruct an axial image are approximately a quarter of the gantry rotation time (75 ms), or half the time required for a single-source canner.

Comment

The shortest time that a standard CT imaging can acquire 1000 projections during a 360-degree x-ray tube rotations is currently close to 0.3 seconds. Temporal resolution the time required to produce an image, is thus close to 300 ms. This may be compared with the 300 seconds it took the first clinical EMI scanner to acquire the data needed to reconstruct a CT image. Temporal resolution in routine clinical CT scanning was approximately 3 seconds in the 1980s and 1 second in the 1990s, but it is unlikely to undergo dramatic improvements in the foreseeable future due to the forces generated by spinning the gantry at such a high speed (Table 6.2).

After an x-ray tube rotates approximately 180 degrees, it is possible to synthesize a set of projections pertaining to x-ray tube angles 180 to 360 degrees. This process is known as rebinning and permits a CT image to be obtained when the x-ray tube rotates 180 degrees. Although using 180-degree rotation provides the best temporal resolution for a single-source CT scanner, the image quality will be inferior because of reduced spatial sampling. Rays at 0 and 180 degrees are normally offset by half a detector width, which increases spatial resolution. This benefit is lost when only half of the projections are used to create an image. Furthermore, the use of only half of the x-ray photons in 180-degree rotation will increase the amount of mottle. If motion artifact is a serious concern, then using 180-degree rotation may be clinically beneficial, even though it has inferior resolution and noise characteristics.

Dual-source CT scanners acquire image data (projections) at double the rate of single-source scanners. This in turn can improve temporal resolution performance by a factor of two because the required projections are obtained in half the time. Temporal resolution of current dual-source CT scanner is approximately one-quarter of the x-ray tube rotation time (Fig. 6.21). In cardiac imaging, it is often more important to be able to create an image quickly than achieving the best resolution and the lowest noise. Inferior resolution plus increased mottle in the resultant images can still be adequate for cardiac imaging tasks.

References

Huda W. CT II. In: *Review of Radiological Physics*. 4th ed. Philadelphia: Wolters Kluwer; 2016:167.

Lin E, Alessio A. What are the basic concepts of temporal, contrast, and spatial resolution in cardiac CT? *J Cardiovasc Comput Tomogr*. 2009;3(6):403–408.

Table 6.2 NOMINAL X-RAY TUBE ROTATION TIMES FOR REPRESENTATIVE CT SCANNERS THROUGH THE FIRST FOUR DECADES OF THIS IMAGING MODALITY

Nominal Decades	X-Ray Rotation Time Seconds	Comment
1970s	300	EMI scanner
1980s	3	Engineering improvements
1990s	1	Spiral computed tomography scanning
2000s	0.3	Multislice scanning

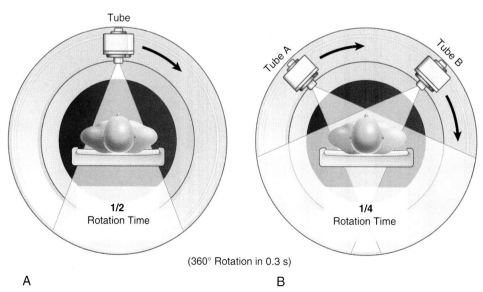

(360° Rotation in 0.3 s)

A   B

Fig. 6.21 The best possible temporal resolution that is achievable on a single-source CT scanner is approximately half of the x-ray tube rotation time (A), whereas on a dual-source CT scanner it is approximately a quarter of the x-ray tube rotation time (B).

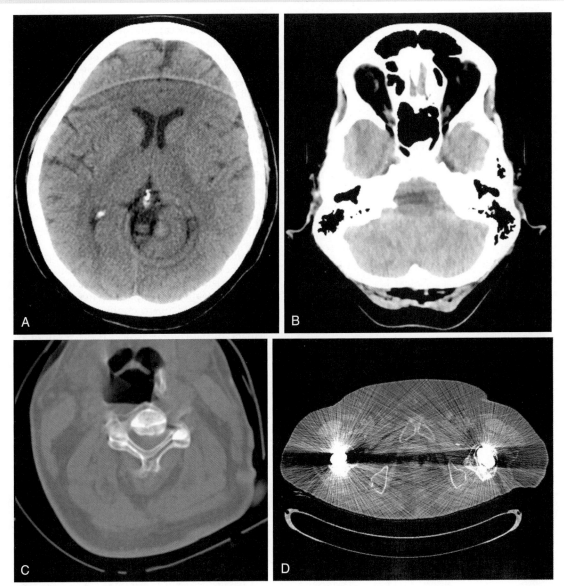

Fig. 6.22

1. What is the most likely cause of the artifact depicted in Fig. 6.22A?
   A. Miscalibrated detector
   B. Gibbs artifact
   C. Motion artifact
   D. Photon starvation

2. What is the most likely cause of the artifact depicted in Fig. 6.22B?
   A. Beam hardening
   B. Motion artifact
   C. Stair-step artifact
   D. Improper CT protocol

3. What is the most likely cause of the artifact depicted in Fig. 6.22C?
   A. Cervical spine fracture
   B. Volume averaging
   C. Ghosting
   D. Motion artifact

4. What is the most likely cause of the artifact depicted in Fig. 6.22D?
   A. Motion
   B. Metal artifact
   C. Ghosting
   D. Ba spill

CASE 6.11

Artifacts

Fig. 6.22 (A) Miscalibrated detector. (B) Beam hardening. (C) Motion artifact. (D) Metal artifact.

1. **A.** A CT scanner assumes equal amounts of radiation will produce equal outputs by each detector element. If the detector elements are not calibrated properly, they will send out high and low signals for the same receptor dose. Since the detector rotates around the patient, the pattern is circular.

2. **A.** The CT scanner assumes the beam energy does not change as it moves through the patient. Because bone preferentially attenuates low x-ray energies (due to the photoelectric effect), an artifact may arise if enough bone is present to substantially harden the beam spectrum.

3. **D.** The CT scanner assumes a patient is stationary so it sees each part of the anatomy from all angles. If the patient moves during scanning, the projections no longer overlap correctly, causing artifacts in the reconstructed image.

4. **B.** Metal causes a very distinctive artifact. This artifact is the result of the metal blocking most of the photons from reaching the detector (photon starvation).

Comment

One or more ring artifacts may arise in the image if the detector is not properly calibrated. This occurs because the system assumes each detector element will give the same signal when hit with the same amount of radiation. Miscalibration causes the signal to be too high or too low and to vary across detector elements. It is in a ring shape because the detectors rotate around the patient. Recalibration by the technologist or service engineer usually solves the problem. A single bright ring may be seen when a detector element is broken. This usually requires a new detector element.

Aliasing artifacts can occur when edges are imaged because of inadequate data sampling (i.e., too few detectors in each centimeter of the detector arc). Unfortunately, there is seldom a change that can be made to eliminate aliasing. Metal implants and long path lengths through highly attenuating materials can result in no photons reaching the detector (photon starvation). This appears as streaks of high noise across the image. Scanning a patient with their arms above their heads, avoiding the metal implant, iterative reconstruction, and increased kilovolts may reduce starvation artifacts.

Patient motion (e.g., heart beating or patient breathing) produces ghosts or streaks in the direction of motion. Due to the axial acquisition plane in CT, motion artifact is seen as an abrupt displacement of anatomy when viewing sagittal and coronal reconstructions. Sedation, restricting patient movement, and decreasing scan time can all reduce motion artifact. Partial volume artifact is the result of averaging the linear attenuation coefficient in a voxel that contains two different tissues. In Fig. 6.23, partial volume averaging of the right major fissure (see Fig. 6.23A) with the more superior right middle lobe (see Fig. 6.23C) may appear to represent a pulmonary nodule (see Fig. 6.23B). Thinner slices and smaller detectors help to reduce partial volume artifact in images.

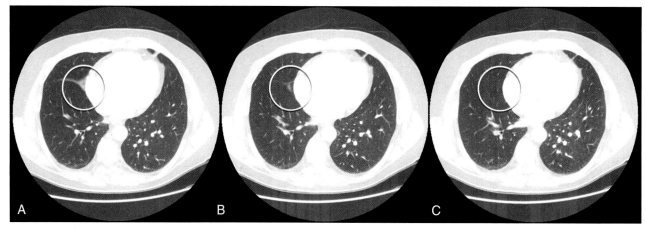

Fig. 6.23 Partial volume artifact. This is caused by two types of tissue being present in the same image voxel. In B, both soft tissue from the major fissure (A showing the slice below) and lung (C showing the slice above) are present in the fissure *(circle)*. As a result, the attenuation for that pixel will be an average of the attenuation for lung and the attenuation for soft tissue. Looking at B alone, this could appear to represent a pulmonary nodule.

Beam-hardening artifacts are caused by lower-energy photons being preferentially absorbed, resulting in a more penetrating (higher-energy) beam hitting the detector. When the energy increases, the computed HU values are lowered (darker), showing darker areas behind strong absorbers (e.g., bone). Software algorithms that incorporate prior knowledge of the patient (e.g., skull in head CT scan) have been developed to reduce beam-hardening artifacts.

References

Barrett JF, Keat N. Artifacts in CT: recognition and avoidance. *Radiographics.* 2004;24(6):1679–1691.

Huda W. CT II. In: *Review of Radiological Physics.* 4th ed. Philadelphia: Wolters Kluwer; 2016:168.

CASE 7.1

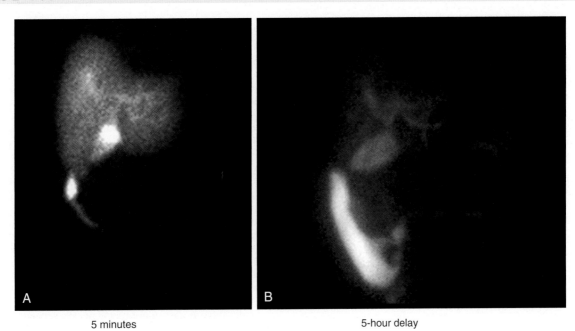

5 minutes   5-hour delay

Fig. 7.1

1. How many neutrons are there in carbon-14 ($Z = 6$)?
   A. 6
   B. 8
   C. 14
   D. 20

2. Which is an isobar of carbon-13?
   A. Carbon-12
   B. Nitrogen-13
   C. Oxygen-13
   D. B and C

3. Which is (are) radioactive?
   A. Carbon 11
   B. Carbon 13
   C. Carbon 14
   D. A and C

4. What is the most likely kinetic energy of an alpha particle emitted by an unstable radionuclide (e.g., radon)?
   A. Approximately electron volt
   B. Approximately kiloelectron volt
   C. Approximately megaelectron volt
   D. A, B, or C

CASE 7.1

Nuclear Structure

Fig. 7.1 (A and B) Hepatobiliary scan in a patient with bile leak following cholecystectomy. Note the pooling of radiotracer on the delayed image (B). The unstable nucleus of Tc-99m releases a photon that is detected to make these images.

1. **B.** The mass number of a nucleus is the total number of nucleons, which are the protons and neutrons. Carbon-14 has a total of 14 nucleons and must have six protons because the atomic number of carbon is six (by definition). Accordingly, carbon-14 must have eight neutrons (14 minus 6).

2. **D.** Two nuclei are deemed to be isobars when the total number of nucleons is the same (i.e., sum of the number of protons and neutrons). Carbon-13 has a total of 13 nucleons, which is the same as nitrogen-13 and oxygen-13.

3. **D.** Carbon-11 is a short-lived positron emitter that has great research potential in biomedical research (positron emission tomography). Carbon-14 is a low-energy beta minus (β−) emitter that is widely used in biomedical research to study various metabolic processes.

4. **C.** Most nuclear transformations such as the ejection of an alpha particle will have energies that are a few megaelectron volts. Outer shell electrons have binding energies of a few electron volts, and inner shell electrons have binding energies of "keV" to "tens keV."

Comment

Nuclei contain protons and neutrons, often referred to as nucleons. The mass number (A) is the number of protons (Z) added to the number of neutrons (N). Iodine-127 (also written ^{127}I or $_{53}{}^{127}I$) has an atomic number Z of 53 and 127 nucleons (53 protons and 74 neutrons). Two nuclides having the same mass number (A) are called isobars, and two with the same atomic number Z are called isotopes (Fig. 7.2). Stable nuclides with low mass number such as oxygen-16 (^{16}O) have approximately equal numbers of neutrons and protons, but in heavy nuclides such as lead this becomes approximately 1.5:1 (e.g., ^{208}Pb has 82 protons and 126 neutrons).

Radionuclides are unstable isotopes that emit radiation in their attempt to reach a stable (i.e., lower-energy) nuclear state. Transformation of an unstable nuclide is referred to as

Table 7.1 EXAMPLES OF THE PRODUCTION OF RADIONUCLIDES USED IN NUCLEAR MEDICINE

Production Mode	Radionuclides	Decay Mode
Cyclotron	^{67}Ga, ^{111}In, ^{123}I, ^{18}F	Positron or electron capture
Fission	^{131}I, ^{133}Xe, ^{99}Mo	Beta minus
Neutron activation	^{125}I, ^{89}Sr	Beta minus
Generator	^{99m}Tc, ^{82}Rb, ^{68}Ga	Isomeric and positron

radioactive decay. The original nuclide is referred to as the "parent," and the nuclide produced after any nuclear transformation is referred to as a "daughter." A transmutation occurs (i.e., a new element is formed) whenever there is a change in the number of protons. A large variety of parent radionuclides can be stored in generators where their daughters are extracted and used in imaging. For example, Mo-99 decays to Tc-99m, which we use extensively in imaging.

Radionuclides can be produced using several different methods (Table 7.1). They can be by-products of fission (i.e., when heavy nuclides break up) in nuclear reactors. Fission products include Mo-99, I-131, and Xe-133. Fission-produced radionuclides generally decay by a β− process (i.e., an electron emitted from the nucleus). Radionuclides are also produced in nuclear reactors, where neutrons smash into a target material. As the neutrons hit the target, some get incorporated into the nucleus and create a radioactive atom in a process called neutron activation. Note that neutrons activate only a portion of the total target material, and radioactive atoms cannot be chemically separated from the nonradioactive atoms. Examples include Sr-89 and I-125.

Finally, a cyclotron can be used to bombard a material with protons moving at very high speeds. Protons are absorbed in a fraction of the atoms they bombard. These atoms then become new elements (because the Z changes) and are radioactive. Cyclotron-produced materials can be chemically separated and supplied as pure materials (i.e., carrier free). These typically decay by positron decay or electron capture (EC). Examples include F-18, In-111, and Ga-67.

References

Huda W. Nuclear medicine I. In: *Review of Radiological Physics*. 4th ed. Philadelphia: Wolters Kluwer; 2016:177–180.

Mettler Jr FA, Guiberteau MJ. Radioactivity, radionuclides, and radiopharmaceuticals. In: *Essentials of Nuclear Medicine Imaging*. 6th ed. Philadelphia: Elsevier; 2012:1–7.

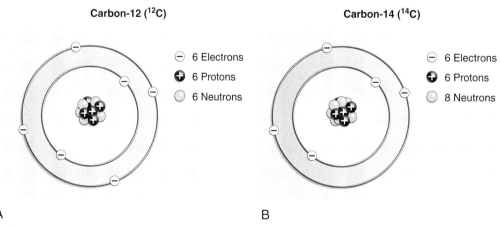

Fig. 7.2 Stable carbon-12 (A) has six protons and six neutrons in its nucleus and is stable. Carbon-14 (B) also has six protons but eight neutrons and is an unstable isotope. Carbon-14 decays by emitting a beta minus particle and has a half-life of 5700 years.

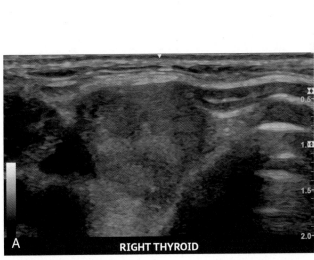

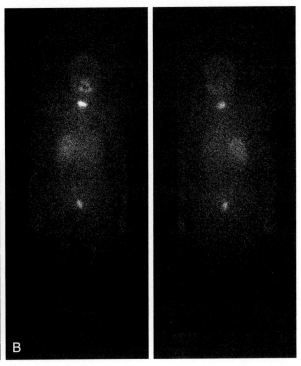

A   RIGHT THYROID

B

Fig. 7.3

1. What type of radionuclide is produced when a stable nucleus absorbs a neutron in a nuclear reactor?
   A. β− emitter
   B. Positron emitter
   C. EC
   D. Alpha emitter

2. What best characterizes radionuclides that are produced when a fissile nucleus splits into two daughter nuclei?
   A. Positron emitters
   B. Alpha emitters
   C. Long lived
   D. Carrier free

3. What best characterizes radionuclides that are produced in cyclotrons?
   A. Positron emitters
   B. Short lived
   C. Carrier free
   D. A, B, and C

4. How far do beta particles most likely travel in soft tissues (mm)?
   A. <1
   B. 1
   C. 10
   D. 100

CASE 7.2

Beta and Alpha Decay

Fig. 7.3 Iodine 131 (I-131) treatment (B) in a patient with thyroid cancer detected on ultrasound (A). I-131 emits beta minus particles to destroy neoplastic tissue.

1. **A.** When a nucleus absorbs neutrons, it likely has too many neutrons to be stable. Too many neutrons can also be expressed as having too few protons. Accordingly, nuclei created in reactors by absorbing neutrons are usually β– emitters, where the daughter nuclei have an atomic number ($Z + 1$) that is one higher than the unstable parents (Z).

2. **D.** Fission products are typically β– emitters. Because fissile materials are different chemicals, fission products are "carrier free." Every nucleus of a fission product (e.g., ^{99}Mo) is radioactive, whereas ^{99}Mo produced by neutron activation in a reactor will be mixed up with stable ^{98}Mo.

3. **D.** Cyclotrons accelerate positive particles (e.g., protons). When positive charge(s) are added to a target nucleus, the resultant radionuclides have too much positive charge, resulting in decay by positron emission and/or EC. These are typically short-lived (1 to 100 minutes) and are carrier free because the addition of a proton creates a new element (e.g., O → F).

4. **A.** The range of beta particles in soft tissues is less than 1 mm. When a beta emitter such as ^{131}I is taken up by the thyroid gland, virtually all the kinetic energy will be deposited within this organ, which is an important consideration for therapeutic applications in nuclear medicine (NM).

Comment

β– decay involves a neutron being converted into a proton, which occurs in nuclei with too few protons (or too many neutrons) (Table 7.2). Energy is released during this nuclear transformation in the form of an energetic electron, which is referred to as a beta particle because the process is known as beta decay (Fig. 7.4B). An example of a β– emitter is ^{131}I, which is used to treat patients with thyroid cancer. Because the beta particles typically travel less than 1 mm in soft tissues, virtually all of their kinetic energy will be deposited in the organ that takes up this radionuclide.

Beta plus (β+) decay involves a proton being converted into a neutron, which occurs in nuclei with too many protons (i.e., neutron deficient). Energy is released in this nuclear transformation in the form of a positron, which is a positively charged electron. After an energetic positron has lost all its kinetic energy, it will annihilate with an electron, producing two 511-keV photons emitted in opposite directions (i.e., 180 degrees apart). The most popular positron emitter used in positron emission tomography (PET) imaging is ^{18}F. In EC, a proton captures an atomic electron and is converted into a neutron. EC occurs in nuclei with too many protons (i.e., neutrons deficient) and therefore competes with positron emission (see Fig. 7.4A).

Heavy radionuclides with a high mass number can emit an alpha particle consisting of two neutrons and two protons (and no electrons). A common alpha emitter is radium (Ra) 226, which decays to radon (Rn) 222, which is also an alpha emitter. When ingested, inhaled, or injected, alpha particles can pose a high risk. Domestic radioactive radon (and its daughters) is believed by the Environmental Protection Agency to be responsible for up to 15% of lung cancers in the United States.

References

Cherry SR, Sorenson JA, Phelps ME. Modes of radioactive decay. In: *Physics in Nuclear Medicine*. 4th ed. Philadelphia: Elsevier; 2012:19–26.

Huda W. Nuclear medicine I. In: *Review of Radiological Physics*. 4th ed. Philadelphia: Wolters Kluwer; 2016:177–179.

Table 7.2 RADIOACTIVE DECAY MODES FOR AN UNSTABLE PARENT CONTAINING *Z* PROTONS AND WITH A MASS NUMBER A

Decay Mode	DAUGHTER NUCLEUS	
	Mass Number	**Atomic Number**
Isomeric	A	Z
Beta minus	A	$Z + 1$
Positron	A	$Z – 1$
Electron capture	A	$Z – 1$
Alpha	$A – 4$	$Z – 2$

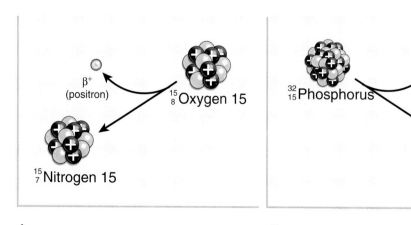

A   B

Fig. 7.4 Oxygen-15 (A) is a positron emitter, which decays into nitrogen-15 with one fewer proton (7 vs. 8) and thus moves to the left (with the x-axis representing atomic number). (B) Phosphorus-32, which is a beta minus emitter, decays to sulfur-32, which has one more proton (16 vs. 15) and thus moves to the right. In both cases the parent is "higher" than the daughter, as the vertical direction indicates the overall energy level.

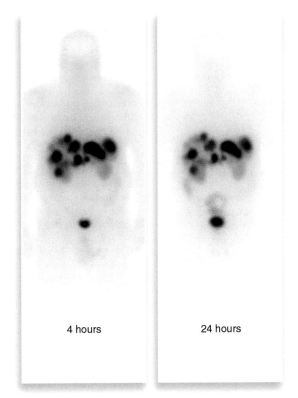

4 hours   24 hours

Fig. 7.5

1. Which is most likely to change when an unstable radionuclide undergoes an isomeric transition?
   A. Mass number
   B. Atomic number
   C. A and B
   D. Neither A nor B

2. How much activity remains after 10 half-lives (%)?
   A. 10
   B. 1
   C. 0.1
   D. 0.01

3. How will increasing the half-life of a radionuclide affect the amount of activity?
   A. Increase
   B. Decrease
   C. Unchanged

4. If the physical half-life of ^{131}I is 8 days and the biological half-life in the thyroid gland is also 8 days, what is the effective half-life (days)?
   A. 4
   B. 8
   C. 16
   D. 64

CASE 7.3

Isomeric Transitions and Activity

Fig. 7.5 Octreotide scan in a patient with metastatic carcinoid tumor. Multiple areas of increased uptake are seen throughout the liver at 4 and 24 hours. Normal renal and bladder uptake. The half-life of indium-111 is 2.8 days, which easily allows for 24-hour imaging.

1. **D.** In an isomeric transition, an excited nucleus emits excess energy in the form of electromagnetic radiation (gamma ray). The parent and daughter nuclei have the same mass number A and the same atomic number Z.

2. **C.** After one half-life, half of the initial activity (50%) remains, and after two half-lives, the activity that remains is ¼ (i.e., ½ × ½). After 10 half-lives, the activity is down to $(½)^{10}$, or 1/1024, which is 0.1%.

3. **B.** When the half-life increases, the decay constant $(λ = 1/T_{½})$ is reduced. Because activity is the product of the decay constant and the number of nuclei (i.e., $N × λ$), a very long half-life implies negligible activity.

4. **A.** There is a formula that can be used to compute the effective half-life, namely $1/T_{effective} = 1/T_{physical} + 1/T_{biological}$. In this example, the effective half-life is 4 days and is used to estimate the thyroid dose when ^{131}I is administered to patients therapeutically.

Comment

The stable nuclear configuration is called the ground state, and higher energy configurations are unstable isomeric states. When an isomeric state transforms into any lower configuration, gamma rays are emitted. The definition of a gamma ray is electromagnetic radiation that originates in a nuclear process, whereas x-rays are also electromagnetic radiation but are produced by electrons. Parent and daughter nuclei have the same mass number and atomic number after an isomeric transition.

^{99m}Tc is 140 keV higher than ^{99}Tc, and transitions between these energy levels will result in the emission of 140-keV gamma rays. This photon energy (i.e., 140 keV) is high enough to escape from a patient but low enough to be easily detected. Nuclides such as ^{99m}Tc and ^{123}I (160 keV) are thus ideal for use in NM. Isomeric states with long lifetimes are called metastable, and a metastable state is denoted by a lowercase m as in ^{99m}Tc or alternatively technetium-99m.

A radionuclide's activity is defined as the number of transformations (decays) that occur each second. The United States uses non-SI units of millicuries (mCi), where 1 mCi is $3.7 × 10^7$ nuclear transformations per second. The SI unit of activity is the becquerel (Bq), where 1 Bq is 1 nuclear transformation per second. Accordingly, 1 mCi is 37 MBq. Radioactivity decays exponentially and is characterized by the decay constant λ, which is related to the half-life $(T_{½})$ by $T_{½} = 0.69/λ$. Because nuclear radioactivity activity is inversely proportional to the decay constant λ, long-lived radionuclides have a very small λ and thus very little activity. This can be illustrated by considering ^{99}Tc, the daughter of ^{99m}T, which is unstable. The half-life of ^{99}Tc is very long (100,000 years), and therefore the activity of ^{99}Tc in a patient who has undergone a ^{99m}Tc scan will be negligible.

In one physical half-life $(T_{½})$, half of the radionuclides will decay. Half the initial activity remains after one half-life, a quarter remains after two half-lives, and an eighth remains after three half-lives. A useful benchmark is that after 10 half-lives, only 0.1% of the initial activity remains (Fig. 7.6). Contaminated clothing and objects are typically stored for 10 half-lives, after which time the remaining activity would be deemed to be "negligible" (i.e., the same as natural background). The effective half-life (T_e) of a radionuclide encompasses both radioactive (physical) decay (T_p) and biological clearance (T_b) in any organ. Effective half-lives are always shorter than both physical and biological half-lives and may be obtained using the formula $1/T_e = 1 T_b + 1 T_p$.

References

Bushberg JT, Seibert JA, Leidholdt Jr EM, Boone JM. Radioactivity and nuclear transformation. In: *The Essential Physics of Medical Imaging*. 3rd ed. Philadelphia: Wolters Kluwer; 2012:579–594.

Huda W. Nuclear medicine I. In: *Review of Radiological Physics*. 4th ed. Philadelphia: Wolters Kluwer; 2016:177–179.

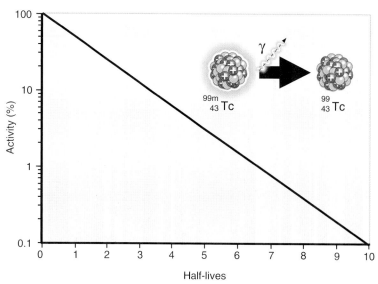

Fig. 7.6 There is an exponential decay of any radionuclide, as shown on this semi-log graph, where after 10 half-lives only 0.1% of the initial activity remains.

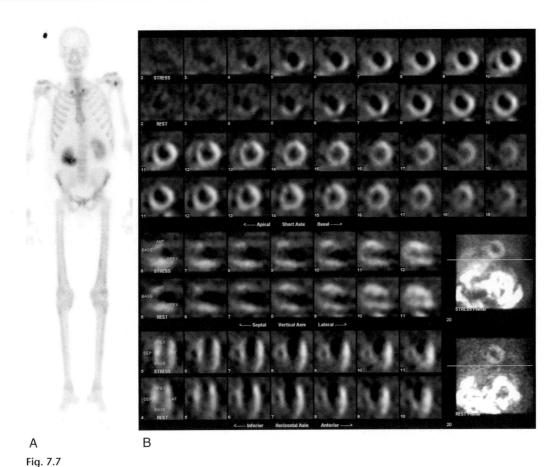

A   B

Fig. 7.7

1. Which parameters influence the amount of activity that can be eluted from a radionuclide generator?
   A. Initial activity
   B. Parent half-life
   C. Daughter half-life
   D. A, B, and C

2. If long-lived radionuclide has a daughter with a half-life of 1 hour, how long will it take for equilibrium to be established (hours)?
   A. 1
   B. 2
   C. 4
   D. 8

3. How long will a ^{99}Mo/^{99m}Tc generator most likely be useful in an NM department?
   A. Day
   B. Week
   C. Month
   D. Year

4. Which of the following radionuclides produced by generators are positron emitters?
   A. Rubidium-82
   B. Gallium-68
   C. Indium-113m
   D. A and B

CASE 7.4

Radionuclide Generators

Fig. 7.7 ^{99m}Tc methyl diphosphonate (MDP) bone scan (A) demonstrating hypertrophic osteoarthropathy. Myocardial perfusion scan (B) demonstrating a large fixed anterior wall defect. The ^{99m}Tc used in both scans is drawn from a generator.

1. **D.** The maximum activity that can be initially eluted from a generator is the initial parent activity. At any later time, the maximum activity will be determined by the parent activity which decays according the parent half-life. At any time, the actual amount of daughter activity depends on the daughter half-life.

2. **C.** For a long-lived radionuclide, equilibrium is established after four daughter half-lives, and this is referred to as "secular." For a short-lived parent, equilibrium is also established after four daughter half-lives. In the case of a short-lived parent, however, the equilibrium is generally referred to as "transient."

3. **B.** The half-life of ^{99}Mo is 67 hours so that after a week the activity is less than 25% of the initial activity. ^{99}Mo/^{99m}Tc generators are generally replaced every week.

4. **D.** Rubidium-82 is a positron emitter used in cardiac PET imaging and is produced when strontium-82 decays. Gallium-68 is a positron emitter utilized in tumor imaging and is produced when germanium-68 decays. Indium-113m is a gamma emitter that used to be used for ventilation imaging and is produced when tin-113 decays.

Comment

Radionuclide generators continuously produce a useful daughter radionuclide from the radioactive decay of a longer-lived parent (Fig. 7.8). Technetium generators have an aluminum column that is loaded with ^{99}Mo (i.e., parent) and which decays to ^{99m}Tc (i.e., daughter). When saline is passed through the column, it elutes the ^{99m}Tc that has been produced. ^{99}Mo is not soluble in saline and will therefore remain on the column, whereas the ^{99m}Tc is eluted as sodium pertechnetate. Approximately 90% of diagnostic NM studies make use of ^{99m}Tc-labeled pharmaceuticals, with new agents added every year.

A generator starting with approximately 100 GBq of ^{99}Mo initially has no ^{99m}Tc present. As ^{99}Mo decays, ^{99m}Tc activity is produced and ^{99m}Tc activity (daughter) increases until equilibrium is reached *after four daughter half-lives*. Because ^{99m}Tc has a half-life of 6 hours, a ^{99}Mo generator requires 24 hours to reach equilibrium. Activities of parent and daughter are always taken to be approximately equal at equilibrium. Consequently, a 100-GBq ^{99}Mo generator also has approximately 100 GBq ^{99m}Tc at equilibrium. The half-life of ^{99}Mo is 66 hours, so these generators generally are used for approximately 5 working days, which corresponds to approximately two parent half-lives. After 5 days, yield of a ^{99m}Tc generator would be less than 25% of initial activity, at which time it is returned to the vendor and replaced by a new generator.

When the parent radionuclide is approximately 10 times longer lived than the daughter, generators are said to reach transient equilibrium, but when the parent is greater than 100 times longer lived than the daughter, the generator is said to reach secular equilibrium. The important point is that both secular and transient equilibrium occur after four daughter half-lives. In secular equilibrium, parent and daughter activities are taken to be equal, whereas in transient equilibrium, parent and daughter activities are only approximately equal. For clinical purposes, the differences between secular and transient equilibrium are of negligible importance.

For cardiac imaging procedures, ^{82}Rb can be obtained from ^{82}Sr, which is a positron emitter. ^{82}Rb is rapidly taken up by the myocardium and is used to assess cardiac perfusion. The very short half-life of ^{82}Rb (1.25 minutes) permits sequential studies every 5 minutes. Additional isotopes produced by generators for NM include ^{113m}In, which is a gamma emitter, and ^{68}Ga, which is a positron emitter. ^{113m}In was previously used for ventilation imaging and has been proposed for liver imaging in developing countries. ^{68}Ga is used in neuroendocrine tumor imaging, where it is combined with somatostatin analog.

References

Huda W. Nuclear medicine I. In: *Review of Radiological Physics*. 4th ed. Philadelphia: Wolters Kluwer; 2016:180–181.

Society for Nuclear Medicine and Molecular Imaging. Targeted cancer treatment with nuclear medicine therapy. www.snmmi.org/therapyinfographic. Accessed March 2, 2018.

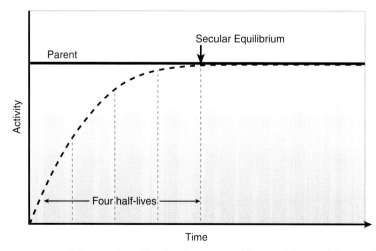

Fig. 7.8 A long-lived parent has constant activity over time. The daughter starts with no activity, reaching equilibrium where parent/daughter activities are equal after four daughter half-lives. Because the parent is long lived, this would be referred to as secular equilibrium.

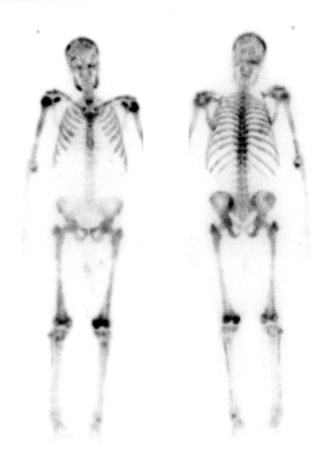

Fig. 7.9

1. What percentage of 140-keV photons are likely absorbed (photoelectric effect) in the sodium iodide crystal on a clinical gamma camera?
   A. >90
   B. 60
   C. 30
   D. <10

2. What's the most likely energy resolution of a sodium iodide scintillator at a photon energy of 140 keV (%)?
   A. 5
   B. 10
   C. 20
   D. 50

3. Which "energy pulse(s)" will likely be accepted by a 20% window for Tc-99m settings (keV)?
   A. 130
   B. 140
   C. 150
   D. A, B, and C

4. Which radionuclide from contaminated shoes is most likely to generate a photopeak at an energy of 511 keV?
   A. ^{18}F
   B. ^{68}Ga
   C. ^{82}Rb
   D. A, B, and C

CASE 7.5

Scintillators

Fig. 7.9 The 140-keV photons emitted from a patient injected with ^{99m}Tc methyl diphosphonate (MDP) interact in the scintillators and form the image. This is an example of a "superscan," where there is diffuse osseous radiotracer uptake secondary to metastases. Bone uptake is so avid that the kidneys are not visualized.

1. **A.** The thickness of a sodium iodide crystal in a clinical gamma camera is approximately 10 mm. Most of the incident photons with an energy of 140 keV (>90%) will be absorbed in the crystal due to the photoelectric effect.

2. **B.** The energy resolution of a radiation detector is the width of the photopeak measured as a full width half maximum (FWHM). For a gamma camera, the width of the technetium-99m photopeak is approximately 14 keV (FWHM), which is expressed as 10% of the photopeak energy (140 keV).

3. **D.** The pulse height analyzer (PHA) settings on a typical gamma camera is 20%. The PHA will thus accept photon energies in the range of 126 to 154 keV, which will capture most of the photoelectric events that occur in the sodium iodide crystal.

4. **D.** Gallium-68, fluoroine-18, and rubidium-82 are all positron emitters, which result in the emission of 511 keV photons. Differentiating between these three positron emitters could be achieved by estimating the half-life by making a repeat measurement later.

Comment

After a gamma ray has been absorbed in a sodium iodide scintillator, many light photons are produced, with the total amount of light directly proportional to the energy deposited. The total light produced by the scintillator is typically detected, turned into an electrical signal, and amplified by a photomultiplier tube (PMT), as shown in Fig. 7.10. The detected signals appear as an energy spectrum, where the vertical axis is the number of counts at a given energy, and the corresponding horizontal axis is the photon energy. A typical spectrum consists of photopeak events, as well as scatter events that appear below the photopeak energy. When two photons are detected at the same time, this is referred to as a coincidence, which will generally have an energy higher than the photopeak.

Detection efficiency is the percentage of incident gamma rays totally absorbed in the scintillator, via the photoelectric effect. At 140 keV, a 10-mm-thick sodium iodide crystal will absorb virtually all the photons, whereas at 500 keV, it detects very few photons because the photoelectric effect is proportional to $1/E^3$. Each absorbed gamma ray will produce a slightly different amount of light because of statistical fluctuations. This causes a "broadening of the photopeak" (i.e., instead of all photons detected at the true energy, they will have a small range about the true energy). This spread is measured by the FWHM and is expressed as the energy resolution, which is the percentage of the photopeak energy. For example, the FWHM of a ^{99m}Tc photopeak (140 keV mean peak) in a 10-mm sodium iodide crystal is approximately 14 keV, so the corresponding energy resolution is 10%.

Pulse height analysis (PHA) is used to determine which part of any detected spectrum will be used to create the image. PHA sets off a lower signal value (S_-) and an upper signal value (S_+) for pulses that are detected by the scintillator and PMT detector. Signals lower than S_- or higher than S_+ are excluded from subsequent analysis. Energy signals lower than the photopeak (i.e., Compton scatter events in a patient) and higher than the photopeak (i.e., coincidences) do not contain imaging information and their inclusion would markedly degrade image quality.

The PHA window width (i.e., $S_+ - S_-$) is generally set to twice the energy resolution. If the photopeak window is not centered on the peak energy (i.e., symmetrical), a flood image will depict artifacts in the form of "hot spots." For a sodium iodide crystal, using a PHA window width that is 20% of the photopeak energy will likely accept most of the photopeak events, as well as rejecting most of the scatter for common radionuclides such as techentium-99m and I-123. When the PHA window is increased, both the number of photopeak and scatter events will increase, and vice versa. Accordingly, with a wide window we have too many scatter events, and with a narrow window too few photopeak events, so an optimized balance is required.

References

Huda W. Nuclear medicine I. In: *Review of Radiological Physics.* 4th ed. Philadelphia: Wolters Kluwer; 2016:182–184.

Radiological Society of North America. RSNA/AAPM physics modules. Nuclear medicine: 1. Radiation detection instrumentation in nuclear medicine practice. https://www.rsna.org/RSNA/AAPM_Online_Physics_Modules_.aspx. Accessed March 2, 2018.

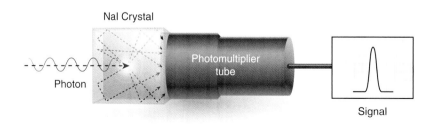

Fig. 7.10 Schematic representation of a scintillator detector where the sodium iodide *(NaI)* crystal absorbs a photon, releasing an amount of light that is proportional to the energy deposited in the crystal. The light is detected by a photomultiplier tube that produces a signal (pulse) whose intensity is proportional to the energy deposited in the crystal.

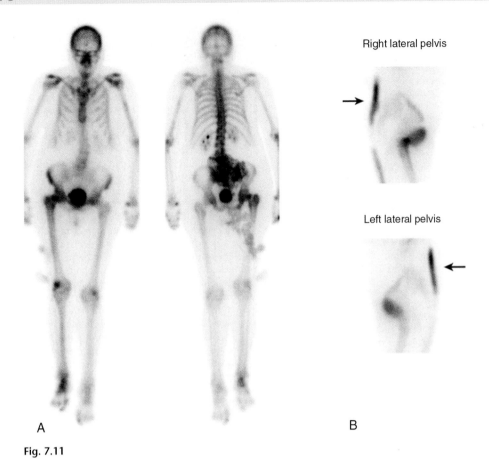

Right lateral pelvis

Left lateral pelvis

A   B

Fig. 7.11

1. When a contaminated swipe with 1 kBq of ^{99m}Tc is placed into an efficient well counter, how may "counts" are most likely in the 140-keV photopeak every second?
   A. 1
   B. 1000
   C. 1,000,000

2. Which type of contamination can be detected by a Geiger-Müller (G-M) detector?
   A. Beta
   B. X-ray
   C. Gamma
   D. A, B, and C

3. How many photons must be detected to produce a detection event in a G-M detector?
   A. 1
   B. 10
   C. 100
   D. >100

4. When 1 MBq of ^{99m}Tc and 1 MBq of ^{18}F are measured in a dose calibrator, which will likely produce the "biggest signal"?
   A. ^{99m}Tc
   B. ^{18}F
   C. Same

CASE 7.6

Detectors in Nuclear Medicine

Fig. 7.11 ^{99m}Tc methyl diphosphonate (MDP) bone scan demonstrating irregular uptake along the posterior pelvic area (A). Lateral views confirm that this artifact is due to contaminated clothing posterior to the patient (B). Gamma camera detectors are designed to detect almost all ^{99m}Tc photons to give acceptable image quality at a reasonable patient dose.

1. **B.** An activity of 1 kBq will result in 1000 nuclear transformations every second, so the number of gamma rays emitted will also be close to 1000 per second. A well counter will detect a high percentage of these.

2. **D.** In a G-M probe, an incident photon or a beta particle interacts with gas atoms and an electron is ejected and accelerated by a high voltage. These electrons subsequently knock out additional electrons in an "electron avalanche" (Fig. 7.12B).

3. **A.** A G-M detector can detect a single photon that interacts with the air in the chamber. One interaction results in an electron avalanche (see Question 2) and can result in an audible "click." G-M detectors are thus extremely sensitive and can detect a single photon.

4. **B.** The signal from 1 MBq of ^{18}F will likely be three times higher than the signal from 1 MBq of ^{99m}Tc. The (physics) reason for this is there is more energy released from an ^{18}F nuclear disintegration event than from a ^{99m}Tc event.

Comment

When detector sensitivity is important, samples of radioactivity can be placed into a "well" within a large NaI crystal. This well counter has a high sensitivity because it absorbs most emitted gamma rays. The output from a well counter is used to identify radionuclides from the measured photopeak energy. For example, a photopeak of 140 keV would thereby identify the radionuclide as ^{99m}Tc, whereas a photopeak at 365 keV would thereby identify this as ^{131}I. Well counters have excellent efficiency and are used to quantify the amount (MBq) of activity present. A contaminated swab with of 1 kBq of activity results in up to approximately 1000 detected counts per second in the well counter.

Uptake of radioiodine in patients can be measured using a gamma camera or an uptake probe that permits comparisons of patient activity (uptake) with a known amount in a neck phantom. Thyroid uptakes using the probe are usually obtained 24 hours after administration of the radioiodine and are measured at a standard distance from the scintillator crystal in the probe (e.g., 25 cm). Neck counts are corrected for background activity and then compared with a standard in a neck phantom. Uptake probes are also used to monitor radioactive iodine (e.g., ^{131}I) in workers who have handled this nuclide to ensure their uptake is within regulatory safety limits.

A dose calibrator consists of an ionization chamber (see Fig. 7.12A) that can accurately measure the activity of a radioisotope dose. Dose calibrators have a response that varies with radionuclide, and so each radionuclide has a unique electronic setting selected by the user. Accurate measurements require the dose calibrator to be set up with an appropriate nuclide specific calibration factor. Activity is measured prior to injection into a patient to ensure proper dosage. In a Geiger counter (see Fig. 7.12B), when a photon or beta particle interacts with an atom and ejects an electron, this triggers an avalanche resulting in a very large signal. Geiger counters are very sensitive and thus ideal for detecting very small amounts of radiation (contamination). However, they are also very crude devices and cannot accurately measure dose or dose rate.

Nuclear Regulatory Commission regulations require administered doses to be ±20% of the dose prescribed for a given diagnostic study. All dose calibrators must undergo detailed quality control measurements. The daily constancy is checked by measuring a ^{137}Cs source which is "constant" because it has a half-life of 30 years with interdaily variations less than 5%. Commercially available calibrated sources check the accuracy at installation and annually. Linearity means that when the activity doubles, so does the calibrator response. It is typically checked quarterly by measuring the decay of a vial of ^{99m}Tc over several days. Geiger counters are checked on a regular basis by testing their response using a small sealed source (typically Cs-137) to ensure that the readings do not drift over time or miss contamination. It is also important to check the battery before each use using a button on the meter. If the battery is low, the reading can be incorrect.

References

Huda W. Nuclear medicine I. In: *Review of Radiological Physics.* 4th ed. Philadelphia: Wolters Kluwer; 2016:182–184.

Radiological Society of North America. RSNA/AAPM physics modules. Nuclear medicine: 1. Radiation detection instrumentation in nuclear medicine practice. https://www.rsna.org/RSNA/AAPM_Online_Physics_Modules_.aspx. Accessed March 2, 2018.

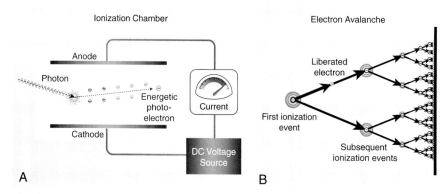

Fig. 7.12 An ionization chamber (A) simply collects all the electrons produced following an ionization event and subsequent ionizations caused by the energetic photoelectron. This can be compared with a Geiger counter (B), where each electron results in an electron avalanche, resulting in a very large signal. Ionization chambers are thus insensitive but accurate, whereas Geiger counters are sensitive but crude, making them good for contamination detection.

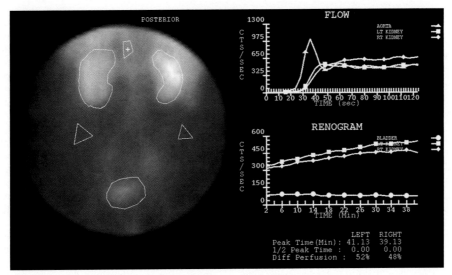

Fig. 7.13

1. During a detection event, what is the sequential order of the key components of a gamma camera (collimator; crystal; PMTs)?
   A. Collimator; PMT; crystal
   B. Crystal; collimator; PMT
   C. Collimator; crystal; PMT
   D. PMT; collimator; crystal

2. How many of the incident gamma rays are transmitted through the collimator of a typical gamma camera (%)?
   A. 10
   B. 1
   C. 0.1
   D. 0.01

3. What is the best descriptor of a crystal detector used in a gamma camera?
   A. Scintillator
   B. Photoconductor
   C. Photostimulable phosphor
   D. Ionization chamber

4. What does the PMT array in a gamma camera measure?
   A. Total scintillator light
   B. Pattern of light distribution
   C. A and B
   D. Neither A nor B

CASE 7.7

Gamma Cameras

Fig. 7.13 99mTc Mag 3 renogram demonstrating retention of radiotracer in both kidneys in this patient with acute tubular necrosis.

1. **C.** On a gamma camera, the incident gamma rays have to first pass through a collimator, so only those that are traveling perpendicular to the collimator plane get through to the detector/scintillator (sodium iodide). After being absorbed by the detector crystal, some of the energy is converted into light and detected by PMTs (light detectors).

2. **D.** Only approximately 0.01% of the incident photons will be allowed to pass onto the detector in a gamma camera. The fundamental problem of gamma cameras is the low number of counts in the resultant image, which results very high levels of image mottle.

3. **A.** Gamma cameras use a scintillator as a detector, generally sodium iodide crystals that absorb photons and convert a fraction of this absorbed energy into visible light. Photoconductors and photostimulable phosphors are currently used in radiography. A dose calibrator is an ionization chamber.

4. **C.** The total amount of light that is generated is collected by the PMT, converted to a signal, and subjected to PHA. The pattern light causes a pattern of PMT signals across the detector, which is used to identify the x and y coordinates where the incident gamma ray was absorbed and to place photopeak events (only) into the correct image pixel.

Comment

Gamma cameras use a collimator to determine the general location that gamma rays originate from in the patient. Gamma rays originating within the patient pass through a hole in a lead collimator. To reach the scintillator, photons must be traveling parallel to the collimator holes. Collimators provide directional and coarse spatial information relating to the in-plane location of each single gamma ray interaction. When a collimator is used, we obtain a projection image that is analogous to a chest x-ray. Gamma rays that pass through the collimator are absorbed by the photoelectric effect by a sodium iodide scintillator crystal. The thickness of a typical crystal is 10 mm, which is sufficient to capture most of the incident photons for Tc-99m (140 keV) and I-123 (160 keV).

Scintillators absorb incident gamma photons and produce many light photons. The amount of light is directly proportional to the absorbed energy, so that the total amount of light can be used to identify the photopeak interactions that will be used to create the NM image. Adding the signals from all the PMTs (typically approximately 37) yields a cumulative electrical signal that reflects the energy of the photon that was absorbed in the crystal (Fig. 7.14). PHA counts only photopeak events, which will be used to create an image. Imaging of radionuclides that emit several very different energies use multiple windows to maximize efficiency. For example, 111In uses two windows for its gamma rays—173 and 247 keV.

The point at which the photon is absorbed will have the highest light intensity, so that the light pattern provides spatial information of the interaction site. Location of the gamma ray interaction is determined by a pulse arithmetic circuit. Along each x and y projection, the pattern of light is analyzed to identify where interaction occurred. The highest signal will be registered by the nearest PMT. Based on the relative strength of signals from each PMT, the pixel location (x and y) can be determined and an event added to this pixel.

Gamma cameras produce projection images of the distribution of radioactivity in patients, built up one gamma ray (i.e., count) at a time. A typical gamma camera image (e.g., lung scan) will likely have approximately 500,000 counts. However, for a 64 × 64 matrix size, this is only slightly more than 100 photons/pixel, which will thus exhibit random interpixel fluctuations (noise) of approximately 10%. The reason for this is that a collimator is essential to obtain spatial information, but very few photons are transmitted through those currently used in NM. A good rule of thumb is only 1 in 10,000 (0.01%) or fewer, so that the resultant images have very few counts.

References

Huda W. Nuclear medicine I. In: *Review of Radiological Physics.* 4th ed. Philadelphia: Wolters Kluwer; 2016:184–187.

Mettler Jr FA, Guiberteau MJ. Instrumentation and quality control. In: *Essentials of Nuclear Medicine Imaging.* 6th ed. Philadelphia: Elsevier; 2012:28–34.

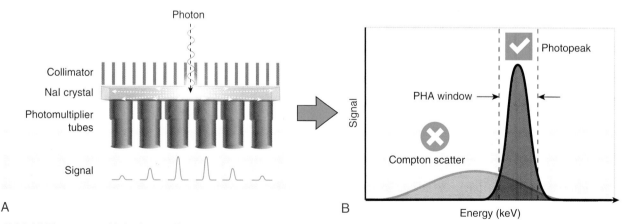

Fig. 7.14 (A) The pattern of light detected by the array of photomultiplier tubes, which permits identification of the location (x, y) of the photon interaction. The total amount of light collected can be used to generate a spectrum (B) where pulse height analysis can be performed to accept most of the photopeak events and reject most of the scatter events.

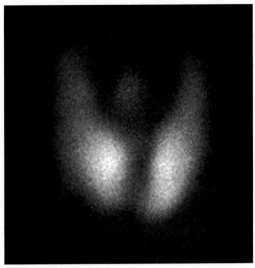

Fig. 7.15

1. Which radionuclide would require the use of a collimator with the most lead (i.e., be the heaviest)?
   A. ^{67}Ga
   B. ^{111}In
   C. ^{123}I
   D. A and B

2. How will increasing the sensitivity of a gamma camera collimator influence the corresponding collimator spatial resolution performance?
   A. Deteriorate
   B. No effect
   C. Improve

3. Which component most likely dominates the system resolution of a gamma camera?
   A. Collimator assembly
   B. Scintillator thickness
   C. PMT array
   D. A, B, and C are similar

4. Which type of collimator likely has the worst spatial resolution performance for any given organ?
   A. Parallel hole
   B. Diverging
   C. Converging
   D. Pinhole

CASE 7.8

Collimators

Fig. 7.15 Pinhole collimator image of the thyroid in a patient with Graves disease demonstrating diffuse increased uptake of iodine 123.

1. **D.** Medium-energy collimators are longer and also have thicker septa to reduce septal penetration. Low-energy collimators are used with ^{99m}Tc and ^{123}I, whereas the higher energies of photons emitted by ^{111}In and ^{67}Ga require the use of medium energy collimators that also have (much) more lead.

2. **A.** If the hole size is made larger, it will transmit more photons, which increases the sensitivity. However, when holes are large, there is an increase in uncertainty as to the gamma event location, so resolution will deteriorate. Collimators in NM trade "sensitivity" with "resolution performance" depending on the construction.

3. **A.** System resolution is a combination of the FWHM of the collimator and detector components, such as scintillator thickness and PMT array (i.e., intrinsic). A typical collimator has a resolution of approximately 8 mm, much larger than the intrinsic resolution of 3 mm.

4. **B.** A parallel hole collimator generates an image that reflects that actual patient size. Pinhole and converging collimators generate a magnified image. By contrast, a diverging collimator might be used to minify the image of a patient, although resolution (i.e., detail visibility) will worsen.

Comment

Parallel-hole collimators are usually made of lead and contain multiple holes that are perpendicular to the collimator plane. The lead strips between the holes are called septa, whose thickness can be varied to adjust the collimator sensitivity and resolution. Parallel-hole collimators always project the same object size onto the camera and have a field of view (FOV) that is independent of the distance between the collimator and patient. Low-energy collimators are thin and can be manufactured from bent foils. These collimators are used with low-energy gamma ray emitters up to approximately 160 keV (e.g., ^{99m}Tc and ^{123}I).

Medium-energy collimators are used with between approximately 160 and 300 keV for radionuclides such as ^{67}Ga and ^{111}In that uses higher-energy photons of 185 and 300 keV. These are both thicker to minimize penetration, as well as having thicker septa (Fig. 7.16). High-energy collimators (>300 keV) have the greatest amount of lead for use with ^{131}I that has the highest photon energy of 365 keV. These collimators are very heavy and are usually manufactured using casting techniques (not foil). When images are obtained of high-energy nuclides using a low-energy collimator, artifacts will be present because many gamma rays are transmitted through the collimator septae.

Converging collimators produce a magnified image and are useful for small anatomy and infant imaging. Diverging collimators project an image size that is smaller than the object size and could be used to increase the FOV to image larger patients and capture the whole organ (e.g., lung) on a single image. Pinhole collimators are cone shaped with a single hole at the apex and result in images that are normally magnified and will also be inverted. As a rule, resolution will improve as the amount of geometrical magnification increases. Accordingly, pinhole and converging collimators should exhibit more detail than parallel-hole collimators, whereas diverging collimators exhibit less.

Collimator sensitivity is defined as the number of counts detected per MBq in a patient being imaged. Parallel-hole collimator sensitivity is independent of distance from the collimator because each hole sees a larger area of the patient as it moves away but also loses photons due to the inverse square law. However, parallel-hole collimator resolution falls off rapidly with distance; thus it is best to place the detector as close to the patient as possible. High-sensitivity collimators have large holes and are thinner. High-resolution collimators have small holes and are thicker to better localize the activity. A high-sensitivity collimator will typically transmit three times more photons than a high-resolution collimator but have half the resolution at a distance of 10 cm (i.e., 16 vs. 8 mm FWHM).

References

Cherry SR, Sorenson JA, Phelps M: The gamma camera: basic principles. In: *Physics in Nuclear Medicine*. 4th ed. Philadelphia: Elsevier; 2012:201–204.

Huda W. Nuclear medicine I. In: *Review of Radiological Physics*. 4th ed. Philadelphia: Wolters Kluwer; 2016:185–186.

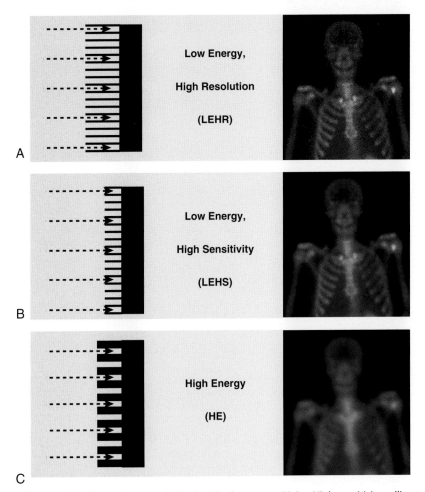

Fig. 7.16 A high-resolution collimator (A) offers the best resolution but the lowest sensitivity. High-sensitivity collimators (B) have reduced resolution performance. High-energy collimators (C) generally have reduced resolution and sensitivity in comparison with low-energy collimators, as shown in the simulated images on the right.

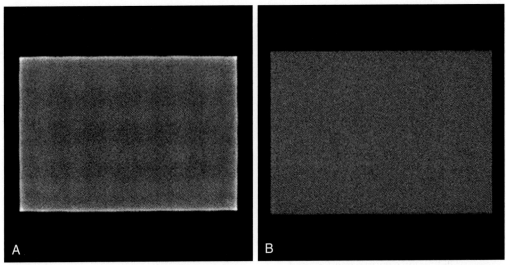

Fig. 7.17

1. How frequently are extrinsic uniformity flood images typically acquired in an NM department?
   A. Daily
   B. Weekly
   C. Monthly
   D. Quarterly

2. Which uniformity value would (just) fail the uniformity test and render the gamma camera to be unacceptable for clinical use (%)?
   A. <1
   B. 1 to 2
   C. 2 to 3
   D. >3

3. Which would most likely be used to determine the extrinsic uniformity of a gamma camera?
   A. Point source of ^{99m}Tc
   B. Point source of ^{57}Co
   C. Flood source of ^{99m}Tc
   D. Flood source ^{57}Co

4. How many million (M) counts would likely be acquired in a flood image that is used to generate sensitivity correction factors?
   A. 0.1
   B. 1
   C. 10
   D. 100

CASE 7.9

Gamma Camera Quality Control

Fig. 7.17 Uniformity testing with the raw uncorrected flood image (A) and corrected image (B).

1. **A.** Gamma camera extrinsic uniformity is typically tested daily by imaging a flood source. Extrinsic means that the test is performed with the collimator on and therefore requires an area source. A practical area source is a plate of ^{57}Co, which has a photon energy of 120 keV that is close to that of ^{99m}Tc but a much longer half-life.

2. **D.** Gamma camera uniformity is the standard deviation of the observed count values expressed as a percentage of the mean number of pixel counts in the image. Thus lower values indicate greater uniformity. Typical values of uniformity are in the range of 2% to 3%. Uniformities in excess of 3% need to be corrected before the gamma camera is used.

3. **D.** Extrinsic uniformity is typically tested by imaging an area source of uniform activity. The collimator blocks a large amount of the photons from point sources, and thus the acquisition time would be too long for daily testing. There are liquid area sources that one can add Tc-99 to for testing, but the technologist dose, difficulty attaining proper uniformity, and relatively short half-life mean that Co-57 sources are typically used instead.

4. **D.** Daily uniformity images are usually composed of approximately 1 million counts, which gives low enough natural fluctuation to assess uniformity but also can be acquired in a reasonable amount of time. Uniformity correction maps are actually used to improve clinical images and require a much higher degree of accuracy (lower natural fluctuations) to ensure good image quality. As a result, 100 million counts is typically used and is performed only monthly due to the extended acquisition time.

Comment

Gamma camera quality control is essential for detecting hardware and software problems that degrade image quality. Although it may differ slightly from scanner to scanner, quality control typically generally follows a standard schedule (Table 7.3). The word "intrinsic" in a quality control test's name indicates that the measurement is made without the collimator and is used to assess the detector and electronics (e.g., count rate). "Extrinsic" measurements use the collimator and test the total system performance. Often different equipment is needed to test intrinsic and extrinsic properties. For example, intrinsic uniformity tests typically use a point source of ^{99m}Tc, whereas extrinsic uniformity tests use a large sheet made of ^{57}Co ($T_{1/2}$ 270 days and 120 keV gamma rays).

Table 7.3 EXAMPLES OF REQUIRED GAMMA CAMERAS TESTS AND THEIR FREQUENCY

Test	Frequency	Comments
Spectrum display	Daily	Ensures peak center and spread are reasonable
Extrinsic uniformity	Daily	Collimator on; also called system uniformity
Spatial resolution	Weekly	Typically extrinsic with a bar phantom
Linearity	Weekly	Spatial resolution image tests this as well
Sensitivity	Annual	Registered counts per MBq of activity
Energy resolution	Annual	Determines window width setting

Uniformity images (floods) are obtained daily before any patient has been scanned. Because it is the whole imaging chain that needs to be assessed, these are usually extrinsic flood acquisitions acquired with approximately 1 million counts. Uniformity is the standard deviation expressed as percentage of mean counts in a flood image. Gamma cameras are expected to have a uniformity of 2% to 3% at most, although some newer cameras yield acceptable images even at 6%. Nonuniformities are unacceptable for clinical imaging because hot or cold spots may mimic pathology and degrade quantitative measurements. It is important to check with the vendor for the specified acceptable uniformity. Uniformity correction maps that are used to correct the planar images are usually derived from a uniform flood image of 100 million counts and acquired monthly.

Photopeak window of the PHA is evaluated daily using a source that irradiates the whole crystal. Typically, approximately 3 million counts are necessary to assess the peak. If the photopeak is incorrectly centered, there will be "hot spots" seen at the location of each PMT, and the total number of counts accepted by the PHA will also be markedly reduced. "Hot spots" can also be seen in flood images when the wrong energy correction factors are applied, such as ^{67}Ga instead of ^{99m}Tc. However, in this case the flood image contains the expected number of counts, whereas an incorrectly centered PHA will show a reduced number of counts.

Resolution and linearity are typically tested with a bar phantom. Quadrant bar phantoms have four sets of parallel bars, with each rotated through 90 degrees. The most common dimensions are 3.5, 3.0, 2.5, and 2.0 mm (Fig. 7.18). Bar pattern phantoms also check for linearity (i.e., ability to image straight lines), performed weekly.

References

Huda W. Nuclear medicine I. In: *Review of Radiological Physics.* 4th ed. Philadelphia: Wolters Kluwer; 2016:187–188.

Mettler Jr FA, Guiberteau MJ. Instrumentation and quality control. In: *Essentials of Nuclear Medicine Imaging.* 6th ed. Philadelphia: Elsevier; 2012:55–61.

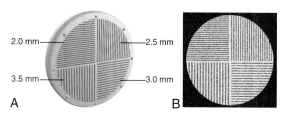

Fig. 7.18 (A) A phantom that can be used to assess the linearity and resolution performance of a gamma camera. (B) The intrinsic resolution performance generated with a flood source. The scanner successfully reproduces all line patterns.

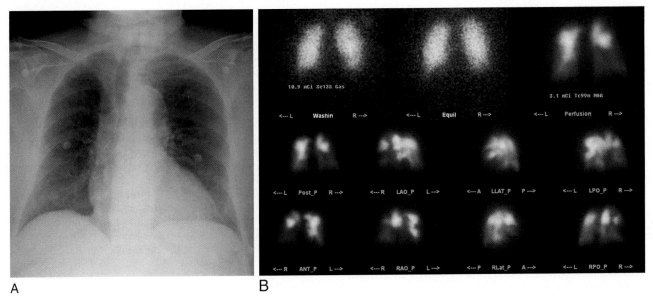

A

B

Fig. 7.19

1. Which type of lesion is most likely to be detected in a planar gamma camera image?
   A. Hot spot
   B. Cold lesion
   C. A and B similar

2. How can the resolution of a gamma camera collimator be improved?
   A. Smaller holes
   B. Thicker septa
   C. Increased thickness
   D. A, B, and C

3. Which is most likely to improve when a high-resolution collimator is replaced by a high-sensitivity collimator?
   A. Mottle
   B. Artifacts
   C. Contrast
   D. A, B, and C

4. How could the number of counts in a gamma camera image be increased?
   A. Administer more activity
   B. Count for longer time
   C. Use a high-sensitivity collimator
   D. A, B, and C

CASE 7.10

Gamma Camera Image Quality

Fig. 7.19 Chest radiograph (A) of a patient with shortness of breath and a ventilation/perfusion scan (B) in the same patient demonstrating high probability of pulmonary embolism. Although less detail is present, the functional information provided by the nuclear medicine scan is crucial for diagnosis.

1. **C.** If the lesion does not take up activity, this results in a cold spot. On the other hand, if the lesion preferentially takes up activity, this is a hot spot. To achieve good contrast, it is not important whether the lesion is hot or cold; the most important aspect is that the lesion is different (i.e., high contrast).

2. **D.** The resolution of a gamma camera collimator can be increased by making the holes smaller, increasing the septa thickness, and making the collimator thicker (which will make the holes longer). In general, the more lead a collimator has, the better it will be able to localize the origin of gamma rays within a patient but the lower sensitivity.

3. **A.** A high-sensitivity collimator will generate more counts and thereby reduce mottle for a given imaging time. Artifacts and contrast are unlikely to be affected by switching between high resolution and high sensitivity.

4. **D.** One way to get more counts are to use more activity, which would increase the patient dose. Counting for a longer time is also technically possible, but this would increase motion artifacts and might not be well received by patients. Use of high-sensitivity collimators would increase counts but at the price of reduced resolution performance.

Comment

In NM, contrast is defined as the difference in count density (counts per pixel) in an abnormality compared with the corresponding count density in the surrounding anatomy (normal background). Contrast will generally be degraded by septal penetration, as well as Compton scatter events that are (unfortunately) accepted by the PHA. Contrast in NM images occurs either when radiopharmaceuticals localize in the organ of interest ("hot spot") or when there is an absence of uptake ("cold spot"). Radioactivity in other normal tissues will produce undesirable background counts, degrading contrast, which may be quantified in a relative manner as the target-to-background ratio.

Gamma camera lung images contain only approximately 10 photons in each square millimeter, whereas a chest radiograph is likely to have more than 10,000 x-ray photons in each square millimeter. The low number of counts causes the poor (i.e., mottled) appearance in NM images, as shown in Fig. 7.20. Although there are ways of increasing NM image counts, these have an important downside. If the administered activity is increased, the patient dose will go up. If the imaging time is increased, this may not be acceptable to patients, and motion artifacts will increase. If a higher-sensitivity collimator is used, the spatial resolution performance will be degraded.

With a low-energy high-resolution (LEHR) collimator, a gamma camera will have an FWHM of approximately 8 mm at an imaging distance of 10 cm. The gamma camera intrinsic resolution is typically 3 mm, which is very small compared with the collimator. Resolutions in NM are normally expressed as an FWHM because this can be readily measured with electronic calipers. However, to compare NM with computed tomography (CT), we must convert FWHM to line pairs (lp)/mm using the formula of FWHM = 1/2 (lp/mm). Thus an FWHM of 8 mm corresponds to a limiting spatial resolution of 0.06 lp/mm. Gamma camera resolution is nearly 10 times worse than CT (0.7 lp/mm). Note that in CT, a resolution of 0.7 lp/mm would correspond to an FWHM of 0.7 mm, which is far too small to be measured directly.

References

Huda W. Nuclear medicine I. In: *Review of Radiological Physics*. 4th ed. Philadelphia: Wolters Kluwer; 2016:192–193.

Radiological Society of North America. RSNA/AAPM physics modules. Nuclear medicine: 2. Gamma cameras/image quality. https://www.rsna.org/RSNA/AAPM_Online_Physics_Modules_.aspx. Accessed March 2, 2018.

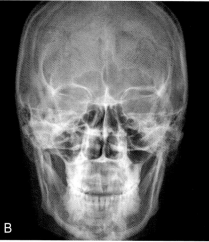
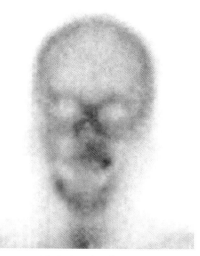

10^8 photons/mm^2 10^4 photons/mm^2 10 photons/mm^2

Fig. 7.20 The least amount of mottle is seen in the photograph (A) where the highest number of photons are used to create the image. The gamma camera image (C) has the highest level of mottle because these images are created using the lowest number of photons. X-rays (B) are intermediate between these two extremes.

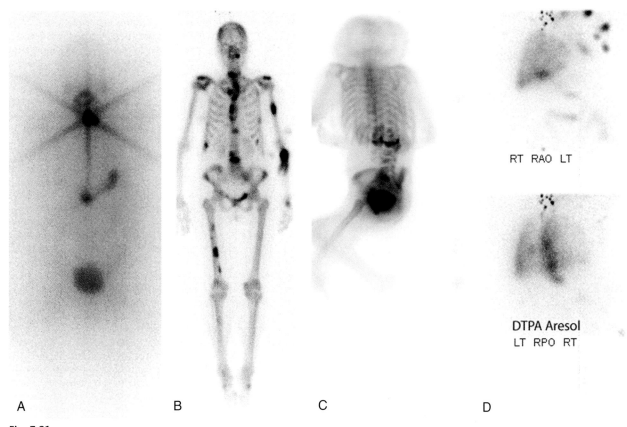

A   B   C   D

Fig. 7.21

1. What is the cause of the artifact shown in Fig. 7.21A?
   A. Cracked crystal
   B. Damaged collimator
   C. Off-peak imaging
   D. Septal penetration

2. What artifact is shown in Fig. 7.21B?
   A. Extravasation
   B. Ghosting
   C. Ringing
   D. Reverberation

3. What is the cause of the artifact shown in Fig. 7.21C?
   A. Incorrect energy corrections
   B. Patient or camera motion
   C. Broken photomultiplier tube
   D. Incorrect corrections (energy)

4. What is the cause of the artifact shown in Fig. 7.21D?
   A. Patient contamination
   B. Photon starvation
   C. Incorrect center of rotation
   D. Poor uniformity correction

CASE 7.11

Gamma Camera Artifacts

Fig. 7.21 (A) Star artifact due to septal penetration during an I-131 scan. (B) Intravenous infiltration of radiotracer (left arm) during bone scan. (C) Motion artifact during bone scan. (D) Sneeze during the aerosol portion of a ventilation perfusion scan resulting in contamination.

1. **D.** Darker pixels indicate more radiation hitting the detector. If a high-energy radioisotope (e.g., I-131) is imaged using a low-energy collimator, most of the photons will penetrate the septae. The most photons penetrate if they hit the collimator perpendicular to the septae; thus the previous hexagonal septae creates a six-point star pattern.

2. **A.** If the injection is not performed correctly, radioisotope pools at the injection site creating a "hot spot." This will also cause fewer counts to collect in the desired regions because less radioisotope collects in those regions.

3. **B.** NM scans have long acquisition times. If the patient moves during these times, then counts will begin to accumulate from the new positions. This will result in blurring and ghosts in the image that often make localization and diagnosis difficult or impossible.

4. **A.** In this ventilation study the patient sneezed shortly after inhaling the radioisotope. Because mucus retains some of the isotope, the hot spots seen around the neck and head region are contamination from the sneeze (likely present on the patient's skin and clothing).

Comment

Reduced resolution and ghosting will occur if the patient moves during image acquisition. Due to the relatively long acquisition times of NM (tens of minutes) compared with x-ray (<1 second), motion is a common problem in NM. Patients may also inadvertently degrade image quality by contributing to contamination. Radioactivity is often present in bodily secretions, so sneezing (see Fig. 7.21D), coughing, drooling, etc. may contaminate patient clothing, the table, or even the detector itself. This is most prevalent when imaging the mentally incapacitated, children, and patients with incontinence.

Loss of image uniformity is a common problem in NM. A damaged collimator (from being dropped or having something dropped on it) will typically result in a line of either increased or decreased signal, depending on whether the septae were bent open or closed. Similarly, physical damage to the scintillator will often result in a cracked crystal (Fig. 7.22), which is characterized by a signal void (where the crack is located and not converting gamma to light) surrounded by a signal increase (due to refracted light collecting adjacent to the crack). The PMT may also cause circular or hexagonal signal voids if it fails or becomes separated (decoupled) from the scintillator.

Septal penetration (see Fig. 7.21A) is visualized as a six-pointed star and occurs when a low-energy collimator is used to image higher-energy photons (typically I-131). The pattern is due to the fact that there is more penetration when the high-energy photons hit perpendicular to the septal wall (most collimators have a hexagonal pattern so the star artifact has six points). Metal objects worn by the patient or implanted in the patient will cause a signal void in an image because they block most gamma rays from reaching the detector due to their high attenuation coefficient. For this reason, we typically request that patients wear only the hospital gown when being scanned.

Technical errors also contribute to the creation of artifacts. When the PHA window is not centered correctly on the photopeak (mistuned or detuned), the resultant image will incorrectly use Compton scattered photons to make the image. This appears as a haze and generally results in a loss of contrast. The uniformity correction will also be incorrect, and as a result the location of individual PMTs will be visible on the image. If the technologist does not inject the radiopharmaceutical properly, then extravasation may occur (see Fig. 7.21B). If the injection site is in the FOV, a hot spot will be visible where the radiopharmaceutical pooled. If the injection site is not in the FOV, it can be difficult to confirm extravasation, as the only other indication is typically an overall decrease in the number of counts because all the radiopharmaceutical did not arrive at the region of interest.

References

Huda W. Nuclear medicine I. In: *Review of Radiological Physics*. 4th ed. Philadelphia: Wolters Kluwer; 2016:187–188.

Mettler FA Jr, Guiberteau MJ. Instrumentation and quality control. In: *Essentials of Nuclear Medicine Imaging*. 6th ed. Philadelphia: Elsevier; 2012:62–69.

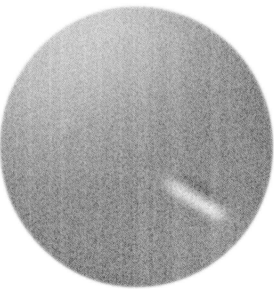

Fig. 7.22 Cracked crystal. Note the signal void in an adjacent area of increased signal.

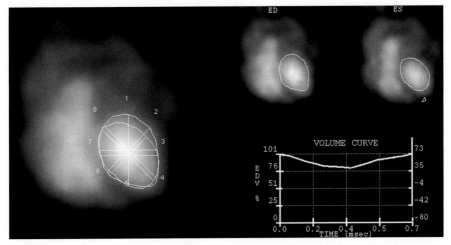

Fig. 7.23

1. How much activity is most likely administered to a patient undergoing a diagnostic NM examination (MBq)?
   A. 5
   B. 50
   C. 500
   D. 5000

2. Which type of collimator is most often used in a diagnostic NM examination?
   A. Parallel
   B. Converging
   C. Diverging
   D. Pinhole

3. Which best characterizes thyroid uptake of radioiodine?
   A. Active transport
   B. Capillary blockage
   C. Compartmental localization
   D. Compartmental leakage

4. Which term best characterizes uptake of sulfur colloid by the liver and spleen?
   A. Diffusion
   B. Phagocytosis
   C. Cell sequestration
   D. Receptor binding

CASE 7.12

Gamma Camera Protocols

Fig. 7.23 Multigated acquisition (MUGA) scan performed using 99mTc pertechnetate--labeled red blood cells demonstrates an ejection fraction of 20%.

1. **C.** Radioisotope dosage (activity) is a trade-off between patient dose and image mottle; 500 MBq is typically used because it gives an acceptable patient dose and acceptable number of counts during the scan time (typically tens of minutes). Activities of 5 and 50 MBq are much too low, and 5000 MBq is much too high.

2. **A.** Parallel-hole collimators are the workhorse of NM imaging, having an acceptable FOV, good trade-off between sensitivity and resolution, and no magnification. Converging collimators may be used for imaging small organs and in children where magnification is useful. Pinhole collimators are typically seen only in thyroid imaging (and even here parallel-hole collimators are often used). Diverging collimators are used only when an FOV larger than the detector is needed (which is rare).

3. **A.** Uptake of radioiodine into the thyroid is accomplished by active transport that is mediated by the sodium/iodide symporter. In a normal thyroid this process can concentrate radioiodine 20 to 40 times and disease processes, such as hyperthyroidism, can make this concentration even higher.

4. **B.** Radiolabeled sulfur colloid beads are 0.1 to 1 μm in diameter, making them small enough to interact with individual cells in the body. They are extracted by the Kupffer cells of the liver through phagocytosis and are then fixed into the tissue. Other organs with phagocytes can also extract sulfur colloid (although to a much lesser extent).

Comment

Evaluation of function, not anatomy, sets NM studies apart from most other modalities. In general, anatomical detail is poor or nonexistent in NM images; however, because the radioisotope is combined with molecules that are analogues for those found in the human body, functional data can be determined by the flow and accumulation of the radioisotope. In some instances, quantitative functional data can be calculated by determining the amount of isotope collecting in a region of interest and comparing it to the background and injected activities (e.g., cardiac ejection fraction). Examples of a variety of studies designed to study physiology using different radionucleotides can be found in Table 7.4.

Most clinical NM imaging is performed using ^{99m}Tc labeled radiopharmaceuticals. The parallel-hole collimator is almost universally used except in very task-specific situations. Although it varies by task, approximately 500 MBq of Tc99m is typically used for an adult patient to ensure sufficient counts for a diagnostic image. ^{99m}Tc imaging will generally use a 20% window centered on the photopeak of 140 keV. NM matrix sizes are generally 64 × 64 or 128 × 128, with low count densities and high levels of mottle (±10%). By contrast, the large matrix sizes (512 × 512) and radiation intensities in CT result in typical interpixel fluctuations (mottle) of approximately ±3 Hounsfield units (0.3% differences in attenuation coefficient).

Table 7.4 COMMON NUCLEAR MEDICINE STUDIES ARE LISTED WITH THEIR RESPECTIVE RADIOTRACER AND PHYSIOLOGIC MECHANISM

Study	Radionucleotide	Physiology/Function
Bone (pyrophosphate) scan	^{99m}Tc	Physicochemical adsorption
Cholescintigraphy	^{99m}Tc	Hepatobiliary function (secretion)
Macroaggregated albumin	^{99m}Tc	Lung perfusion (capillary blockade)
Renal DTPA	^{99m}Tc	Kidney filtration; diffusion
Sulfur colloid	^{99m}Tc	Phagocytosis; liver and spleen
Tagged red blood cell	^{99m}Tc	Vascular compartmental localization; gastrointestinal bleed detection
Tagged white blood cell	^{111}In	Cell sequestration
Octreotide	^{111}In	Neuroendocrine tumor (somatostatin receptors)
Thyroid uptake scan	^{123}I	Active transport
Ventilation	^{133}Xe	Lung ventilation (compartmental localization)
Ga-67 citrate	^{67}Ga	Inflammation and infection

DTPA, Diethylenetriamine pentaacetic acid.

Gamma cameras apply several corrections to improve image quality. A linearity correction is needed to ensure removal of a spatial distortion. Without this correction, straight lines appear as wavy. Nonuniformities in detector response are also needed. Each PMT will produce slightly different outputs even if the same amount of radiation hits the scintillator above it. Furthermore, the output varies depending on how far from the center of the PMT the radiation hit the scintillator. Nonuniformity correction factors can be generated using flood images obtained with an exceptionally high number of counts (e.g., >100 million) that minimize random noise.

The primary shortcoming of NM is mottle, which is a direct consequence of the small number of counts. This problem can be overcome only when there are more photons accepted, which is very difficult to engineer, and changes made in the protocol adversely impact other areas of image quality. Increasing activity would result in higher patient doses, and changing the collimator to improve sensitivity would require a corresponding loss of spatial resolution. Increasing the PHA window width will always increase the number of counts used to create the image, but this will occur at the cost of accepting more scatter, which will generally decrease contrast. Scanning for a longer time allows more counts to accumulate but increases the already long acquisition times and leads to more motion artifacts.

References

American College of Radiology. ACR–SPR practice parameter for the performance of skeletal scintigraphy (bone scan). Resolution 28. Revised 2017. https://www.acr.org/-/media/ACR/Files/Practice-Parameters/skeletal-scint.pdf?la=en. Accessed March 2, 2018.

American College of Radiology. ACR–AAPM practice parameter for reference levels and achievable administered activity for nuclear medicine and molecular imaging. Resolution 53; 2015. https://www.acr.org/-/media/ACR/Files/Practice-Parameters/reflevels-nucmed.pdf?la=en. Accessed March 2, 2018.

Huda W. Nuclear medicine I. In: *Review of Radiological Physics*. 4th ed. Philadelphia: Wolters Kluwer; 2016:186–187.

CASE 8.1

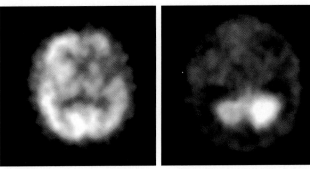

Fig. 8.1

1. What is the principal advantage of using single-photon emission computed tomography (SPECT) over planar imaging in nuclear medicine (NM)?
   A. Improved resolution
   B. Tomographic slices
   C. Lower patient doses
   D. Shorter imaging times

2. Which gamma camera characteristic *must* be modified to perform SPECT imaging?
   A. More photomultiplier tubes (PMT)
   B. Thinner crystal
   C. Thicker collimators
   D. No modifications needed

3. What type of collimator could be used to improve the spatial resolution performance in neuro SPECT imaging?
   A. One-dimensional (1D) converging
   B. Two-dimensional (2D) converging
   C. 1D diverging
   D. 2D diverging

4. Which radionuclides would require the use of a medium-energy collimator?
   A. Gallium-67
   B. Indium-111
   C. Iodine-123
   D. A and B

CASE 8.1

Single-Photon Emission Computed Tomography and Image Acquisition

Fig. 8.1 Brain single-photon emission computed tomography demonstrating crossed cerebellar diaschisis in a patient with left middle cerebral artery infarction.

1. **B.** SPECT stands for single-photon emission computed tomography and will generate tomographs (slices) of the distribution of radioactivity in a patient. A tomographic slice eliminates the anatomical clutter of overlying and underlying radioactivity and thus increases lesion contrast. This is analogous to a chest CT compared with a chest x-ray.

2. **D.** SPECT uses conventional projection images obtained by rotating the gamma camera around the patient. No major modifications are required for the detector or collimator.

3. **A.** A 1D converging collimator (AKA fan beam) would result in magnified projections within the image plane and improve spatial resolution while keeping adequate coverage of the brain. A 2D converging (AKA cone beam) collimator would result in sampling artifacts. Diverging collimators will minify images, which will degrade resolution.

4. **D.** Both gallium-67 and indium-111 are considered medium-energy gamma emitters and would both require the use of a medium-energy collimator. Technetium-99m and iodine-123 are low-energy gamma emitters; both would use low-energy collimator in SPECT imaging.

Comment

Tomographic views of the radioisotope distribution within a patient can be obtained using a SPECT system. The term *single photon* differentiates SPECT from the two photons detected after a positron annihilation event in positron emission tomography (PET). Emission (radiation originates in the patient) differentiates SPECT from transmission (originates outside the patient) of x-ray beams used in CT. SPECT generates tomographs (slices) that require the use of computers to reconstruct the acquired projections. SPECT is to gamma camera imaging as CT is to radiographs.

A conventional gamma camera is used to acquire projection images, which are reconstructed into tomographic images.

SPECT systems use one or more gamma camera heads (detectors) that rotate around an axis of rotation (Fig. 8.2). The radius of rotation is the distance from the camera (detector) face to central axis. Parallel-hole collimators are commonly used for SPECT imaging. For small body regions such as the head, fan beam collimators with a limited field of view may be used to improve resolution in one direction (*x*-axis). Fan beam collimators are a hybrid of parallel holes in one direction (*y*-axis along the patient long axis) and converging collimator in the perpendicular (*x*-axis) direction.

A low-energy collimator would be used for technetium-99m and iodine-123, which are usually made of lead foil. Medium-energy collimators would be used for gallium-67 and indium-111, which are usually much heavier and manufactured using casting techniques. A single-head gamma camera rotates 360 degrees around the patient, using either 60 or 120 stops. Typically this yields either 60 or 120 low-quality planar images in total. Each stop is designed to allow for sufficient counts to accumulate from that angle (projection). When 60 projection images are obtained at 20 seconds each, the scanning time is 20 minutes. In cardiac imaging, 180-degree rotations are frequently used, with images acquired with the smaller 64 × 64 matrix in an attempt to increase acquisition speed.

As in planar NM scanning, technetium-99m is the workhorse of SPECT imaging. Technetium-99m SPECT is combined with hexamethyl propyleneamine oxime (HMPAO) to examine cerebral perfusion and in epilepsy imaging. Technetium-99m with tilmanocept is used to identify sentinel lymph nodes (lymphoscintigraphy). Technetium-99m, when combined with sestamibi, is used for parathyroid imaging and for determining myocardial perfusion. Indium-111 is combined with octreotide to identify tumors with somatostatin receptors (e.g., neuroendocrine tumors). Iodine-123 is used to image thyroid abnormalities. Gallium-67 may be used to image chronic inflammation and infection. Finally, myocardial viability can be assessed using Tl-201 (in addition to technetium-99m sestimibi). Many other radioisotopes are also used in SPECT imaging.

References

Huda W. Nuclear medicine II. In: *Review of Radiological Physics*. 4th ed. Philadelphia: Wolters Kluwer; 2016:191–192.

Radiological Society of North America. RSNA/AAPM physics modules. Nuclear medicine: 4. SPECT/SPECT - CT/image quality. https://www.rsna.org/RSNA/AAPM_Online_Physics_Modules_.aspx. Accessed March 2, 2018.

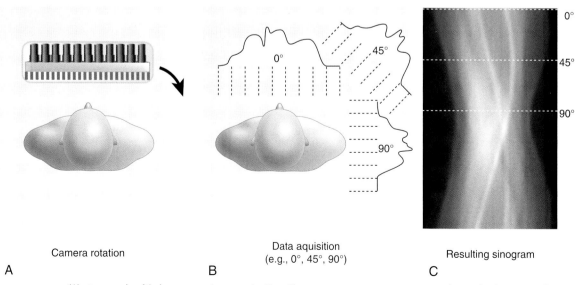

Camera rotation	Data aquisition (e.g., 0°, 45°, 90°)	Resulting sinogram
A	B	C

Fig. 8.2 A gamma camera (A) at an angle of 0 degree acquires a projection. The gamma camera rotates to acquire projections at angles ranging from 0 to 180 degrees (B), which are stacked sequentially in a sinogram (C).

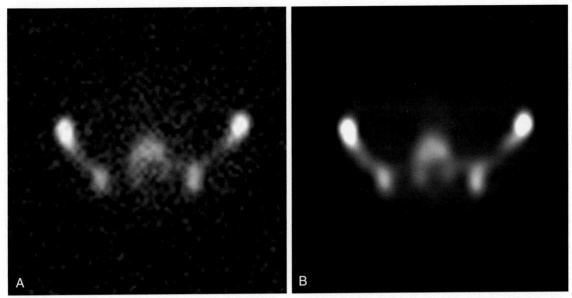

Fig. 8.3

1. What is the most likely SPECT matrix size in cardiac SPECT?
   A. 32 × 32
   B. 64 × 64
   C. 128 × 128
   D. 256 × 256

2. What is the most likely total image acquisition time in typical cardiac SPECT examination (minutes)?
   A. 1
   B. 2
   C. 5
   D. 15

3. What is normally observed when SPECT images are reconstructed using filtered back projection (FBP) algorithms?
   A. Streaking
   B. High noise
   C. Ring artifacts
   D. A and B

4. What kind of image processing is most likely used to improve the appearance of SPECT images?
   A. Frequency filtering
   B. Unsharp mask enhancement
   C. Image inversion
   D. Histogram equalization

CASE 8.2

SPECT Image Reconstruction

Fig. 8.3 Filtered back projection (FBP) iterative reconstruction (IR)–simulated single-photon emission computed tomography images of the pelvis using FBP (A) and IR (B).

1. **B.** Cardiac SPECT uses a 64 × 64 matrix size. The size of the heart is relatively small, and there is little advantage to having a large reconstruction matrix. For a 20-cm field of view, the pixel size will be 200/64 or approximately 3 mm, which is substantially less than the typical SPECT resolution of 8 mm.

2. **D.** To maximize the acquired counts, we use an image acquisition time that is clinically acceptable of approximately 15 minutes. Much shorter times would result in too much mottle, and much longer times would increase motion artifacts and unlikely be tolerated by many patients.

3. **D.** The number of counts in each point in any gamma camera projection will be low, and this will give rise to streaking artifacts, as well a very high image mottle when tomographs are reconstructed using FBP. When the number of counts in any part of a projection is low, iterative reconstruction (IR) algorithms result in better-quality images.

4. **A.** SPECT images are often processed using frequency filtering to smooth the image. When images are made smoother, there will be a corresponding loss of spatial resolution. The other options are processing steps sometimes used in radiography and/or fluoroscopy.

Comment

FBP can be used to obtain tomographs from any type of projection imaging, including CT, SPECT, and PET. In the early days of SPECT imaging, FBP was used because this algorithm was readily available and required minimal computing time. In all NM images, the average number of counts per pixel is low, with some pixels always having "very low" counts. Furthermore, comparatively few projections are created (i.e., ~100 in SPECT vs. ~1000 in CT). The low number of counts combined with the low number of projections gives a poor reconstruction. As a result, FBP reconstruction from gamma camera projections will generally result in streaky tomographic images with elevated levels of noise/mottle.

In current clinical SPECT imaging, IR algorithms are now used to generate tomographs. Empirical evidence clearly demonstrates that IR algorithms are more accurate and minimize artifacts when processing images that have low counts (Fig. 8.4). Because NM images are relatively small (e.g., 64 × 64), image reconstruction times using IR algorithms are relatively low and much shorter than in CT where the amount of data is up to two orders of magnitude higher (i.e., 512 × 512 matrix sizes).

When a 64 × 64 matrix is used to acquire projection images, the SPECT system will generate an isotropic volume data set that has 64 slices, each with a 64 × 64 matrix. With this volume data set, this will permit transverse, sagittal, and coronal views to be generated. As with other 3D imaging modalities such as CT and MR, it is also possible to generate any desired imaging plane reconstruction. In cardiac imaging, the acquired 3D volume set can be used to produce a "bull's eye" plot consisting of a display of data in the short-axis slices of the left ventricle. These slices are then mapped in larger rings from apex to base, facilitating an objective evaluation of this type of tomographic examination.

Quality of SPECT tomographic images can be improved using "frequency filtering." These "frequency filters" can either enhance or remove selected spatial frequencies to generate a final (processed) image that has the desired characteristics. Low spatial frequencies in all imaging relate to large objects and will pertain to the overall image contrast. High spatial frequencies refer to small details and edges and will pertain to spatial resolution performance. Use of low-pass filters will remove high frequencies, resulting in smoother (i.e., less noisy) images. The price of using a low-pass filter is that there will be a loss of spatial resolution, and the image will show less distinct edges. Use of a high-pass filter would result in sharper images that have enhanced edge definition but at the cost of increasing image noise.

References

Cherry SR, Sorenson JA, Phelps M. Tomographic reconstruction in nuclear medicine. In: *Physics in Nuclear Medicine*. 4th ed. Philadelphia: Elsevier; 2012:270–277.

Huda W. Nuclear medicine II. In: *Review of Radiological Physics*. 4th ed. Philadelphia: Wolters Kluwer; 2016:191–192.

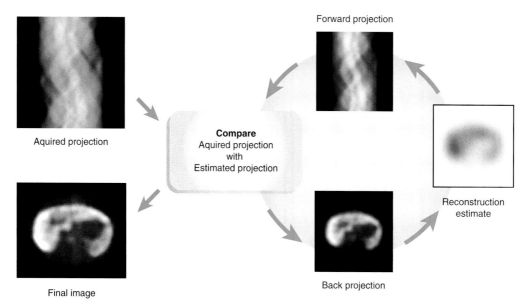

Fig. 8.4 Schematic representation of iterative reconstruction. An estimate of the image can be used to simulate projection data (sinogram), which is compared with the acquired sinogram. The differences between the two sets of data are used to improve the reconstructed image, and this process is repeated until the differences between the two sets of sinograms are "low."

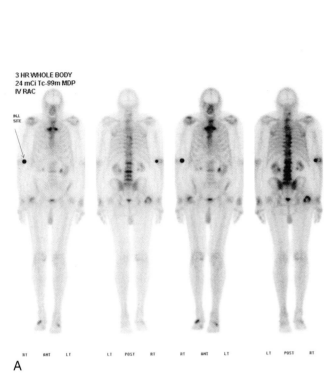

3 HR WHOLE BODY
24 mCi Tc-99m MDP
IV RAC

INJ.
SITE

RT   ANT   LT   LT   POST   RT   RT   ANT   LT   LT   POST   RT

A

B

Fig. 8.5

1. What is the principal benefit of using contour patient rotation in SPECT imaging?
   A. Improved resolution
   B. Higher contrast
   C. Reduced mottle
   D. Fewer artifacts

2. Which radionuclides can be used to perform SPECT imaging?
   A. Gallium-67
   B. Indium-111
   C. Iodine-123
   D. A, B, and C

3. Which protocol parameters are representative of a technetium 99m SPECT bone study?
   A. 128 × 128 acquisition matrix
   B. 360-degree rotation with 60 views
   C. 30 seconds acquisition/view
   D. A, B, and C

4. Which radiopharmaceuticals can be used for cardiac SPECT imaging?
   A. Technetium 99m–labeled sestamibi
   B. Technetium 99m–labeled tetrofosmin
   C. ^{201}Tl chloride
   D. A, B, and C

CASE 8.3

SPECT Protocols

Fig. 8.5 Planar images (A) compared with reconstructed single-photon emission computed tomography images (B) of a bone scan in a patient with multilevel severe degenerative changes of the spine and asymmetric hip uptake.

1. **A.** Spatial resolution performance of all parallel-hole collimators will drop very rapidly with increasing distance from the collimator. Use of contour patient rotation will minimize the distance between the patient and detector, and thereby improve spatial resolution performance.

2. **D.** Examples of SPECT examinations typically performed in clinical practice include scans using iodine-123, infection scans using gallium-67, and white blood cells studies that are labeled with indium-111.

3. **D.** A 128 × 128 acquisition matrix, a 360-degree rotation with 60 views, and a 30-second acquisition/view are representative choices for protocol parameters in a SPECT technetium-99m bone study.

4. **D.** Historically, ^{201}Tl was used to perform cardiac SPECT studies; however, its relatively long half-life (73 hours) results in relatively high patient doses (e.g., 20 mSv). By using technetium-99m–labeled sestamibi and tetrofosmin, the shorter half-lives of 6 hours generally result in a substantial reduction in patient dose (e.g., 50%).

Comment

SPECT can be added to planar bone scintigraphy to improve lesion localization and is often performed to evaluate spine lesions. Modern scanners use range finders to detect the body contour and minimize the detector to body distance. This improves resolution by reducing source-to-detector distance. Bone scan SPECT is acquired with 800 MBq technetium-99m–labeled phosphonates, a 128 × 128 acquisition matrix, and 60 views over 360 degrees with 30 seconds per view. As with the other scans mentioned later, the SPECT image quality is improved when the acquisition is performed using a contoured orbit rather than a circular orbit (Fig. 8.6). A DaTscan (iodine-123 ioflupane) is a native SPECT brain scan that uses 150 MBq of iodine-123 to aid in the diagnosis of Parkinson disease. Perfusion imaging is also a SPECT native scan used for several indications including localizing epilepsy, diagnosing encephalitis, and corroborating brain death. A dosage of 800 MBq of technetium-99m–labeled bicisate ethyl cysteinate dimer (ECD) is injected and crosses the blood-brain barrier for imaging. Brain scans require high resolution and so are typically acquired with 128 × 128 matrix and a low-energy high-resolution (LEHR) collimator.

Cardiac SPECT systems likely use two or three gamma cameras in an L or similar shape, which will reduce the total time needed to acquire projection data. The patient is positioned with the arms above the head, if possible, to avoid attenuation artifacts. Myocardial perfusion imaging currently uses either technetium-99m–labeled sestamibi or technetium-99m–labeled tetrofosmin. Due to its relatively high patient dose, ^{201}Tl is typically used only if there is a shortage of technetium-99m, which may happen when generators go down for repair. For 1-day rest/stress testing, 400 MBq will be used at rest and 1200 MBq will be used during stress. The larger amount (three times) is necessary to avoid activity from rest confounding the stress read. Cardiac scans are typically acquired with a LEHR collimator for technetium-99m and 64 projections over 180 degrees at 20 s per projection, with a 64 × 64 matrix size.

Cardiac SPECT acquisitions are generally electrocardiogram gated to reduce motion artifact. Data are reprocessed into orientations specific to the heart: horizontal long axis, vertical long axis, and short axis. Polar maps are also standard to display the heart segments and corresponding defects. The strength of SPECT myocardial perfusion imaging is the ability to quantify function. Typical metrics include defect size and severity, left ventricle ejection fraction (LVEF), and left ventricular volume. LVEF is the ratio of diastolic minus systolic activity to systolic activity. LVEF is calculated for both rest and stress series and compared. It is important to note that LVEF may become inaccurate if extensive motion artifact exists or if gating is not accurate due to arrhythmias.

Medium-energy radionuclides are also used for SPECT imaging but require the use of an appropriate medium-energy collimator. An indium-111 white blood cell scan uses white blood cells (neutrophils) removed from the patient. These are tagged and reinjected intravenously into the patient where they localize to areas of infection. A dosage of 20 MBq is typically used and a dual-head camera is needed for indium-111 SPECT due to the low count rate. Gallium-67 citrate localizes to infection but through a different route than indium-111. Instead of binding to leukocytes, it binds to plasma transferrin and crosses the capillary epithelium at the infection site. Unlike indium-111–tagged white blood cells, gallium-67 will also localize in lymphomas, lung carcinomas, and other tumors. A dosage of 200 MBq is typically administered. Both indium-111 and gallium-67 SPECT use a 64 × 64 matrix, 360-degree rotation, medium-energy collimators, and 25 seconds per angle.

References

Anagnostopoulos C, Harbinson M, Kelion A, et al. Procedure guidelines for radionuclide myocardial perfusion imaging. *Heart*. 2004;90(Suppl 1):i1–i10.

Huda W. Nuclear medicine II. In: *Review of Radiological Physics*. 4th ed. Philadelphia: Wolters Kluwer; 2016:192–198.

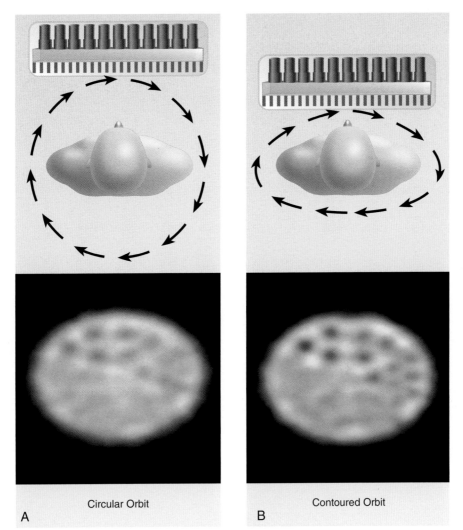

Circular Orbit

A

Contoured Orbit

B

Fig. 8.6 During single-photon emission computed tomography, the detector assembly moves in an orbit around the patient and pauses to acquire counts at each angle before continuing the rotation. A circular orbit (A) results in inferior count localization to a contoured orbit (B) as seen with this simulated elliptical phantom.

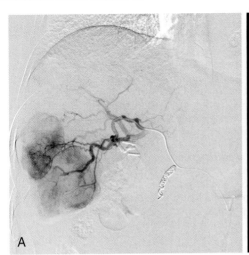
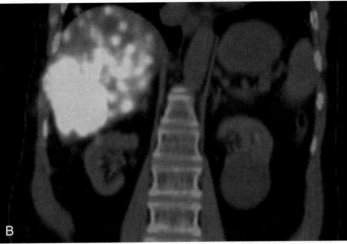

Fig. 8.7

1. What is the average CT x-ray photon energy (keV) in a typical CT/SPECT acquisition that uses a technetium-99m–labeled pharmaceutical?
   A. 30
   B. 60
   C. 90
   D. >90

2. How many technetium-99m gamma rays are transmitted through 1 cm of soft tissue that has an estimated attenuation coefficient of 0.15 cm^{-1} at 140 keV (%)?
   A. 95
   B. 85
   C. 15
   D. 5

3. For which type of material will there be the greatest uncertainty in estimating the attenuation at 140 keV from a Hounsfield unit (HU) measurement obtained at an x-ray tube voltage of 120 kV?
   A. Lung
   B. Tissue
   C. Bone
   D. Pacemaker

4. What is the benefit of adding CT to a SPECT study?
   A. Attenuation correction
   B. Anatomical data
   C. Fewer artifacts
   D. A and B

ANSWERS

CASE 8.4

CT and SPECT

Fig. 8.7 Bremsstrahlung single-photon emission computed tomography image (B) in a patient who received yttrium-90 treatment to the right hepatic lobe during angiography (A).

1. **B.** Most CT examinations in SPECT/CT are performed at a voltage of 120 kV, with an average x-ray spectrum energy of approximately 60 keV. This energy is much lower than the technetium-99m energy of 140 keV, so that attenuation correction factors must be extrapolated from the measured attenuation at 60 keV to the applied attenuation at an energy of 140 keV.

2. **B.** By definition, an attenuation coefficient of 0.15 cm⁻¹ will attenuate approximately 15%, or a fraction of 0.15, of the incident photons. If 15% is lost in 1 cm, then it follows that 85% is transmitted, because the lost and transmitted percentages must add to 100%.

3. **D.** Attenuation coefficients of lung, tissue, and bone measured at one energy are straightforward to extrapolate to any other energy. Attenuation by a pacemaker will depend critically on the pacemaker atomic number, which is unknown, and frequently result in attenuation correction artifacts adjacent to such materials.

4. **D.** Adding a CT scan to a traditional SPECT imaging examination can correct for attenuation of gamma rays that originate within the patient. Distribution of the radioactivity that provides invaluable physiological information can be shown superimposed on the exquisite anatomical data that are provided by the CT scan. CT images may cause attenuation correction artifacts near any metallic implants.

Comment

Coregistration of NM images with CT improves lesion localization, and CT attenuation information (HU) can be used to perform attenuation corrections (Fig. 8.8). Sixteen-slice CT scanners are adequate for most applications. Spiral CT scanning of the whole body, taken from the eyes to upper thigh, can be performed in less than 30 seconds. Representative CT image acquisition parameters are 120 kV tube voltage, 512 × 512 matrix size, 5-mm reconstructed image slice thickness, and 1.5 CT pitch. Because CT scan times are much shorter than SPECT

and PET acquisition times, patient motion between the scans can result in misregistration artifacts.

CT images are obtained at an average energy of 60 keV, which is much lower than photon energies used in SPECT, including 140 keV for technetium-99m and 160 keV for iodine-123. The acquired attenuation data from 60 keV CT are extrapolated to estimate what the attenuation would be at the radioisotope's energy. For soft tissue, lung, and bone, this extrapolation can be performed in a straightforward manner. However, attenuation extrapolation is very problematic for nonstandard materials such as metallic implants. Attenuation in SPECT imaging can be substantial. For technetium-99m, 5 cm of tissue attenuates half of the activity and 10 cm reduces the detected activity to approximately a quarter, requiring attenuation correction factors of approximately 2 and 4 for activity at soft tissue depths of 5 cm and 10 cm, respectively.

When images of a uniform water phantom are initially acquired, the measured activity concentrations in the initial SPECT image will be less in the phantom center than at the periphery. The measured attenuation coefficients of the water phantom can then be used to correct for gamma ray attenuation. After the attenuation correction has been performed, there will be the expected uniform pattern of activity in this water phantom. For cardiac studies, attenuation by the diaphragm and female breast are specific concerns because they can mimic defects. It should be noted that CT attenuation correction also offers better quantification of radiotracer uptake.

When CT scans are performed for attenuation correction only, very low doses are used, where the CT radiation intensity (CTDI$_{vol}$) is approximately 1 mGy and the corresponding patient effective dose is approximately 1 mSv. For nondiagnostic CT scans where images are used to localize uptake against patient anatomy, the CT radiation intensity would likely be increased to approximately 4 mGy, and the corresponding effective doses would be approximately 4 mSv. To generate diagnostic images, substantially more radiation would be used. For these diagnostic scans, the radiation intensity would likely be a CTDI$_{vol}$ of 16 mGy, which results in a "high" effective dose of approximately 16 mSv.

References

Bushberg JT, Seibert JA, Leidholdt EM Jr, Boone JM. Nuclear imaging—emission tomography. In: *The Essential Physics of Medical Imaging*. 3rd ed. Philadelphia: Wolters Kluwer; 2012:735–742.

Huda W. Nuclear medicine II. In: *Review of Radiological Physics*. 4th ed., Philadelphia: Wolters Kluwer; 2016:196–198.

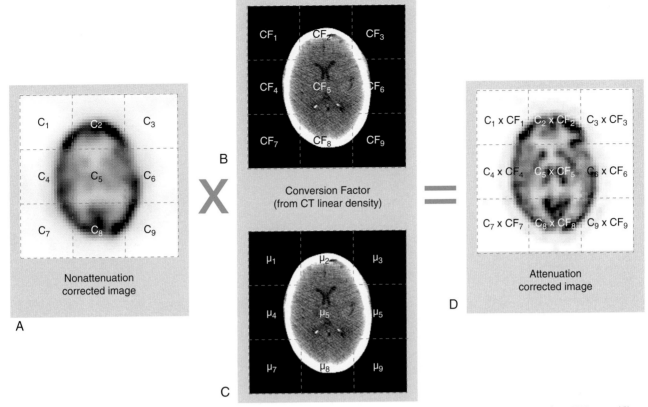

Fig. 8.8 Schematic overview of the attenuation correction process. (A) Measured concentrations of activity in each pixel. A CT image (C) provides data on tissue attenuation, which can be used to generate correction factors for activity at each pixel (B). The attenuation correction image (D) is obtained as the product of the concentration in each pixel (C) and the corresponding pixel correction factor (CF).

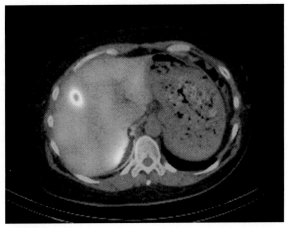

Fig. 8.9

1. Which of the following is most likely to improve spatial resolution performance in SPECT imaging?
   A. Contoured orbits
   B. High-sensitivity collimators
   C. Longer imaging times
   D. B and C

2. Which imaging metric is improved the most when SPECT images are compared with the corresponding planar gamma camera images?
   A. Lesion contrast
   B. Spatial resolution
   C. Temporal resolution
   D. Lower dose

3. Which artifact appears only in a SPECT image?
   A. Center of rotation
   B. Septal penetration
   C. Cracked crystal
   D. Detuned photopeak

4. What will most likely be visible in SPECT images when the acquired planar images have very poor uniformity (e.g., >3%)?
   A. Streak artifacts
   B. Ring artifacts
   C. Hot spots
   D. Cold spots

CASE 8.5

SPECT Image Quality

Fig. 8.9 Single-photon emission computed tomography/CT image from an octreotide scan in a patient with a metastatic carcinoid tumor.

1. **A.** A contoured orbit around the patient will markedly improve spatial resolution because the distance(s) between the patient and the gamma camera collimator is minimized. With a circular orbit, the gap between the collimator and the "thin" body regions will degrade the spatial resolution performance. A higher-sensitivity collimator implies a poorer resolution, and longer imaging times increase the chance of motion blur.

2. **A.** Lesion contrast, the difference in lesion counts compared with the adjacent background tissues, will improve markedly in SPECT. The reason for this is that the overlying and underlying tissues, which all contain radioactivity, have been "removed." SPECT resolution cannot be better than the underlying gamma camera resolution and takes much longer than gamma camera imaging (i.e., worse temporal resolution). Patient doses in a gamma camera and SPECT will be similar.

3. **A.** Common gamma camera artifacts include septal penetration, cracked crystals, and detuned photopeak. However, a center of rotation artifact is possible only when the gamma camera is rotated around an incorrect isocenter, which will result in artifacts (e.g., blur and halos) in reconstructed SPECT images.

4. **B.** When the gamma camera has a poor uniformity or uses uniformity correction factors that have excessive noise, the resultant images will have ring artifacts (sometimes called bull's eye artifacts). This is like CT, where detectors that have not been correctly calibrated (i.e., are noisy) can cause multiple rings to form in the image.

Comment

A medical physicist will normally assess the overall performance of a SPECT system prior to its acceptance by the radiology department. This also establishes the system baseline against which future quality control (QC) measurements are compared. System spatial resolution is assessed using a point or line source with data acquired "free in air." Commercially available SPECT phantoms (e.g., Jaszczak) are used to semiquantitatively assess spatial resolution, uniformity, and image contrast at acceptance and then every quarter, to ensure that there has been no drop in system performance of key image quality metrics (Fig. 8.10).

QC is very important in SPECT and more important than in planar NM imaging. Small amounts of noise and nonuniformities are generally better tolerated in planar NM imaging but less so in SPECT, where reconstruction is used to generate tomographs. The SPECT image reconstruction algorithm amplifies image noise and nonuniformities, resulting in poor image quality. Nonuniformities will appear as ring artifacts (AKA "bull's eye") in SPECT images. In multihead SPECT systems, partial rings will be produced if only one of the heads has poor uniformity.

SPECT QC is performed by a technologist and uses the Jaszczak phantom. This phantom is filled with radioisotope and has a uniform section to test uniformity and look for artifacts, a rod section to assess resolution, and a sphere section to assess detectability and contrast. An accurately aligned center of rotation can be checked using point sources or line source once every month. The gamma camera head tilt angle can be checked using a bubble level and is generally performed every quarter. Table 8.1 shows key tests and typical frequency for their performance.

Quantum mottle (noise) is the most important factor in SPECT imaging. As in all NM examinations, this occurs because of the low number of photons used to reconstruct each voxel. Noise is especially high in SPECT imaging because only a limited amount of time can be spent collecting counts for each projection. Spatial resolution of SPECT will also be worse than in planar imaging because the planar images make the tomographs and their resolution is degraded by the reconstruction algorithm. The major benefit of SPECT is the improved contrast that results from the elimination of overlapping structures and activity from those structures.

TABLE 8.1 SINGLE-PHOTON EMISSION COMPUTED TOMOGRAPHY QUALITY CONTROL

Test	Frequency	Comments
Center of Rotation	Monthly	Failure results in decreased resolution and circular artifacts
Uniformity Correction Map	Monthly	Eliminates ring and other artifacts
System Performance	Quarterly	Jaszczak phantom test using technetium-99m

References

Huda W. Nuclear medicine II. In: *Review of Radiological Physics*. 4th ed. Philadelphia: Wolters Kluwer; 2016:192–193.

Mettler FA Jr, Guiberteau MJ. Instrumentation and quality control. In: *Essentials of Nuclear Medicine Imaging*. 6th ed. Philadelphia: Elsevier; 2012:59–61.

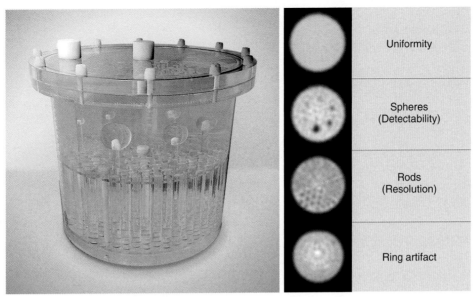

Fig. 8.10 The Jaszczak phantom is filled with radioisotope (technetium-99m) and imaged in the single-photon emission computed tomography scanner. Each of the three sections (uniformity, spheres, and rods) are compared with baseline measurements made during acceptance testing to look for degradation. The uniformity section is also inspected for artifacts such as the bull's eye (AKA ring) artifact shown.

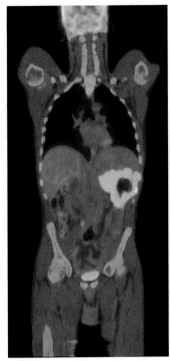 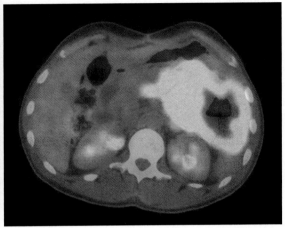

Fig. 8.11

1. How are positron emitters typically manufactured?
   A. Nuclear reactors
   B. Fission products
   C. Cyclotrons
   D. A, B, and C

2. What type of detector does PET imaging use?
   A. Scintillator
   B. Photoconductor
   C. Photostimulable phosphor
   D. Ionization chamber

3. Increasing which detector characteristic would likely improve the detection of 511 keV photons?
   A. Atomic number
   B. Physical density
   C. Detector depth
   D. A, B, and C

4. What detector property should be increased for a detector used in a PET scanner?
   A. Detection efficiency
   B. Light output
   C. Photopeak full width half maximum (FWHM)
   D. A and B

CASE 8.6

PET Detectors

Fig. 8.11 Positron emission tomography/CT demonstrating a large hypermetabolic mass in the left upper quadrant of the abdomen confirmed to be a gastrointestinal stromal tumor following biopsy.

1. **C.** Positron emitters are always made in cyclotrons. A cyclotron accelerates positively charged nuclei (e.g., protons) to high energies so that the radioactivity created has "too much positive" charge. Nuclei with too many protons therefore decay by emitting positrons or by electron capture, both of which will reduce the positive charge of the resultant daughter nuclei (e.g., fluorine 18 [$Z = 9$] to oxygen 18 [$Z = 8$]).

2. **A.** PET detectors are scintillators with PMTs connected to the crystal. Annihilation photons are converted to light by the scintillator, which is then detected by a PMT, which is a very sensitive and quantitative light detector.

3. **D.** Increasing atomic number, physical density, and detector depth would all most likely improve the detection of 511-keV photons. Higher atomic number means more photoelectric interactions, higher physical density gives more interactions per length, and deeper depth gives a longer path for the annihilation photons to interact along.

4. **D.** Detection efficiency and light output should be maximized because this will ensure as many annihilation photons as possible interact and are correctly counted by the system. However, the photopeak should be as narrow as possible (low FWHM) to allow for a smaller energy window and thus less counted scatter.

Comment

PET requires that the radionuclide be a positron emitter. These isotopes are created in a cyclotron by accelerating protons to hit a target with one fewer protons than the desired radioisotope. When the proton collides with the target, the proton is, for our purposes, absorbed and a new radioactive atom with too many protons is formed. Common isotopes are fluorine-18 ($T_{1/2}$ = 110 min), carbon-11 ($T_{1/2}$ = 20 min), and rubidium-82 ($T_{1/2}$ = 1.3 min). The short rubidium-82 half-life means the parent strontium-82 generator achieves equilibrium in approximately 5 minutes (four daughter half-lives). On-site cyclotrons are required to generate short-lived radionuclides, which include carbon-11, nitrogen-13 ($T_{1/2}$ = 10 min), and oxygen-15 ($T_{1/2}$ = 2 min). Virtually any chemical of interest in medicine can be labeled using carbon, nitrogen, and oxygen, so these short-lived radionuclides are valuable research tools.

By far, fluorine-18 is the most used radioisotope in PET and is typically coupled to a glucose analog called fluorodeoxyglucose (FDG). The average energy of rubidium-82 (1.4 MeV) results in a mean range over 5 mm. This range is much higher than the other common positron emitters (fluorine-18, carbon-11, oxygen-15) that all have mean ranges of approximately 1 mm. Consequently, images obtained using rubidium-82 will be markedly blurrier than those obtained with lower-energy positron emitters such as fluorine-18. Compared with soft tissue, the distance traveled by positrons in lung will be markedly greater because of the low density of lung, and the distance traveled in bone will be much shorter in bone because of its higher density. When a positron interacts with an electron in the body, it produces two 511-keV photons moving in exactly opposite directions. This is called an annihilation interaction, with the annihilation photons used to create the image.

PET imaging systems make use of scintillators to detect annihilation 511-keV photons. The most common detector materials that are currently used include bismuth germanate (BGO), lutetium oxyorthosilicate (LSO and LYSO), and gadolinium oxyorthosilicate (GSO). Important scintillator properties include the photoelectric absorption efficiency, the scintillator energy resolution, and the time that it takes for the generated light to decay (Table 8.2). Best absorption characteristics are in BGO, but this scintillator also has the worst energy resolution. Inorganic scintillators (i.e., GSO, LSO, and LYSO) emit more light, which generally improves their energy resolution. These inorganic scintillators also have shorter decay times, which provide improved performance at the high count rates encountered in PET imaging.

Crystals can be arranged in blocks, consisting of either 6 × 6 or 8 × 8 arrays of detector elements. Each block thus has 36 or 64 "crystal elements" that are coupled to an array of PMTs (Fig. 8.12). As in gamma cameras, the PMTs will provide positional information, as well as offering pulse height analysis (PHA). Photons within the acceptable energy range, typically between 450 and 550 keV, are counted as valid 511-keV annihilation photons. Detector blocks are arranged in rings, and several rings are placed adjacent to each other to provide multislice capability. Multiring PET scanners can easily have up to 20,000 individual "crystal elements" in a total of 400 blocks.

TABLE 8.2 COMPARISON OF SODIUM IODIDE (NAI), BISMUTH GERMANATE (BGO), LUTETIUM OXYORTHOSILICATE (LSO), AND LUTETIUM-YTTRIUM OXYORTHOSILICATE (LYSO) SCINTILLATORS

Characteristic	NaI	BGO	LSO	LYSO
Density (g/cm³)	3.7	7.1	7.4	7.1
Atomic number (Z)	51	75	66	64
Decay time (ns)	230	300	40	45
Energy resolution (%)	7	13	11	9
Light photons per keV	30	9	26	30

References

Huda W. Nuclear medicine II. In: *Review of Radiological Physics*. 4th ed. Philadelphia: Wolters Kluwer; 2016:193–194.

Radiological Society of North America. RSNA/AAPM physics modules. Nuclear medicine: 4. PET/PET - CT/image quality. https://www.rsna.org/RSNA/AAPM_Online_Physics_Modules_.aspx. Accessed March 2, 2018.

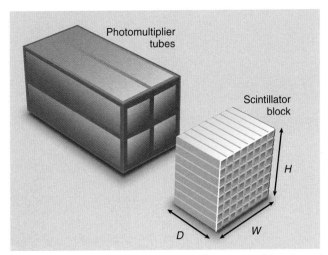

Fig. 8.12 Scintillator block with a depth *(D)*, width *(W)*, and height *(H)* made from a material such as bismuth germanate or lutetium oxyorthosilicate. Immediately behind the scintillator is an array of photomultiplier tubes where the total light produced is used to identify photopeak events (511 keV) using pulse height analysis, and the light pattern is used to identify the location of the photon interaction.

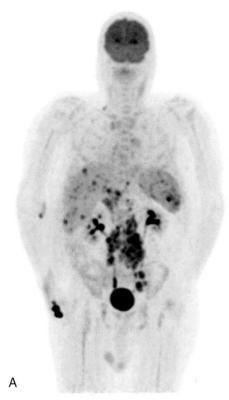

A

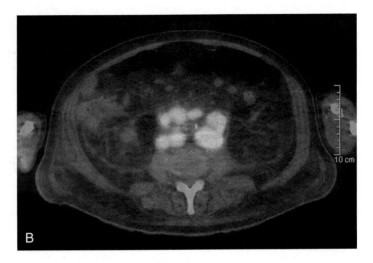

B

Fig. 8.13

1. Which of the following has the highest-energy photon used in PET imaging?
   A. Fluorine-18
   B. Carbon-11
   C. Rubidium-82
   D. All have same energy

2. How do PET systems identify the amount of energy deposited in a detector following a photon interaction?
   A. PHA
   B. Coincidence photon detection
   C. Analyze light pattern
   D. A, B, and C

3. How do PET imaging systems obtain spatial information?
   A. Low-energy collimators
   B. Medium-energy collimators
   C. High-energy collimators
   D. Coincidence photon detection

4. Which types of events are generally recorded by a PET imaging system?
   A. Line of response (LOR)
   B. Coincidence event
   C. Scatter event
   D. A, B, and C

CASE 8.7

Image Acquisition

Fig. 8.13 Positron emission tomography/CT in a patient with lymphoma demonstrating numerous hypermetabolic lymph nodes on coronal maximum intensity projection images (A) and axial fused CT images (B).

1. **D.** The higher the positron energy, the farther it will travel before losing all of its kinetic energy and undergoing annihilation. When a positron has lost of all its kinetic energy and meets an electron, two 511-keV photons are emitted 180 degrees apart, irrespective of positron initial kinetic energy.

2. **A.** The total amount of light produced following a photon interaction in a PET scintillator is collected and analyzed using PHA. The PHA system would eliminate lower-energy scatter events and higher-energy events, where two interactions occurred in the same crystal at exactly the same time (pulse pile-up).

3. **D.** PET imaging systems do not use a collimator to obtain spatial information as occurs with planar imaging performed using a gamma camera. PET systems obtain spatial information by detecting two 511-keV annihilation events at the same time, which produces an LOR along which the annihilation event is assumed to have occurred.

4. **D.** When two events are detected at the same time (within the temporal window) and have the appropriate amount of energy according to PHA, this corresponds to an LOR. A random coincidence event occurs when one annihilation photon from one positron decay is detected at the same time as an annihilation from a second positron event. A scatter event occurs when a scattered photon is detected at the same time as an annihilation photon and is nonetheless accepted by the PHA window.

Comment

PET scanners use coincidence detection to help eliminate unwanted counts and to localize where activity is located. As a result, PET does not use a collimator like SPECT. Coincidence detection is the detection of both 511-keV annihilation photons created by a positron interaction. Detection of both photons must occur within a specified time interval, called the coincidence timing window. Furthermore, a coincidence detection is counted only if the energy of each detected photon is near 511 keV using PHA. The superior image quality of PET compared with SPECT is in no small part due to PET's ability to localize activity without a collimator.

When a coincidence is detected, the machine knows that the source of the photons (i.e., the radioisotope) must be located in the patient somewhere along the line connecting the two detectors that the annihilation photons hit. This line is called the LOR and is illustrated in Fig. 8.14A. Very few of the detected photons, typically less than 1%, would be accepted by the coincidence circuitry of the PET system. The system is able to localize the radioisotope position by considering all of the LORs that occur at various angles. Combining all the LOR data at a selected angle permits the generation of the corresponding "projection" at that angle (see Fig. 8.14B) similar to projections seen in SPECT.

If both annihilation photons escape the patient without scattering and are detected in the scintillator, then this is called a "true" event. This is the desired event and what is used to create the image. The coincidence timing window eliminates many unwanted counts but not all. A "random" event occurs when one photon from one annihilation event and a second photon from a second annihilation event are accidentally detected within the coincidence timing window. In general, the result of incorrectly counting random events is a reduction in image contrast and the artifactual appearance of activity outside the patient. Only true events contain useful data for creating the image, whereas "randoms" degrade image quality.

A "scatter" event occurs when one or both annihilation photons are scattered before detection. Much of the scatter is eliminated because, when a photon scatters by Compton interaction, it loses some of its energy and thus will often fall outside the PHA energy window. However, sometimes (especially at smaller scattering angles) the photon retains enough energy to be incorrectly counted as a true event. In general, the result of incorrectly counting scatter as true is increased blur because the LOR is shifted away from its true position. As a result, "scatter" events, like "random" events, degrade image quality, and their occurrences should be minimized to optimize imaging performances in PET.

References

Bushberg JT, Seibert JA, Leidholdt EM Jr, Boone JM. Nuclear imaging—emission tomography. In: *The Essential Physics of Medical Imaging.* 3rd ed. Philadelphia: Wolters Kluwer; 2012:722–731.

Huda W. Nuclear medicine II. In: *Review of Radiological Physics.* 4th ed. Philadelphia: Wolters Kluwer; 2016:194–195.

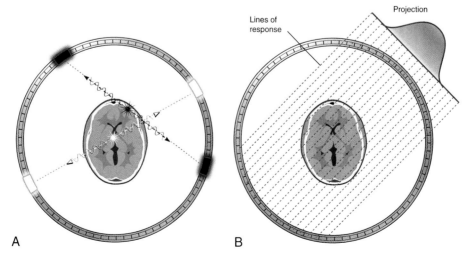

A   B

Fig. 8.14 Two lines of response (LORs) occurring at different times where two detectors identify 511-keV photons simultaneously (A). Positron emission tomography obtains spatial information by coincidence detection, which is much more efficient than the collimators used in gamma camera imaging. Multiple parallel LOR at a single angle create a projection (B) similar to that obtained in single-photon emission computed tomography or CT.

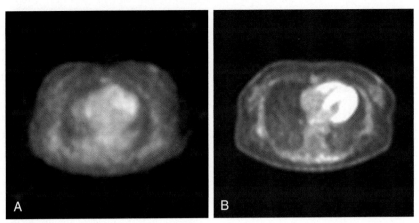

Fig. 8.15

1. What is generally produced from all the acquired LORs by a PET imaging system?
   A. Raw projections
   B. Image sinograms
   C. Attenuation coefficients
   D. A and B

2. Which is the best way to reconstruct PET images from acquired projections?
   A. Back projection
   B. FBP
   C. IR
   D. A, B, and C are similar

3. What is the major benefit of three-dimensional (3D) PET over 2D PET?
   A. Improved sensitivity
   B. Better resolution
   C. Higher contrast
   D. Lower dose

4. What is the major benefit of time of flight (TOF) PET imaging compared with standard PET?
   A. Improved resolution
   B. Reduced noise
   C. Lower dose
   D. B and C

CASE 8.8

Imaging Modes

Fig. 8.15 Positron emission tomography obtained without time of flight (TOF) (A) and with TOF (B) in different patients.

1. **D.** All the LORs at a given angle (say 45 degrees) are used to generate the projection that corresponds to this angle of 45 degrees. A stack of projections from 0 to 180 degrees is the acquired data set that is known as a sinogram and sent for reconstruction into tomographic slices. Attenuation coefficients are determined in CT imaging, not PET.

2. **C.** Back projection refers to allocating the measured amount of activity equally to each pixel along the LOR and is not used clinically because such images are too blurry to be of practical use. FBP is not used because the low number of counts in PET results in images that have too much noise and exhibit streak artifacts. IR reconstruction is typically used because it offers the best image quality.

3. **A.** The benefit of 3D PET is that there is up to an order-of-magnitude improvement in system sensitivity. The trade-off for increased sensitivity is an increase in random and scatter events. Contrast and resolution are unlikely to be improved, and the patient dose will not be affected because the administered activities are similar in 2D and 3D PET.

4. **A.** The use of TOF will help to localize the region of the LOR where an annihilation event actually occurred. As such, this will help to improve the resolution. TOF PET will not impact image noise, and the administered activity would not be changed with a TOF PET, so patient doses are not affected.

Comment

Collecting all LORs for a given gantry angle creates a projection. Arranging these projections as a function of angle will generate a sinogram, similar to that obtained in CT and SPECT. Because of the symmetry of the acquired projections, PET sinograms run only from 0 to 180 degrees, whereas those in CT run from 0 to 360 degrees. Images are reconstructed from sinograms using IR algorithms.

PET systems have detector rings that extend over an axial length up to 22 cm. The multiple detector rings enable several transverse image slices to be acquired at each position as the patient is moved through the scanner. Overlapping images are usually obtained every 15 to 22 cm for several minutes per position. Axial coverage in PET for a single detector position is approximately 20 cm, and images are acquired taking between 2 and 3 minutes at each detector position. A PET scan to cover the body region alone would likely use five detector positions, whereas up to 11 positions are needed for "head to toe" PET imaging, as is done in melanoma patients. Some modern scanners can continuously feed the table through the scanner to speed up imaging (similar to a very slow CT scan).

In 2D scanning, lead shadow shields, also known as septa, are used to define planes and limit scatter from adjacent slices. Septae help to localize activity in the patient to a given ring of the detector. Next, more precise localization is determined by coincidence detection. In addition to limiting scatter, septae help to limit random coincidences and keep activity from outside the field of view from being counted (e.g., photons from fluorine-18 activity in the brain being counted during abdominal scanning). Scanning in 2D is not available in modern machines because technological advances have made 3D acquisition the better option. Older systems found in departments may still offer 2D acquisition as an option.

In 3D mode, septal collimator rings are removed and coincidences are detected among multiple rings of detectors (cross plane). Consequently, when PET scanners are operated in 3D mode, coincidences will be detected in the whole volume that is being imaged (Fig. 8.16). Fourier rebinning, often referred to as FORE, is used to reorganize the acquired 3D data into 2D data sets (sinograms) for subsequent reconstruction. Removal of septa in 3D mode increases PET scanner sensitivity by up to an order of magnitude in comparison to 2D PET. This allows for either better image quality, faster scanning, or lower doses.

References

Cherry SR, Sorenson JA, Phelps M. Tomographic reconstruction in nuclear medicine. In: *Physics in Nuclear Medicine.* 4th ed. Philadelphia: Elsevier; 2012:270–277.

Huda W. Nuclear medicine II. In: *Review of Radiological Physics.* 4th ed. Philadelphia: Wolters Kluwer; 2016:191–195.

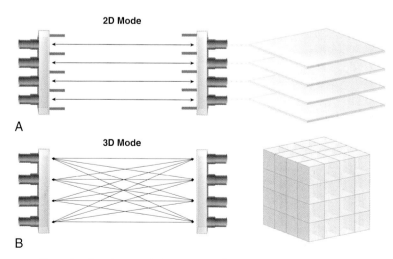

Fig. 8.16 Comparison of how two-dimensional positron emission tomography (PET) works (A), which generates a number of individual slices, with three-dimensional PET (B), which allows detection events that occur between different slices and creates a volume of data, not individual slices.

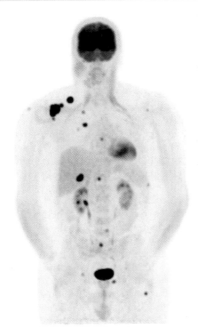 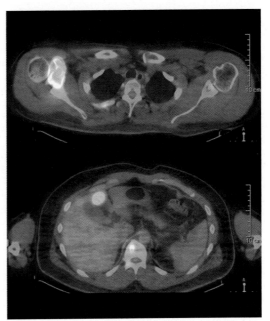

Fig. 8.17

1. Which type of CT scan would likely use the lowest radiation intensity (i.e., $CTDI_{vol}$)?
   A. Attenuation correction
   B. Anatomical localization
   C. Diagnostic image
   D. A, B, and C are similar

2. What is the most likely ratio of the average photon energy in PET imaging compared with that of CT imaging?
   A. 1:10
   B. 1:3
   C. 3:1
   D. 10:1

3. How does the CT scanner in PET/CT differ from a standard CT?
   A. Uses a higher kV
   B. No tube current modulation
   C. Follows noncircular orbit
   D. No differences

4. What are the most likely interactions between 511-keV photons and soft tissue?
   A. Photoelectric
   B. Compton scatter
   C. Coherent scatter
   D. A, B, and C are similar

CASE 8.9

CT in PET

Fig. 8.17 Positron emission tomography/CT in a patient with metastatic melanoma.

1. **A.** The radiation intensity required solely for attenuation correction would likely be an order-of-magnitude lower than that for a diagnostic scan. A CT image generated to identify the anatomy that takes up activity would most likely use a radiation intensity that is intermediate between that of attenuation correction and a full diagnostic scan.

2. **D.** The average photon energy in CT imaging, most likely performed at 120 kV, would be approximately 60 keV. This is nearly 10 times lower than the 511-keV photons that are used to generate LORs in PET imaging.

3. **D.** The CT scanner used with a PET system would be similar to that used for CT imaging. Specifically, CT systems in PET use similar kilovolts, may employ tube current modulation, and make use of circular orbits.

4. **B.** For soft tissues, Compton scatter is more important at energies greater than 25 keV. At 511 keV, Compton scatter will be the only interaction of importance because photoelectric effects will be negligible. Although there are very few PE interactions at 511 keV, there will be a sizeable number of Compton interactions.

Comment

A PET/CT system contains a hybrid of PET and x-ray CT imaging systems that are adjacent to each other. A moving bed is used to pass the patient through the bores of both systems. Most current PET/CT scanners can produce excellent whole-body fused or coregistered PET/CT images in less than 30 minutes. CT images associated with the PET scan may be either for attenuation purposes only (lowest dose and nondiagnostic), localization (reduced dose but still not diagnostic for the CT alone), or a full diagnostic scan (highest dose). As in SPECT/CT, a typical CTDI$_{vol}$ for the CT portion of the imaging would be diagnostic 16 mGy, localization 4 mGy, and attenuation 1 mGy.

The CT available in a PET/CT scanner is essentially the same as a standalone CT scanner. Modern PET/CT scanners include dose reduction technology (e.g., tube current modulation), can have up to 64 detector rows, and may be used as diagnostic CT scanners to accommodate patient overflow from dedicated CT scanners. The average photon energy in CT images is approximately 60 keV, which is much lower than the 511-keV photons used in PET (Fig. 8.18). Attenuation coefficients at PET energies are extrapolated from the measured attenuations obtained in CT imaging (i.e., HU values). During PET image reconstruction, corrections are made for attenuation of 511-keV photons within the patient, based on the estimated attenuation values.

Attenuation correction factors used in PET depend solely on the total thickness of tissue traveled by both annihilation photons. The correction factor for activity in the center of a circular patient would be identical to the correction factor for the same amount of activity at the patient surface. Despite using higher photon energies than SPECT, attenuation correction factors are higher in PET imaging because both photons must make it through the patient and be detected. For a small 20-cm patient, technetium-99m activity at the maximum depth of 10 cm will need a correction factor of 4.5 (μ is 0.15 cm^{-1} and $\mu \times t$ is 1.5). For a PET scan, the correction factor at all depths along this 20-cm path length is approximately 7 (μ is 0.10 cm^{-1} and $\mu \times t$ is 2).

Implanted metal objects may cause "hot" artifacts in CT attenuation-corrected PET images. Similarly, when patients have barium contrast material in the gastrointestinal (GI) tract, adjacent areas may appear anomalous. Using bolus injection of intravenous contrast for a chest CT scan may show foci of apparent increased activity near venous structures, which are just attenuation correction artifacts. Hot spots adjacent to strong attenuators should always be confirmed on non–attenuation-corrected images.

References

American College of Radiology. Image wisely. CT protocol selection in PET-CT imaging; 2012. http://www.imagewisely.org/imaging-modalities/nuclear-medicine/articles/ct-protocol-selection. Accessed March 2, 2018.

Huda W. Nuclear medicine II. In: *Review of Radiological Physics*. 4th ed. Philadelphia: Wolters Kluwer; 2016:198–199.

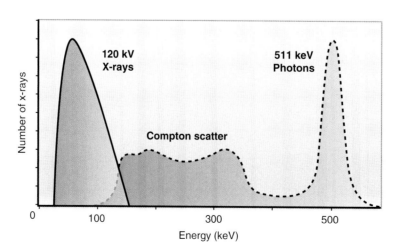

Fig. 8.18 X-ray spectrum (*solid line*) that is used to measure photon attenuation in patients, which have an average energy of approximately 60 keV. The *dashed line* shows the spectrum detected by a positron emission tomography (PET) imaging that includes the photopeak at 511 keV and Compton scatter at lower energies. PET attenuation correction requires an extrapolation of attenuation measurements at 60 keV to much higher application energy (511 keV).

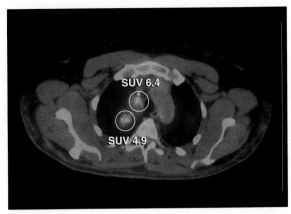

Fig. 8.19

1. Which nuclear medicine imaging modality likely has the worst contrast?
   A. Planar
   B. SPECT
   C. PET
   D. B and C

2. What is the likely ratio of the number of counts in a lung PET scan to that of a perfusion lung gamma camera acquisition?
   A. 100:1
   B. 10:1
   C. 1:10
   D. 1:100

3. What is the most likely resolution (FWHM) of a current clinical PET imaging system (mm)?
   A. 1
   B. 2
   C. 5
   D. 10

4. Which imaging system most likely has the highest sensitivity (i.e., counts per MBq of administered activity)?
   A. Gamma camera
   B. SPECT
   C. PET
   D. A and B

CASE 8.10

PET Image Quality

Fig. 8.19 Positron emission tomography/CT demonstrating increased fluorodeoxyglucose uptake in two right upper lobe pulmonary nodules. *SUV,* Standard uptake value.

1. **A.** Contrast is high in any tomographic imaging modality compared with that of projection imaging. When overlying and underlying activity is added to any slice, the lesion contrast will be markedly reduced so that planar imaging will always result in the worst contrast.

2. **B.** A typical PET acquisition will have 5 million counts per reconstructed slice, which is approximately 10 times higher than a typical planar NM image (500,000 counts). This is the fundamental reason PET images are generally of higher quality than planar images.

3. **C.** The resolution of a PET scanner (FWHM) is approximately 5 mm, which may be compared with the 8 mm of a typical gamma camera operated with a high-resolution collimator. By comparison, the FWHM of a CT scanner would be less than 1 mm.

4. **C.** Gamma cameras used in planar and SPECT imaging have a very low sensitivity due to the collimators. Collimators are not necessary in PET imaging, because spatial information is obtained by detecting annihilation photons "in coincidence." The number of detected photons in PET imaging is approximately 10 times higher than in gamma camera imaging.

Comment

In PET, as in all nuclear medicine, contrast is defined the difference in counts per voxel in a lesion compared with the corresponding counts per voxel in the surrounding anatomy (normal background). Uptake in surrounding tissue due to patient motion, scatter coincidences, and random coincidences all increase background levels and decrease contrast. Similarly, abnormally high glucose levels will cause glucose to compete with FDG and result in less radioactivity pooling to the region of interest and thus much lower contrast. For small lesions, the blurring because of the poor resolution of this imaging modality will also reduce the number of counts in the center of the lesion.

Imaging using fluorine-18–labeled FDG will show hot spots in areas of increased metabolism and is especially useful in tumor imaging. PET allows for semiquantitative measurement of hot spots to aid in diagnosis using a value called the standard uptake value (SUV). SUV is calculated as the activity concentration in the lesion (kBq/mL) divided by the average concentration in the patient. An SUV of approximately 2 is typically used as a threshold differentiating benign from malignant lesions. SUV is not an absolute measurement, meaning that the SUV between two different patients cannot be meaningfully compared; however, if the SUV changes within the same patient, then it may be indicative of a metabolic change. If repeated SUV measurements decrease for a lesion during therapy, this could be indicative of treatment efficacy.

PET scanners are inherently more sensitive than traditional gamma cameras. The fundamental reason for this is that gamma cameras use parallel-hole lead collimators to obtain spatial information, and these transmit only a small number of incident photons (e.g., 0.01%). PET cameras are much more sensitive because spatial information is obtained using coincidence detection, which permits many more data points to be acquired. Replacing 2D with 3D PET imaging has improved sensitivity by another order of magnitude. Consequently, current clinical PET images have high counts and much lower levels of image mottle. PET images use several million counts, which is typically 10 times more than planar images that would likely use 500,000 counts.

Spatial resolution of current commercial PET systems is approximately 5-mm FWHM. This can be assessed by analyzing the blur when imaging a line source of activity. On current PET systems, resolution will get worse as one moves away from the scanner center toward the detector because 511-keV photons may be detected by one of two adjacent detectors (non-coplanar). The spatial resolution is also influenced by the range of the positron, which is approximately 1 mm in soft tissue for fluorine-18. Positrons travel further in lung tissue, resulting in the degradation of spatial resolution performance. In cardiac imaging, rubidium-82 has a much a higher mean positron energy (1.4 MeV) than does fluorine-18 (0.25 MeV). For this reason, the rubidium-82 positrons travel a mean distance of approximately 2.5 mm. Rubidium-82 cardiac images can be differentiated from those of fluorine-18 because they will appear blurrier (Fig. 8.20)

References

Adams MC, Turkington TG, Wilson JM, Wong TZ. A systematic review of the factors affecting accuracy of SUV measurements. *AJR Am J Roentgenol.* 2010;195(2):310–320.

Huda W. Nuclear medicine II. In: *Review of Radiological Physics.* 4th ed. Philadelphia: Wolters Kluwer; 2016:195–196.

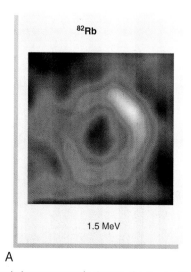

A   B

Fig. 8.20 Cardiac positron emission tomography image obtained with rubidium-82 (A) is much blurrier than the corresponding image obtained using fluorine-18 (B). Positron energy in rubidium-82 is much higher than in fluorine-18, so the positrons travel further before annihilating which increases the resultant amount of image blur.

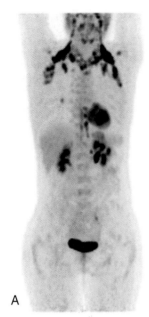

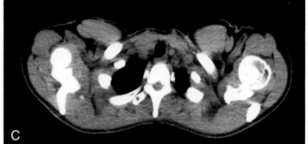

Fig. 8.21

1. Which PET QC test is performed daily?
   A. Blank
   B. Resolution
   C. Sensitivity
   D. Contrast

2. Which of the following could exhibit nonuniformities in a blank image?
   A. One or more broken detectors
   B. Table in gantry during scanning
   C. Faulty uniformity correction factors
   D. A, B, and C

3. Which image quality metrics can be assessed using a modified Jaszczak phantom?
   A. Spatial resolution performance
   B. Low contrast detection
   C. Image uniformity
   D. A, B, and C

4. What is the typical registration error between PET and CT images (mm)?
   A. 0.01
   B. 0.1
   C. 1
   D. 10

CASE 8.11

PET/CT Quality Control

Fig. 8.21 Positron emission tomography/CT images demonstrating hypermetabolic brown fat on coronal maximum intensity projection images (A) and axial attenuation-corrected images (B) that correspond to fat density on CT (C).

1. **A.** PET systems typically require a daily scan with a source that is automatically swept over the detectors to test that their output is uniform and correct. The QC test name "blank" is indicative of the fact that the bore is empty during this test (with the exception of the test source).

2. **D.** A broken detector will fail to register counts and will cause a dark band in the blank image and sinogram. If a foreign object is in the scanner during the blank scan, then it will block the radiation from the source and cause discontinuities in the image. Faulty uniformity correction factors will cause one or more detectors to put out signals that are too high or too low and thus give a bright or dark discontinuity.

3. **D.** The Jaszczak phantom modified for PET use has three separate sections. The uniformity section tests for image uniformity and presence of artifacts. The rods section tests the in-plane resolution, and the SUV/contrast section has several small vials filled with radioisotope that can be used to test low contrast detectability.

4. **C.** All modern PET scanners are PET/CT scanners. To ensure proper localization of PET update, the PET image must overlay exactly on top of the CT image. Typically registration errors using a phantom are approximately 1 mm; however, in clinical practice, breathing, heartbeat, and motion can create errors 10 mm or more.

Comment

On a daily basis a blank scan (scan of a uniform phantom or transmission source moved around the detector ring) is acquired. This is used to check the overall response of the detector's radiation (similar to the uniformity flood scan used for gamma cameras). The sinogram of the PET scan is typically displayed and visually inspected for large variations in signal that would indicate a problem. The software may also quantitatively compare the overall variability with a reference image taken previously. There are several causes of variability, including broken detectors, poor energy calibration, objects in the bore during blank scanning, and poor uniformity correction factors.

At least semiannually (and at acceptance and accreditation) clinical image quality is tested using a modified Jaszczak phantom. This phantom is similar to the one used in SPECT QC, but the spheres are removed and fillable wells from 8 to 25 mm

TABLE 8.3 DESCRIPTION OF EACH SECTION OF THE JASZCZAK PHANTOM AND THE CORRESPONDING IMAGING CHARACTERISTICS THAT ARE EVALUATED

Phantom Section	Properties Tested	PET or SPECT
Uniformity	Nonuniformities, artifacts, and noise	Both
Rods	System resolution	Both
Spheres	Low contrast detectability	SPECT
Cylinders	Attenuation and scatter correction	PET
Vials (fluorine 18)	Detectability, SUV values, and partial volume	PET

PET, Positron emission tomography; *SPECT,* single-photon emission computed tomography; *SUV,* standard uptake value.

are present instead. This still allows for evaluation of contrast but also allows for evaluation of SUV and partial volume. The rods section is used to evaluate the resolution of the scanner, whereas the uniformity section can be used to determine uniformity and to detect artifacts. A description of each section of the phantom can be seen in Table 8.3, together with the imaging characteristics that are evaluated by each section.

On an annual basis, at acceptance, or if there are errors found in the daily scan, new correction factors are determined for the detectors. This is called normalization and is typically an extended version of the blank scan that has an increased number of counts (similar to the uniformity correction map created for gamma cameras). Annual QC also uses a corrected version of this extended blank scan to determine conversion factors from counts to activity per unit volume (Bq/mL). This well-counter (AKA activity concentration) factor is crucial for clinical quantitative and SUV measurements.

The CT portion of CT/PET systems must also undergo QC (Fig. 8.22), which involves the typical QC acquired for standalone CT scanners (e.g., air calibration). If diagnostic CT scans are acquired, then full American College of Radiology accreditation and QC must be performed on the CT portion of the scanner. The registration between PET and CT scans is crucial for proper localization of pathology. To ensure misregistration is not due to scanner errors, it is recommended that a phantom with point sources visible on both CT and PET is imaged at acceptance and after major service. The PET and CT images are overlayed, and the distance between the points on the PET and CT images is measured. In general, this distance is expected to be approximately 1 mm.

References

Huda W. Nuclear medicine II. In: *Review of Radiological Physics.* 4th ed. Philadelphia: Wolters Kluwer; 2016:195–196.

International Atomic Energy Agency (IAEA). *IAEA Human Health Series No. 27: PET/CT Atlas on Quality Control and Image Artefacts.* Vienna, Austria: IAEA; 2014.

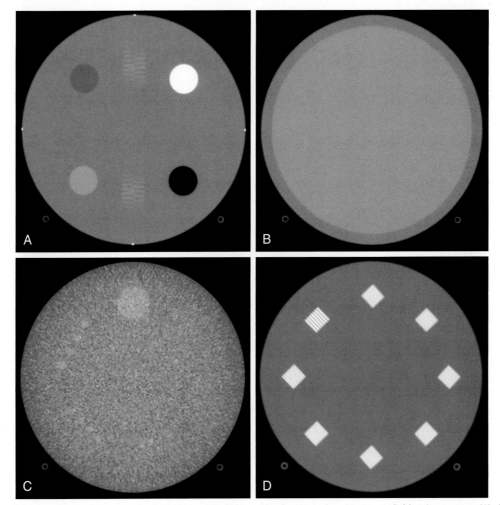

Fig. 8.22 Images of four sections of the American College of Radiology CT phantom showing Hounsfield unit accuracy (A), image uniformity (B), low-contrast lesion detectability (C), and spatial resolution performance (D). This phantom is used to assess the CT portion of a positron emission tomography/CT scanner.

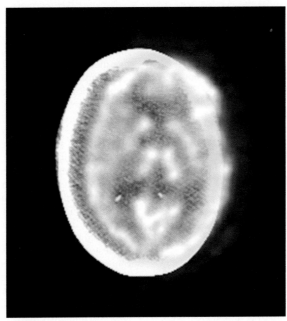

Fig. 8.23

1. Increasing which factor would most likely increase the contrast of a 5-mm lesion in PET?
   A. Administered activity (MBq)
   B. Imaging time (s)
   C. Spatial resolution (lp/mm)
   D. A and B

2. Which of the following could exhibit artifactual hot spots on attenuation corrected PET/CT images?
   A. Hip implant
   B. Metallic stent
   C. Cardiac pacemaker
   D. A, B, and C

3. How would a faulty detector appear on a blank sinogram?
   A. Vertical line
   B. Horizontal line
   C. Angled line
   D. Elliptical shape

4. Which of the following can affect the measured SUV in a PET scan?
   A. Inadequate fasting
   B. Exercise after injection
   C. Partial volume artifact
   D. A, B, and C

CASE 8.12

PET Artifacts

Fig. 8.23 Simulated misregistration artifact during a positron emission tomography (PET)/CT. The patient adjusted his position between the PET and CT portions of the scan.

1. **C.** The lesion diameter must be much bigger that the system's resolution to avoid partial volume artifact. Improving the spatial resolution will therefore increase the contrast of a small lesion by reducing the "blurring out" of the lesion. Reducing the image mottle by increasing the administered activity and/or imaging time will not affect lesion contrast.

2. **D.** When any highly attenuating object is present in the CT scan, it will block a large amount of radiation. When the machine uses that CT scan to correct the PET image for attenuation, the correction factor will be set too high in that region because of errors when extrapolating from measured values at 60 keV to applied values at 511 keV. As a result, these areas are likely to appear anomalously hot.

3. **C.** The sinogram plots the projection angle on the *y*-axis and the detector position along the array along the *x*-axis. When a detector is faulty, it will put out little or no signal and thus appear as a signal void. Because that detector is present in several angles, it creates a line, and that line is diagonal because its position shifts a bit at each projection angle.

4. **D.** When fasting is not adequate, the biodistribution of radiotracer will be radically altered (e.g., pooling in the active digestive tissues) and result in incorrect SUV measurement. Similarly exercise will cause muscles to uptake FDG and substantially alter SUV measurement. Finally, partial volume artifact blurs the lesion, which decreases the maximum lesion uptake value and artifactually lowers SUV in small lesions.

PET artifacts can be categorized by their cause as either machine based or patient based. Machine-based artifacts occur due to a machine error. One typical error is a single detector or block of detectors giving little or no signal. This is easily visualized on the sinogram, where a diagonal black line indicates where no counts are recorded (Fig. 8.24). Although obvious on the sinogram, the reconstructed image will not be degraded perceptibly unless a block or more of detectors are out (though quantitative measurements can still be affected).

Patient-based artifacts occur because of patient anatomy or some action taken by the patient. Attenuation correction can be fooled by the presence of metallic implants or large amounts of contrast in the patient. This is because the HU values of these materials are beyond the standard range that the machine understands. This results in false hot spots near metal implants. Abnormal hot spots can also appear if the patient does not completely rest after injection and before scanning. Muscluar contraction causes the muscles to uptake glucose and thus FDG. This will create unusual hot spots, as well as decreased contrast. Finally, extravasation of the radioisotope can cause a hotspot at injection site, low contrast, and incorrect SUV values.

Partial volume artifact occurs if the physical dimensions of the lesion are comparable with, or smaller than, the system resolution. In this case the resolution causes substantial blurring of the lesion, which will appear to be very faint. If the lesion is small and uptake is not pronounced, the lesion may not be detectable. More often the lesion is detectable, but the lesion size is artificially enlarged and the SUV is artificially decreased by blurring. Partial volume artifacts can be reduced by improving spatial resolution performance, so it will be less evident on modern PET systems that use TOF to improve resolution.

One of the most common causes of PET/CT image artifact is patient motion. If the patient moves between the CT and PET scans, then the two images will not overlay on top of each other correctly. This is called a misregistration artifact and results in incorrect localization of activity. It is especially important near the lung base/liver dome interface, where incorrect localization can alter the diagnosis. If the patient moves during the PET scan, then blurring and ghosting similar to that seen in gamma camera and SPECT imaging will occur. This will degrade image quality and also artificially decrease SUV values.

References

Huda W. Nuclear medicine II. In: *Review of Radiological Physics*. 4th ed. Philadelphia: Wolters Kluwer; 2016:195–196.

Sureshbabu W, Mawlawi O. PET/CT imaging artifacts. *J Nucl Med Technol.* 2005;33(3):156–161.

PET Sinograms

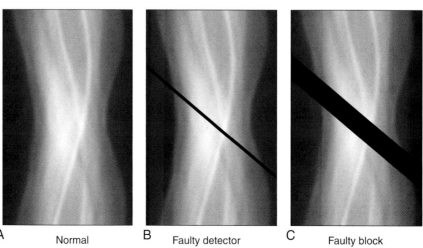

A — Normal B — Faulty detector C — Faulty block

Fig. 8.24 A normal positron emission tomography *(PET)* sinogram (A) shows no obvious defects or anomalies. A faulty detector (B) and faulty block (C) will exhibit obvious defects, which would readily be detected during daily quality control tests where there should be a blank image.

CASE 9.1

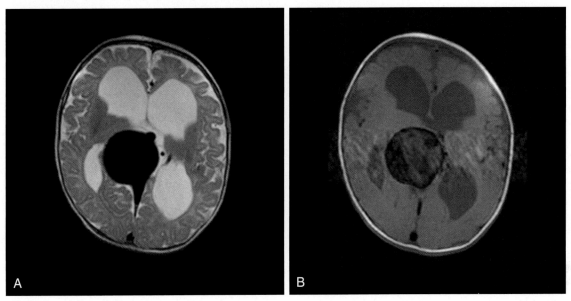

A

B

Fig. 9.1

1. Which of the following nuclei exhibits magnetism?
   A. Helium-4
   B. Carbon-12
   C. Oxygen-16
   D. Phosphorus-31

2. How much weaker is the magnetism of a proton compared with that of an electron?
   A. 2 times
   B. 10 times
   C. 100 times
   D. 1000 times

3. Which of the following has a negative susceptibility?
   A. Molecular oxygen
   B. Deoxyhemoglobin
   C. Iron oxide
   D. Soft tissue

4. Which of the following is paramagnetic?
   A. Calcium
   B. Water (H₂O)
   C. Gadolinium
   D. A and B

CASE 9.1

Magnetism

Fig. 9.1 Vein of Galen aneurysmal malformation in a child. Note the large flow void on T2- (A) and T1- (B) weighted images. Pulsation artifact is also present.

1. **D.** Whenever there is an even number of protons (or neutrons), these will line up in opposite directions, thereby resulting in no net magnetism. Only phosphorus-31 has nuclear magnetism, with 15 protons (an odd number), and is used to perform MR spectroscopy.

2. **D.** The magnetism of a proton is approximately 1000 times weaker than that of an electron. This is one reason why the signal-to-noise ratio in MRI is so low.

3. **D.** Soft tissue has negative susceptibility, which means that the local field is slightly lower than the magnetic field in air. Positive susceptibility would increase the local magnetic field.

4. **C.** A paramagnetic material will modestly increase the local magnetic field. Materials that exhibit paramagnetism include gadolinium, molecular oxygen, and deoxyhemoglobin.

Comment

Magnetism can be produced by moving charges such as electrons flowing in a coil or be an inherent property of particles (e.g., protons and electrons) called spin. Magnetic fields exist as dipoles with a "north pole" that is the origin of the magnetic fields and a "south pole" where the fields return to the object (Fig. 9.2). Magnetic field strength is measured in tesla or gauss, where 1 T is 10,000 gauss. Protons, neutrons, and electrons all have magnetism and will behave like very weak bar magnets. When protons "pair up," their magnetic fields cancel and give zero magnetism. Only nuclei with an odd number of protons and/or an odd number of neutrons are magnetic. By contrast, those nuclei that have an even number of protons, as well as an even number of neutrons, have no nuclear magnetism.

Nuclei that possess magnetism can be depicted by a vector representing the strength and orientation of the "bar magnet."

Magnetic nuclei are sometimes called magnetic moments, dipoles, or most commonly just spins. The high abundance of hydrogen (one unpaired proton) in the body, together with its relatively large nuclear magnetization, makes it the nucleus that is used for most current clinical MRI. When there is no external magnetic field, protons in tissues are randomly oriented so there will be no net nuclear magnetization. It is only the application of an external magnetic field that can result in a net tissue magnetization. The magnetic field associated with electrons is approximately 1000 times stronger than that of a proton. Unpaired electrons such as those found in gadolinium exhibit very strong magnetic fields that can influence adjacent magnetic nuclei.

Magnetic susceptibility is used to characterize how much matter becomes magnetized in an external magnetic field. Local magnetic fields will differ in strength from the external field of atomic electrons in each tissue. The magnitude of this change is quantified as the material susceptibility. Diamagnetic materials have negative susceptibility resulting in decreases of the local magnetic field. Tissue is a classic example of a material that is diamagnetic, with a very small negative susceptibility. However, the important point is that at interfaces there will be *changes* in susceptibility, which in turn result in *changes* in local *magnetic fields*. Changes in magnetic fields at interfaces such as bone and air are very important and can result in MRI signal loss.

In an external magnetic field, paramagnetic materials will increase in the local magnetic field. Paramagnetism results from the presence of unpaired atomic electrons. Gadolinium and deoxyhemaglobin are paramagnetic materials that have a very small positive of susceptibility. A ferromagnetic material will dramatically increase the local magnetic field and thus has a very large positive susceptibility. Steel in some dental devices and implanted medical devices are examples of ferromagnetic materials that can significantly distort acquired signals and result in substantial MR artifacts.

References

Huda W. Magnetic resonance I. In: *Review of Radiological Physics*. 4th ed. Philadelphia: Wolters Kluwer; 2016:205–206.

Pooley RA. AAPM/RSNA physics tutorial for residents: fundamental physics of MR imaging. *Radiographics*. 2005;25:1087–1099.

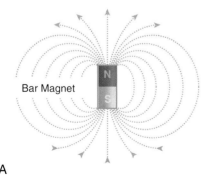
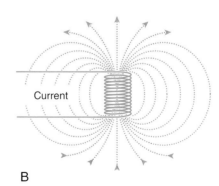

A   B

Fig. 9.2 (A) A bar magnet with lines of force that extend from the north pole to the south pole. Stronger fields have more lines per unit area. (B) The flow of electrons in a coil generates a similar pattern of magnetic lines, showing that magnetism is a property of moving electric charges.

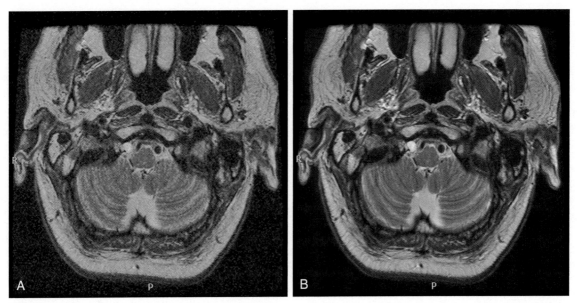

Fig. 9.3

1. What kind of motion do protons exhibit when placed in powerful magnetic fields?
   A. Precession
   B. Oscillation
   C. Rotation
   D. None (stationary)

2. What is the Larmor frequency of protons in a magnetic field of 1.5 T?
   A. 63 Hz
   B. 63 kHz
   C. 63 MHz
   D. 63 GHz

3. When protons are placed in a magnetic field, which are the possible orientations of these protons relative to the external magnetic field?
   A. Parallel
   B. Antiparallel
   C. Perpendicular
   D. A or B

4. How many free protons contribute to the net magnetization when soft tissues are placed in a magnetic field of 1 T (%)?
   A. 0.1
   B. 0.01
   C. 0.001
   D. <0.001

CASE 9.2

Nuclei in Magnetic Fields

Fig. 9.3 Images of a right hypoglossal schwannoma obtained at 0.3 T (A, simulated image) and 3 T (B).

1. **A.** If the magnetic field is directed along the z-axis, the proton's magnetism is aligned at a small fixed angle to the z-axis. The proton moves (precesses) so that the component along the z-axis is fixed and the component in the x-ray plane varies at the Larmor frequency.

2. **C.** The Larmor frequency for protons at 1.5 T is 63 MHz and is directly proportional to the applied magnetic field. For example, the Larmor frequency at 3 would be 126 MHz, which is twice that encountered at 1.5 T.

3. **D.** Protons placed into any magnetic field, aligned along the positive z-axis, will line up into one of two orientations, namely parallel or antiparallel. The parallel orientation is with the proton pointing in the same direction to the external magnetic field and the antiparallel orientation pointing opposite to the external magnetic field.

4. **D.** There are more protons in the parallel orientation than in the antiparallel orientation, which results in a net magnetization aligned along with the external magnetic field (positive z-axis). At 1.5 T, there will be approximately 1,000,003 protons that are parallel and 999,997 antiparallel, so only six protons per million (i.e., <0.001%) contribute to the net magnetization.

Comment

Current MR systems use a solenoid that carries an electrical current to create the main magnetic field. MR scanners generally are superconducting MR magnets to produce this high magnetic field. Superconducting magnets must be kept cold with liquid helium. The electric current that creates the magnetic field is perpetual, so magnetic fields will always be "on." Systems lose their properties should the wire temperature rise and the stored energy is converted to heat (i.e., magnet quench). Ferromagnetic objects such as scissors or oxygen cylinders may be "sucked into" the powerful magnet, which is referred as a "missile effect." Clinical MR magnets are typically 1.5 T, which may be compared with Earth's (very weak) magnetic field of 50 μT.

The ideal MR magnet would have a perfectly uniform magnetic field. High magnetic field uniformity is important to maintain image quality in MRI and essential for performing MR spectroscopy. MR magnets are manufactured to have homogeneity of less than a part per million, or better than 0.001%. Elevators and other nearby ferromagnetic structures can affect magnetic field uniformity, which will generally degrade the quality of acquired MR images. Presence of magnetic inhomogeneities may cause imaging artifacts. Magnetic field inhomogeneities can also influence the detected signals in MRI. Patients with metallic objects are likely to exhibit artifacts because of magnetic field distortions.

When magnetic nuclei are in magnetic fields, they perform a precession motion like a spinning top. The precession rate is known as the Larmor frequency and expressed as a frequency (MHz). One of the most important relationships in MR is that the Larmor frequency for any given magnetic nucleus is directly proportional to the magnetic field. For protons, the Larmor frequency is 63 MHz at 1.5 T. This corresponds to radiofrequencies with a wavelength that is generally comparable to the size of humans (meters). Doubling the magnetic field from 1.5 to 3 T will double the Larmor frequency and halve the corresponding radiofrequency (RF) wavelength. As the RF wavelength is reduced, loss of the RF radiation in the patient (attenuation) will generally increase.

In any magnetic field, protons will point parallel or antiparallel to the applied field. The parallel orientation is a lower energy state, and as such there are a few more protons in the (parallel) orientation. The difference between the parallel and antiparallel magnetization is known as the net tissue magnetization. For any tissue placed into a magnetic field aligned along the z-axis, there will be a very weak *net nuclear magnetization*, called M_z, which is aligned parallel to the direction of the magnetic field (Fig. 9.4). The magnitude of this net tissue magnetization will always be directly proportional to magnetic field so that there is more magnetization at higher magnetic fields. Whenever M_z is displaced, it will always return to this equilibrium orientation, which is the lowest energy state. *The behavior of tissue net nuclear magnetization serves as the starting point for comprehending clinical MRI.*

References

Elster AD. Questions and answers in MRI. Net magnetization (M): what is net magnetization and how does it apply to NMR? 2017. http://mriquestions.com/net-magnetization-m.html. Accessed March 2, 2018.

Huda W. Magnetic resonance I. In: *Review of Radiological Physics*. 4th ed. Philadelphia: Wolters Kluwer; 2016:206–208.

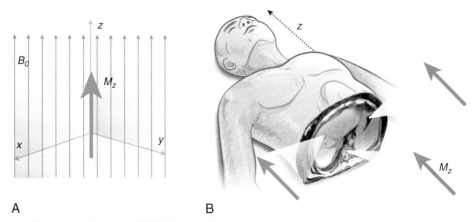

Fig. 9.4 The application of an external magnetic field (B_0) results in a net magnetization that points in the same direction as the external magnetic field (A). In superconducting magnets, the magnetic field is aligned along the long patient axis (B), where the amount of magnetization in different tissues generally varies by less than 10%.

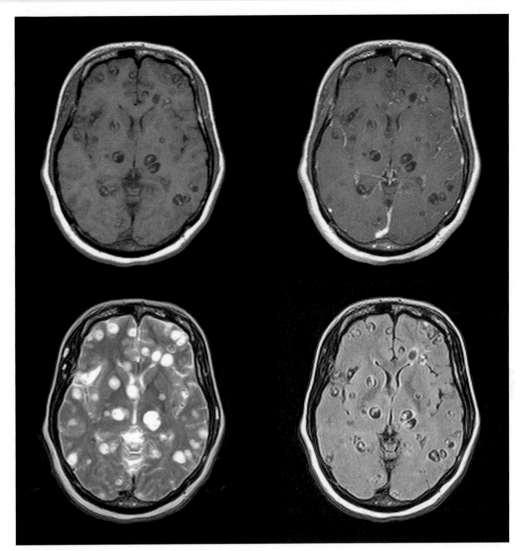

Fig. 9.5

1. What type of electromagnetic radiation can flip the spins in the body during MRI?
   A. RF
   B. Infrared
   C. Microwave
   D. Ultraviolet

2. How will increasing the magnetic field from 1.5 to 3 T affect the Larmor frequency of any magnetic nucleus?
   A. Quartered
   B. Halved
   C. Doubled
   D. Quadrupled

3. Doubling which RF pulse characteristic is likely to double the RF flip angle?
   A. Duration
   B. Intensity
   C. Frequency
   D. A and B

4. Which RF flip angle will have equal longitudinal (M_z) and transverse (M_{xy}) magnetization (°)?
   A. 45
   B. 90
   C. 180
   D. 360

CASE 9.3

Magnetic Resonance

Fig. 9.5 MRI in a patient with neurocysticercosis.

1. **A.** The magnetic field from RF waves oscillates at the correct frequency (Larmor) so it resonates with ^{1}H protons. This adds energy to the proton and causes its magnetic moment to rotate away from the magnetic field direction.

2. **C.** The Larmor frequency is always directly proportional to the applied magnetic field. Although the Larmor frequency is different for every nuclide, when the magnetic field is doubled, so is the Larmor frequency, and this will be the case for every nuclide.

3. **D.** Doubling the duration of any RF pulse will double the flip angle, as will doubling the RF intensity. However, when the pulse frequency increases, it is no longer at the Larmor frequency, and the net magnetization will no longer undergo any rotation.

4. **A.** A 45-degree RF pulse will result in equal magnetizations along the positive z-axis and in the x-y plane. A 90-degree RF pulse rotates the longitudinal magnetization M_z into the x-y plane, a 180-degree RF pulse will rotate it to point along the minus z-axis, and a 360-degree pulse brings the magnetization to its original starting point.

Comment

RF coils are used to emit RF pulses and broadcast at a very narrow range of frequencies known as the *transmit bandwidth*. The RF coils may also be used to detect radio waves from patients. The range of emitted frequencies (i.e., transmit bandwidth) determines the slice thickness. Receiver coils may be different from the transmit RF coil or may be the transmit coil that is electronically switched from transmit mode to receive mode. *Receiver bandwidth*, which is the range of frequencies detected, is directly proportional to the applied gradient strength (mT/m) that defines the range of frequencies emitted in the patient. Because the detected noise is directly proportional to receiver bandwidth, using strong gradients generally increases the amount of detected noise, and thus reduces signal-to-noise ratios.

Head and body coils are designed to transmit and receive uniform RF signal throughout a volume. Head and body coils are generally termed volume coils because these transmit and receive a uniform RF signal throughout a large volume of the patient. RF radiation is not ionizing but will generally deposit energy within the patient, increasing tissue temperatures. Use of surface coils will increase sensitivity of adjacent tissues, but the signal drops off rapidly with increasing distance. Linear coils receive the signal from only one axis (x or y) of rotating transverse magnetization, whereas quadrature coils receive the signal from both axes, which will increase the signal-to-noise ratio (SNR). Modern scanners often use phased array that combines many surface coils around the region of interest.

An applied RF field can interact with the net nuclear magnetization but only when it is at the Larmor frequency. RF fields at the Larmor frequency will make the net magnetization (M_z) rotate away from alignment with the magnetic field at a rate that is directly proportional to the RF intensity. Rotation of the net tissue magnetization will always continue while the RF is applied so that after the RF is switched off, the magnetization will have rotated through some flip angle (θ degree). Doubling the RF pulse duration, or the RF strength, will generally double the flip angle. Two 180-degree RF pulses applied in rapid succession will result in the longitudinal magnetization of *stationary* tissues to return to their initial location. The same sequence of two 180-degree RF pulses applied to moving blood would not result in this return to the initial state because the blood flowing out of the slice does not receive the second signal. This technique could therefore be used to suppress the signal from flowing blood.

After a θ-degree flip, net tissue magnetization can be replaced by two vectors that are directed parallel and perpendicular to the external magnetic field. The parallel vector is called longitudinal magnetization (M_z) and points parallel to the external field. The perpendicular vector is called transverse magnetization (M_{xy}) and is in the plane perpendicular to the external magnetic field (Fig. 9.6). If the flip angle is 90 degrees, all the magnetization is in the transverse plane, and no longitudinal magnetization remains. If the flip angle is 180 degrees, all the magnetization points in the negative longitudinal direction, and no magnetization is present in the transverse plane. At a flip angle of 45 degrees, there will be equal amounts of longitudinal and transverse magnetization.

References

Huda W. Magnetic resonance I. In: *Review of Radiological Physics*. 4th ed. Philadelphia: Wolters Kluwer; 2016:206–208.

Radiological Society of North America. RSNA/AAPM radiology physics educational modules. Basic principles of nuclear magnetic resonance. https://www.rsna.org/RSNA/AAPM_Online_Physics_Modules_.aspx. Accessed March 2, 2018.

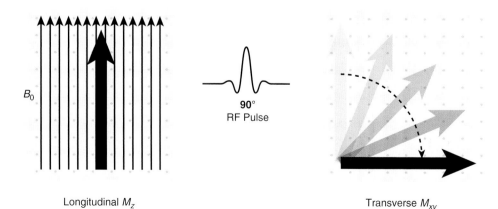

B_0

**90°
RF Pulse**

Longitudinal M_z

Transverse M_{xy}

Fig. 9.6 Initially *(left)* there is only stable equilibrium magnetization pointing along the direction of the external magnetic field B_0, the longitudinal magnetization M_z. The application of a 90-degree radiofrequency *(RF)* pulse at the Larmor frequency rotates this magnetization into the transverse plane, M_{xy}.

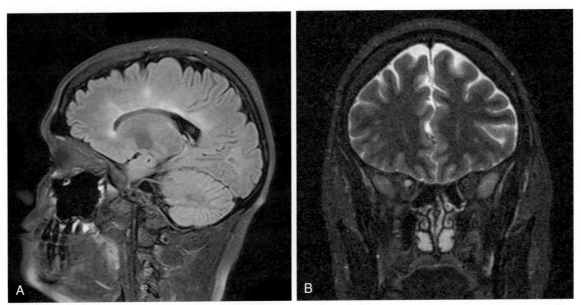

Fig. 9.7

1. Which type of magnetism gives rise to an electrical signal in a coil around a patient placed into a powerful magnet?
   A. Positive longitudinal $(+M_z)$
   B. Negative longitudinal $(-M_z)$
   C. Transverse plane (M_{xy})
   D. A and B

2. What is the most likely frequency of the free induction decay (FID) signal?
   A. $<f_{Larmor}$
   B. f_{Larmor}
   C. $>f_{Larmor}$

3. When the applied magnetic field increases by 1%, how does the frequency of the FID signal change (%)?
   A. No change
   B. <1
   C. 1
   D. >1

4. Which type of coil will most likely result in the largest FID signal versus noise?
   A. Surface
   B. Head
   C. Body
   D. A, B, and C are similar

CASE 9.4

Free Induction Decay

Fig. 9.7 MRI in a patient with multiple sclerosis demonstrating peri-ventricular T2 hyperintensity (A) and right optic neuritis (B).

1. **C.** It is only transverse magnetization, rotating at the Larmor frequency in the x-y plane, that will induce a voltage in an adjacent coil. The induced voltage, known as the FID, will oscillate with the Larmor frequency. No induced signal is obtained from the longitudinal magnetization ($+M_z$ or $-M_z$).

2. **B.** When the longitudinal magnetization is flipped into the transverse plane, it will rotate at precisely the Larmor frequency. For this reason, the FID signal is an induced voltage that also oscillates at the Larmor frequency.

3. **C.** The frequency of the FID signal is always determined by the value of the magnetic field in which the magnetization is rotating. When the applied magnetic field increases by 1%, there is a corresponding linear increase of 1% in the FID frequency.

4. **A.** The largest FID signal will occur with the coil that is closest to the source of signal. Because the surface coil is designed to be placed very close to the body, it will exhibit the highest signal.

Comment

Following a 90-degree RF pulse, all the tissue magnetization is in the transverse plane (M_{xy}) with longitudinal magnetization (M_z) reduced to zero. Transverse magnetization vector (M_{xy}) will rotate about the external magnetic field at the Larmor frequency. If the field strength increases, so will the rotation of M_{xy}, and vice versa. M_{xy} will induce a voltage in a receiver coil wrapped around the tissue RF coils, and which generally generate stronger signals when the coils are placed closer to the source. The magnitude of the transverse magnetization gradually decreases, and this will also result in the corresponding detected signal (voltage) in the RF coils to get progressively weaker.

Voltages detected in the coils are called the FID signals (Fig. 9.8). The FID oscillates at the Larmor frequency and is proportional to M_{xy}. An increase of the local magnetic field will always result in an increased FID signal frequency, and vice versa. It is important to note that it is transverse magnetization (M_{xy}) alone that gives rise to an FID signal. There is no signal from longitudinal magnetization (M_z) itself. It is only after an RF pulse rotates M_z into the x-y plane that a signal can be collected to make the image. It is generally difficult to measure the FID immediately following a 90-degree RF pulse because of the potential for electronic interference from the applied RF pulse. For this reason, MR images are obtained by the creation of echoes of the FID signal and not by the detection of the FID signal itself.

FID signals are very weak because of the small number of nuclei that contribute to the signal. Small number of nuclei contributing to MR signals, together with the overall weakness of all nuclear magnetism, result in low SNRs. Echoes of the FID signal will generally be even weaker, and great ingenuity is being applied to help improve the SNR and generate clinically acceptable images. One way of improving the SNR is to have four images acquired so that the resultant signal will be four times larger with the image acquisition time also being four times longer. Because noise is random, when four images are added the noise only doubles. The resultant SNR is increased by a factor of two, which is the square root of the number of acquired images.

Although large coils receive a high signal, they also receive a large amount of noise and thus smaller coils are typically better. Unfortunately, a coil can only see the anatomy within its loop and only sees to a depth of approximately 75% of its diameter. Thus arrays of small coils are often used to ensure proper coverage of the anatomy of interest. The ideal coil for a given task has coil diameter large enough to see deep structures of interest but small enough to eliminate unwanted noise. The signals from each coil in an array are independently received and then combined to make a single image.

References

Bushberg JT, Seibert JA, Leidholdt Jr EM, Boone JM. Magnetic resonance basics: magnetic fields, nuclear magnetic characteristics, tissue contrast, image acquisition. In: *The Essential Physics of Medical Imaging.* 3rd ed. Philadelphia: Wolters Kluwer; 2012:410–419.

Huda W. Magnetic resonance I. In: *Review of Radiological Physics.* 4th ed. Philadelphia: Wolters Kluwer; 2016:206–208.

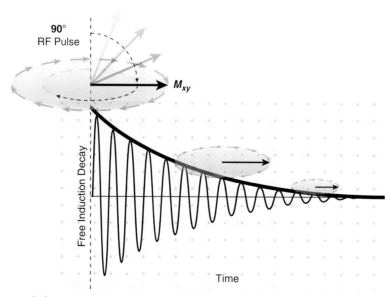

Fig. 9.8 Following a 90-degree radiofrequency *(RF)* pulse at the Larmor frequency, the net tissue magnetization is flipped into the transverse plane, where it rotates at the Larmor frequency inducing an oscillating voltage in an RF receiver coil, known as the free induction decay (FID) signal. The transverse magnetization gets progressively weaker due to spin dephasing, and the resultant FID signal gets progressively weaker.

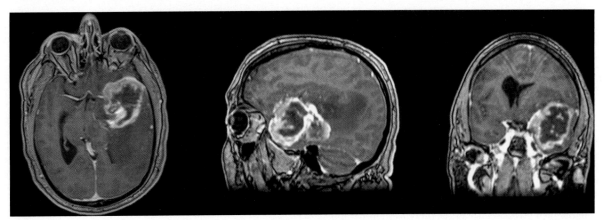

Fig. 9.9

1. What magnetic field frequency can flip a spin into a different orientation?
   A. Zero frequency (stationary)
   B. $<f_{Larmor}$
   C. f_{Larmor}
   D. $>f_{Larmor}$

2. After how many T1 times is longitudinal magnetization fully recovered following a 90-degree RF pulse?
   A. 1
   B. 2
   C. 3
   D. 4

3. Which likely has the shortest T1 time?
   A. Fat
   B. Gray matter
   C. Whiter matter
   D. Cerebrospinal fluid (CSF)

4. What is the difference in T1 times between gray and white matter (%)?
   A. 0.1
   B. 1
   C. 10
   D. 100

CASE 9.5

T1 Relaxation

Fig. 9.9 Glioblastoma multiforme demonstrating avid gadolinium enhancement.

1. **C.** Resonance means that if a magnetic field (like the one that makes up part of an RF pulse) has a frequency equal to the Larmor frequency, its energy will be transferred to the spins and "push them" (typically from the z direction into the x-y plane). Lattice oscillations at the Larmor frequency are also required for flipped spins to revert to their equilibrium location.

2. **D.** Following a 90-degree RF pulse, there is no longitudinal magnetization along the positive z-axis. It will take four T1 (longitudinal relaxation) times to get back to this equilibrium value: 63% has recovered after one T1, 83% after two T1, 95% after three T1, and greater than 99% (i.e., essentially full recovery) after four T1 times.

3. **A.** Fat has a relatively short T1 of approximately 250 ms at 1.5 T. CSF and all fluids have extremely long T1 values, typically 2000 to 3000 ms. Gray and white matter have T1 values of 900 and 800 ms, respectively.

4. **C.** The T1 value of gray matter is approximately 900 ms, which is approximately 10% longer than then T1 of white matter (i.e., 800 ms). This difference is the reason for these two tissues to appear different on T1-weighted images. This is analogous to how the 5 Hounsfield units difference (i.e., 0.5%) between gray and white matter in CT allows for those tissues to be differentiated in CT imaging.

Comment

When protons are placed into a magnetic field, a net longitudinal magnetization is produced (M_z) that is parallel to the external magnetic field. This longitudinal magnetization does not materialize immediately but grows exponentially from zero to M_z. The exponential growth of M_z is characterized by a time constant T1. After time T1, 63% of M_z will have formed, and the maximum (equilibrium) longitudinal magnetization (i.e., M_z) will have established itself after $4 \times$ T1. This equilibrium longitudinal magnetization is the preferred (i.e., lowest energy) orientation, so that if M_z is displaced for any reason it will always revert to this orientation. After any displacement, after four T1 times the initial equilibrium longitudinal magnetization will be fully recovered.

T1 is known as the longitudinal relaxation time whose value is influenced by spin-lattice interactions. When atomic lattice motions are present that produce magnetic fields that oscillate at the Larmor frequency, this will enable spins to be flipped (recover) to the equilibrium orientation. T1 times are generally long in solids, intermediate in tissues, and revert to being long in fluids. Because large slow molecules do not have motions corresponding to magnetic oscillations at the Larmor frequencies, the T1 values are long. Tissues with faster molecular motions have more lattice motions at Larmor frequencies, which will reduce their T1 values. Fluids have extremely fast molecular motions, with a corresponding lack of lattice motion at Larmor frequencies, resulting in long T1.

After an RF pulse displaces the longitudinal magnetization, net tissue magnetization will always revert (recover) to its equilibrium value (i.e., M_z) after $4 \times$ T1 (Fig. 9.10). This occurs for any type of pulse and will be just as true for a 90-degree pulse as for a 180-degree pulse. Because T1 for liver is 500 ms, liver M_z fully recovers after 2000 ms following any RF pulse (e.g., 90 degrees). In tissues such as fat (T1 shorter), recovery occurs more quickly, and in fluids (T1 longer), recovery will take a longer time. Tissue T1 increases with increasing magnetic field strength because there are usually fewer magnetic oscillations at higher frequencies in tissues. For example, quadrupling magnetic fields is generally taken to double tissue T1 longitudinal relaxation time.

Gadolinium atoms are paramagnetic, and adding these to a compound such as diethylenetriaminepentacetate acid is used to create paramagnetic contrast agents. Gadolinium contrast agents may be in extracellular fluid, within the intravascular blood pool, or designed to be organ specific. Gadolinium will interact with local nuclei to behave as a relaxation agent, markedly reducing T1 of surrounding tissue because the increased spin-lattice interactions will reduce T1 times. Increased spin-lattice interactions reduce relaxation times, whereas fewer spin-lattice interactions will result in an increase in the corresponding material relaxation times. Gadolinium-based contrast agents are called positive contrast agents because these reduce T1, which increases image brightness on T1-weighted images.

References

Elster AD. Questions and answers in MRI. T1 relaxation: definition; 2017. http://mriquestions.com/what-is-t1.html. Accessed March 2, 2018.

Huda W. Magnetic resonance I. In: *Review of Radiological Physics.* 4th ed. Philadelphia: Wolters Kluwer; 2016:207–209.

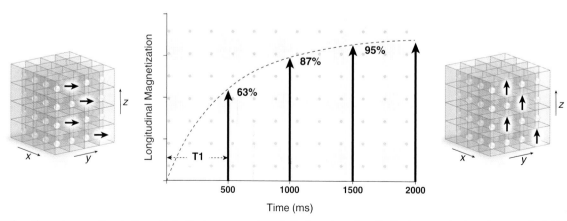

Fig. 9.10 When the net magnetism is flipped into the transverse plane *(left)*, there is no longitudinal magnetization. A magnetic field that oscillates at the Larmor frequency can cause a spin in the transverse plane (M_{xy}) to be rotated back along the equilibrium longitudinal orientation (M_z). For a tissue with a T1 time of 500 ms, the graph shows the growth of longitudinal magnetization has fully reverted to its equilibrium value after four T1, or 2000 ms, with net magnetization fully recovered *(right)*.

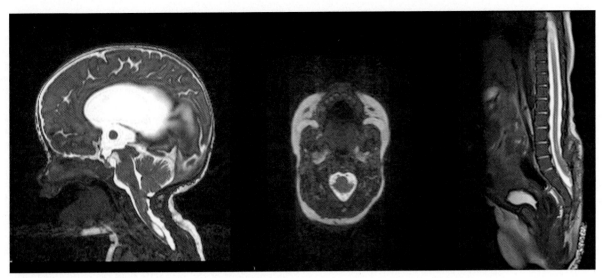

Fig. 9.11

1. If spin-spin interactions increase, what is the effect on T2 times?
   A. Increase
   B. Unchanged
   C. Decrease

2. What is the most likely soft tissue T2 relaxation time (ms)?
   A. 0.5
   B. 5
   C. 50
   D. 500

3. What happens to T2 when the magnetic field strength is increased?
   A. Reduced
   B. Unchanged
   C. Increased

4. How will the introduction of magnetic contrast agents such as super paramagnetic iron oxide (SPIO) affect transverse dephasing, and the corresponding T2 times?
   A. Increased; increased
   B. Increased; decreased
   C. Decreased; increased
   D. Decreased; decreased

CASE 9.6

T2 Relaxation

Fig. 9.11 Chiari II malformation with syringomyelia and myelomeningocele.

1. **C.** When spin-spin interactions increase, there will be more dephasing. With an increase in the amount of transverse spin dephasing, T2 times will be reduced.

2. **C.** As a rule, tissue T2 times are approximately 50 ms or so. This is 10 times longer than a nominal T2* time, which would be 5 ms. For most tissues, T2 times are also 10 times shorter than tissue T1 times, which are typically approximately 500 ms.

3. **B.** As the magnetic field strength changes, empirical data show that the T2 values are left unchanged. This may be contrasted with the fact that an increase in magnetic field will generally increase T1 times.

4. **B.** The addition of a magnetic contrast agent, such as SPIO, will generally increase the amount of dephasing and reduce the T2 time. On T2-weighted spin echo (SE) images, areas that have taken up SPIO will lose their signals and appear to be darker in comparison to the noncontrast images.

Comment

Following a 90-degree RF pulse, the longitudinal magnetization is flipped into the transverse plane and rotates at the Larmor frequency generating an FID signal. This transverse magnetization decays exponentially with a time constant T2, providing the spins are in a perfectly uniform magnetic field. Transverse decay of the FID reflects the fact that the transverse magnetization gets progressively weaker because the spins diphase, meaning they increasingly point in different directions. The FID signal has decayed to 37% of its original value after time T2, and the transverse magnetization will have decayed to zero after $4 \times T2$ (Fig. 9.12). Like T1 time, T2 time is characteristic of the tissue and is responsible for image contrast in T2-weighted images.

Dephasing of the transverse magnetization is a result of spin-spin interactions, which occurs when each spin interacts with the magnetic field of nearby spins. Spins will speed up when they sense a higher magnetic field and will slow down when they sense a lower magnetic field. When spin-spin interactions increase, this will produce a shorter T2, and vice versa. Once spins have dephased so that the FID is zero, the transverse magnetization is irretrievably lost. The addition of contrast agents will increase the amount of dephasing of transverse magnetization and reduce T2 times.

T2 values in soft tissues with intermediate-sized molecules are generally approximately 50 ms. Protons in dense tissue such as bone have an extremely short T2 of less than 0.01 ms and will therefore immediately dephase after a 90-degree RF pulse. It is very difficult to obtain a signal from dense tissue, and bone will generally appear black on an MR image. When the spine is visualized, it is the soft tissues such as bone marrow that is responsible for the signal and not the bone matrix itself. Protons in small, fast, and free molecules like water will have a long T2, generally more than 1 second. Empirical evidence has shown that T2 times are approximately independent of magnetic field strength. This is because spin-spin interactions are independent of the magnetic field strength. This is in marked contrast to spin-lattice interactions (i.e., T1); higher fields would require higher lattice frequencies to flip magnetization back to their equilibrium orientations.

Transverse magnetization (M_{xy}) in tissues with T2 times of approximately 50 ms will decay to zero in approximately 200 ms following a 90-degree RF pulse. Eventually, this will be followed by full longitudinal magnetization (M_z) recovery (i.e., T1), which will occur after approximately 2000 ms for a tissue with T1 equal to 500 ms. It is important to note that T1 relaxation and T2 relaxation times are always independent of each other, as are the processes that create the respective relaxations. For all tissues, T2 must be less than or equal to T1 because transverse magnetization (M_{xy}) cannot be present when longitudinal magnetization (M_z) has fully recovered. If there are few spin-spin interactions, as occurs with water, then the transverse decay is equal to the longitudinal recovery with T1 and T2 exactly equal.

References

Huda W. Magnetic resonance I. In: *Review of Radiological Physics*. 4th ed. Philadelphia: Wolters Kluwer; 2016:207–210.

Poole RA. AAPM/RSNA physics tutorial for residents: fundamental physics of MR imaging. *Radiographics*. 2005;25:1087–1099.

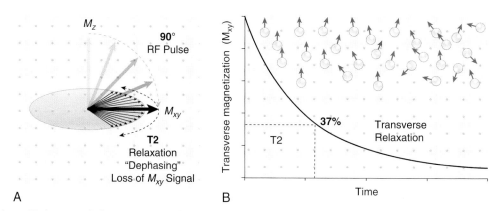

Fig. 9.12 Following a 90-degree radiofrequency *(RF)* pulse (A), tissue magnetization is rotated into the transverse plane *(M_xy)*, leaving no longitudinal magnetization along the z-axis *(M_z)*. The graph in B shows that the spins are initially aligned, resulting in a large transverse magnetization *(M_xy)*. As the spins interact and dephase, the transverse magnetization decays exponentially with a time constant T2.

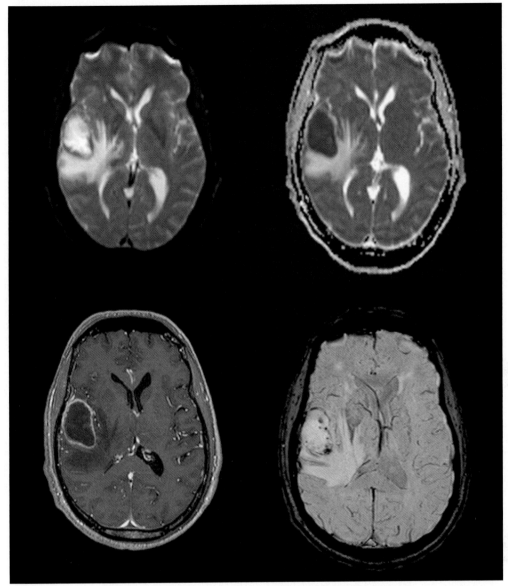

Fig. 9.13

1. What is the most likely homogeneity of a clinical MR magnet (%)?
   A. 0.1
   B. 0.01
   C. 0.001
   D. <0.001

2. Which likely has the greatest magnetic field inhomogeneities that result in the most dephasing of transverse magnetism (M_{xy})?
   A. Cortical bone
   B. Air spaces
   C. Bone/air interface
   D. Soft tissue

3. Which act like magnetic sources in patients?
   A. Hemosiderin
   B. Blood clots
   C. Implants
   D. A, B, and C

4. Which affect T2* dephasing in MRI?
   A. Magnet inhomogeneities
   B. Susceptibility differences
   C. Magnetic sources
   D. A, B, and C

CASE 9.7

T2* relaxation

Fig. 9.13 Cerebral abscess with associated foci of hemorrhage.

1. **D.** In a 10-cm diameter spherical volume (DSV), magnetic field uniformity of an MR magnet is approximately 0.02 parts per million (ppm). If the diameter of the DSV increases to 20 cm, the uniformity is approximately 0.08 ppm and for a 40-cm DSV is approximately 1.5 ppm.

2. **C.** Each tissue has its own susceptibility, and at tissue interfaces, there will be a changing magnetic field at the boundary that will give rise to the dephasing of transverse magnetization (i.e., T2* effect).

3. **D.** Hemosiderin, blood clots, and metallic implants can act as magnetic sources within patients. Because there will be a changing magnetic field with distance from these internal magnetic sources, these changing magnetic fields give rise to the dephasing of transverse magnetization (i.e., T2* effect).

4. **D.** Magnet inhomogeneities, susceptibility differences (interfaces), and magnetic sources (e.g., blood clots) are the three principal sources of T2* effects. Each one of these results in a localized *change* in magnetic field, which will dephase transverse magnetization.

Comment

Measured transverse magnetization (M_{xy}) decays much more rapidly than would be expected from the tissue T2 values. This is because the FID signals decrease exponentially with a decay rate of T2*, where T2* is markedly shorter than the tissue T2 value (approximately 10×). For a tissue with T2* equal to 5 ms, the FID disappears after 20 ms, which corresponds to four times T2*. In the absence of T2* effects, M_{xy} would disappear after 200 ms when the tissue T2 is the expected 50 ms.

The Larmor frequency depends on the magnetic field. If the main magnetic field is *inhomogeneous*, then spins in different positions in the magnet bore will feel different field strengths. These field strength differences will cause the spins to precess at different frequencies and thus to dephase quickly. The impact of magnetic field inhomogeneities is denoted by a "*" in the T2 constant (T2*). Because inhomogeneities can only add to the dephasing, T2* is always shorter than T2. This dephasing results in a loss of transverse magnetization, which causes a rapid reduction in the FID signal.

MR magnets have magnetic field inhomogeneities with differences of a few parts per million (1 ppm = 0.0001%) in a magnetic field at different locations. Near magnetic tissues (e.g., blood clots) there will always be a changing magnetic field that will contribute to T2* effects (Fig. 9.14). Similarly, tissue boundaries demonstrate increased T2* effects (loss of signal) because of differences in susceptibility between the two tissues that are adjacent to the boundary. T2* is the culmination of the dephasing of transverse magnetization from all these three sources of magnetic field inhomogeneity.

Small particles of Fe_3O_4 are termed superparamagnetic and will develop a strong internal magnetization when placed in an external magnetic field. Paramagnetic and ferromagnetic contrast agents influence the local magnetic field homogeneity and are therefore expected to generally shorten both T2 and T2*. Contrast agents that reduce T2 will appear dark on T2-weighted images and are therefore called negative contrast agents. SPIO is composed of nanosized iron oxide crystals and used for liver imaging. Ultrasmall superparamagnetic iron oxide (USPIO) particles can be used to differentiate malignant and benign lesions.

References

Bushberg JT, Seibert JA, Leidholdt Jr EM, Boone JM. Magnetic resonance basics: magnetic fields, nuclear magnetic characteristics, tissue contrast, image acquisition. In: *The Essential Physics of Medical Imaging.* 3rd ed. Philadelphia: Wolters Kluwer; 2012:415–418.

Huda W. Magnetic resonance I. In: *Review of Radiological Physics.* 4th ed. Philadelphia: Wolters Kluwer; 2016:207–210.

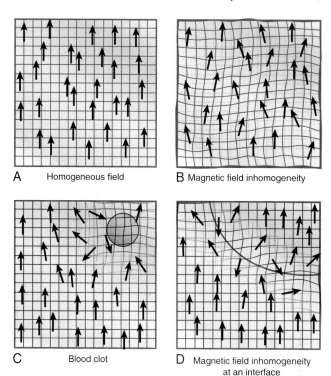

A   Homogeneous field

B   Magnetic field inhomogeneity

C   Blood clot

D   Magnetic field inhomogeneity at an interface

Fig. 9.14 In a homogeneous field (A), all spins stay in phase and point in the same direction, whereas in an inhomogeneous field, they rotate at a different speed to point in different directions, which is referred to as dephasing (B). Inhomogeneous fields occur near magnetic sources such as blood clots (C) and interfaces (D) that will always dephase magnetic spins.

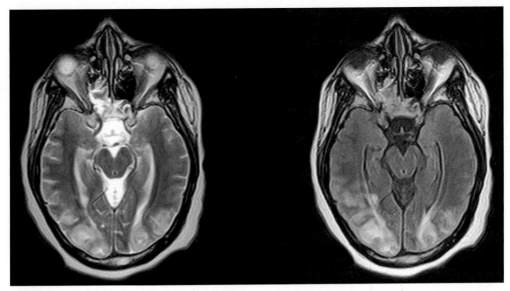

Fig. 9.15

1. If a 180-degree refocusing RF pulse is applied at time echo time (TE)/2, when will the resultant echo most likely be observed?
   A. <TE
   B. TE
   C. >TE

2. Which tissue property will likely affect the intensity of an SE, relative to the initial FID, obtained using a 180-degree RF refocusing pulse?
   A. T1
   B. T2
   C. T2*
   D. ρ

3. What most likely happens to the SNR when TE increases?
   A. Increases
   B. Unchanged
   C. Decreases

4. What is the largest number of SEs that can be obtained following a single 90-degree RF pulse on a commercial MR scanner?
   A. 1
   B. 3
   C. 10
   D. >10

CASE 9.8

Spin Echoes

Fig. 9.15 Posterior reversible encephalopathy syndrome.

1. **B.** When a 180-degree refocusing RF pulse is applied at time TE/2, the resultant echo will be observed at time TE. The choice of TE can be adjusted to any desired value of the operator and is generally used to control the amount of T2 weighting in MR images.

2. **B.** The initial FID signal following a 90-degree RF pulse gets weaker because of T2* and T2 dephasing. The 180-degree refocusing RF pulse will eliminate all T2* effects but cannot recover T2 dephasing due to spin-spin interactions. Spin density and T1 properties do not affect the intensity of the echo *relative* to the FID signal.

3. **C.** As the time to echo (i.e., TE) increases, there will be more dephasing of the transverse magnetization and the signal will get progressively weaker. After just one T2 interval, the echo will be only 37% of the original FID signal. The signal is taken to zero after four T2 intervals.

4. **D.** Fast SE sequences use 180-degree refocusing RF pulses to generate multiple echoes in a single TR interval. Typically, up to 20 echoes could be produced in each TR interval.

Comment

After a 90-degree RF pulse, the net magnetization vector rotates into the transverse plane at the Larmor frequency and produces an FID signal that can be detected in RF coils. This net transverse magnetization (M_{xy}) rapidly dephases primarily because of T2* effects, but also due to T2 effects. When a 180-degree RF refocusing pulse at a time TE/2 is applied, this will generate an SE at time TE (Fig. 9.16). The echo cancels out T2* dephasing effects but is weaker than the FID signal because of T2 effects. Table 9.1 shows how the echo intensity, relative to the initial FID, changes with increasing TE.

An analogy is a row of soldiers who start in phase (lined up), resulting in an FID signal. The speed at which soldiers march is analogous to the Larmor frequency so that a higher local field results in a higher Larmor frequency (spin rotation). Each soldier walks at a different speed and will rapidly be out of step (i.e., dephase), which is analogous to a reduction in the FID signal (T2* effects). Like spins, soldiers always march at a constant speed, and the direction of travel does not affect this marching speed.

At time TE/2, a refocusing 180-degree RF pulse is applied to rotate spins through 180 degrees, which is taken to be analogous to soldiers performing an about-face (180-degree turn). Consequently, at time TE the soldiers will be lined up to generate an echo, which cancels out T2* dephasing. It is notable that some soldiers disappear, which is analogous to true tissue T2 effects so that the echo will be weaker than the initial FID signal (see Table 9.1). Echoes are the signals detected in MRI. Despite the T2 effects, SE imaging as described previously generally gives the strongest signal of available imaging sequences.

One problem with SE imaging as described previously is that it is very slow. A 256 × 256 image requires 256 signals to be acquired, and each signal requires a 90-degree and 180-degree flip along with T1 and T2 relaxation. Fast SE (AKA turbo SE) imaging acquires multiple signals by applying multiple 180-degree pulses after the initial 90-degrees pulse. The number of signals acquired for each 90 degrees is called the echo train length (ETL). Scan time is inversely proportional to ETL. Increasing the ETL to obtain more echoes will always decrease scan time but may also result in a decrease in image quality as later echoes always get progressively weaker.

References

Huda W. Magnetic resonance I. In: *Review of Radiological Physics*. 4th ed. Philadelphia: Wolters Kluwer; 2016:210–211.

Radiological Society of North America. RSNA/AAPM physics modules. MRI: pulse sequences. https://www.rsna.org/RSNA/AAPM_Online_Physics_Modules_.aspx. Accessed March 2, 2018.

TABLE 9.1 RELATIVE INTENSITY OF SPIN ECHOES WITH INCREASING ECHO TIME EXPRESSED IN TERMS OF THE INTRINSIC TISSUE T2 TIME

Time to Echo (TE)	Relative[a] Echo Intensity (%)
T2	37
2 × T2	14
3 × T2	5
4 × T2	2

[a]Compared with free induction decay intensity.

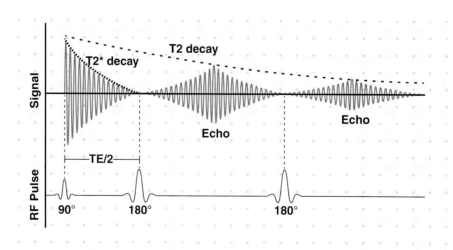

Fig. 9.16 Following an initial 90-degree radiofrequency *(RF)* pulse, the transverse magnetization rapidly decays, with a time constant T2* generating the free induction decay (FID) signal. A 180-degree refocusing pulse at echo time *(TE)*/2 produces an echo at time TE that eliminates all T2* effects but that is weaker than the initial FID because of T2 effects. Additional echoes can be created by applying 180-degree RF pulses that get progressively weaker because of T2 effects. Two of such echoes are shown here (i.e., echo train length = 2).

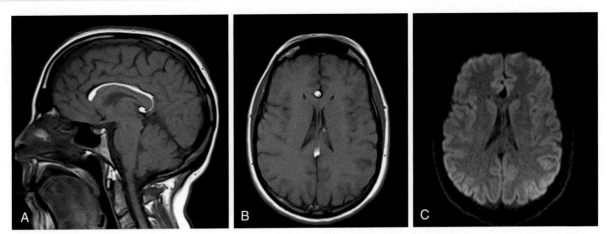

Fig. 9.17

1. What will change along an MR gradient?
   A. Magnetic field
   B. Larmor frequency
   C. RF intensity
   D. A and B

2. How many gradients are required to uniquely identify the location of any given voxel located at an arbitrary point *(x, y, z)*?
   A. 1
   B. 2
   C. 3
   D. 4

3. Which direction is the slice select gradient (SSG) always oriented?
   A. *x* (vertical)
   B. *y* (horizontal and perpendicular to bore axis)
   C. *z* (along bore axis)
   D. Any direction is possible

4. Which gradients must be applied to generate an MR image?
   A. SSG (G_z)
   B. Frequency encode gradient (FEG) (G_x)
   C. Phase encode gradient (PEG) (G_y)
   D. A, B, and C

CASE 9.9

Spatial Localization

Fig. 9.17 Pericallosal lipoma demonstrating hyperintense T1 signal (A and B) and no diffusion restriction (C).

1. **D.** A gradient is a change in magnetic field as a function of position in the scanner. Thus, moving along a gradient, spins will see a progressively larger (or smaller) field. Because the Larmor frequency is proportional to the field strength, the frequency will also be larger or smaller.

2. **C.** With one gradient applied, say, along the x-direction, we can obtain information about where along x the signal is coming from. Applying a second gradient along the y-direction, we can localize along the y-direction, and a third gradient along the z-direction provides the third dimension and therefore the exact location of the signal.

3. **D.** Gradients are vectors, which means that if two equal gradients are applied at 90 degrees to each other, the resultant vector will be at 45 degrees to both vectors. The angle in this plane can be modified to any value by adjusting the strengths of the two gradients. This principle can be extended to three dimensions by using a third gradient so that a gradient can be readily created along "any direction."

4. **D.** When the RF pulses are applied, an SSG is switched on along the selected z-axis. When the echo is generated, an FEG is applied along the selected x-axis. In between these RF pulses and echo, a PEG is applied along the selected y-axis.

Comment

In any uniform magnetic field, the detected MR signals in the form of echoes will originate from the entire patient. As a result, there is no way to localize strong or weak signals to specific tissues and parts of the body (i.e., no spatial encoding). Although the signal cannot be localized in a uniform field, the overall signal can be weakened due to nonuniformities and inhomogeneities. Nonuniformities typically arise from the different local chemical environments (e.g., fat vs. fluid) and interfaces such as air/tissue or bone/tissue. Inhomogeneities are small imperfections in the field uniformity that cause spins to dephase and reduce signal.

Localizing any signal will require the application of a magnetic field gradient that changes the magnetic field at different locations (Fig. 9.18). This gradient will change the Larmor frequency along the gradient direction. As a result, the identification of the detected Larmor frequency will permit the location along this the gradient direction to be determined. Fourier analysis of the MR signal is the method that is used to enable the frequency information to be extracted. Detected signals in MR (i.e., echoes) are analogous to CT projections obtained at different x-ray tube angles. In CT, projections are acquired at different x-ray tube angles, whereas in MR, echoes are acquired using different PEGs.

Activated gradients superimpose a linear gradient along the selected direction. MR scanners have three magnetic field gradient coils that are oriented in the x, y, and z directions. Combination of these three orthogonal gradients will permit the gradients to be oriented in any arbitrary direction (i.e., native acquisition of oblique slices). This is very different than CT, where data are always acquired axially and then postprocessing creates oblique views. Gradients are quantified in terms of mT/m, and a typical gradient will have a strength of 30 mT/m. When one moves a distance of 1 m along a gradient, the magnetic field increases by 30 mT for a positive gradient and decreases by 30 mT for a negative gradient.

Locating a voxel that is responsible for a signal will use three gradients. These are known as the SSG applied along the z-axis, the FEG applied along the x-axis, and the PEG applied along the y-axis. Each detected echo will be obtained in conjunction with a unique set of combinations of slice select, frequency encode, and PEGs. In MRI, when the RF pulses are applied, an SSG is applied (z-axis). When the echo is generated, an FEG is applied (x-axis). In between these RF pulses and echo, a PEG is applied (y-axis). When subsequent echoes are generated, the SSG and FEG are kept the same, but the values of the PEG are changed.

References

Huda W. Magnetic resonance I. In: *Review of Radiological Physics*. 4th ed. Philadelphia: Wolters Kluwer; 2016:211–213.

Radiological Society of North America. RSNA/AAPM physics modules. MRI: image formation. https://www.rsna.org/RSNA/AAPM_Online_Physics_Modules_.aspx. Accessed March 2, 2018.

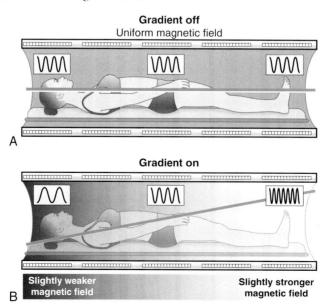

Gradient off
Uniform magnetic field

A

Gradient on

Slightly weaker magnetic field

Slightly stronger magnetic field

B

Fig. 9.18 In a uniform magnetic field (A), the Larmor frequency is the same everywhere, with no spatial localization. When a gradient is applied (B), lower frequencies originate from the left part of the patient where the magnetic field is lower and higher frequencies originate from the right where the magnetic field is higher.

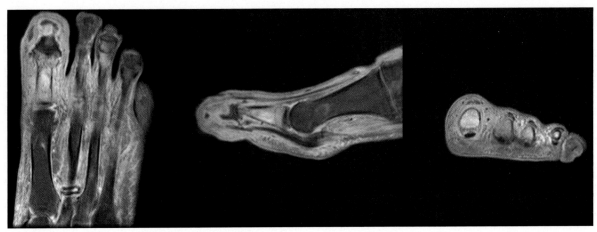

Fig. 9.19

1. Which can be modified to adjust the slice thickness in any MR image?
   A. Transmit bandwidth
   B. Gradient intensity
   C. Magnetic field
   D. A and B

2. If a digitized echo has 196 data points, what will be the matrix size in the reconstructed MR image along the frequency encode direction?
   A. <196
   B. 196
   C. >196

3. If 128 echoes are acquired, what will be the matrix size in the reconstructed MR image along the phase encode direction?
   A. <128
   B. 128
   C. >128

4. How long will it take to acquire 128 × 128 MR image data when the TR time is 1 second and one echo is obtained in each TR interval (s)?
   A. 64
   B. 128
   C. 256
   D. Depends on TE

CASE 9.10

Data Acquisition

Fig. 9.19 Osteomyelitis of the great toe.

1. **D.** For a fixed SSG along the *z*-axis, when the transmit RF bandwidth is increased, this will result in a thicker image slice, and vice versa. If the RF transmit bandwidth is fixed, a strong gradient will result in a narrow slice.

2. **B.** The number of data points sampling the echo will determine the number of pixels along the frequency encode direction (*x*-axis). For a 196 matrix size, the echo must be sampled 196 times to generate a row of *k*-space with 196 discrete numbers.

3. **B.** A separate echo, obtained with a unique PEG, is required for each pixel along the PEG direction. With 128 echoes, there will be exactly 128 pixels along the PEG direction.

4. **B.** If one echo is obtained in each TR interval, then it must take 128 echoes to create a 128 × 128 image. For conventional SE imaging, the total imaging time is TR multiplied by the number of echoes, which thus takes 128 seconds (i.e., approximately 2 minutes).

Comment

To select an MRI slice, a gradient is applied in the desired chosen z-direction whenever an RF pulse is applied (Table 9.2). When an RF pulse is applied, there will be "one location" along the *z*-axis where longitudinal magnetization will be rotated because there is only one location where the Larmor frequency will correspond to the chosen RF pulse. Where the magnetic field is lower, the corresponding Larmor frequencies are less than the applied RF. Where the magnetic fields are higher, the Larmor frequencies will be higher. Each RF pulse will contain a (narrow) range of frequencies that determines the slice thickness and is referred to as the transmit bandwidth. Increasing the transmit RF bandwidth will generate thicker MR slices, and vice versa. If the RF bandwidth is fixed, increasing the SSG results in thinner slices, and vice versa.

FEGs are always applied when echoes are generated (Fig. 9.20). The frequency encode direction (*x*-axis) must be perpendicular to the slice select (*z*-axis) direction. Pixels with lower magnetic fields have signals from this region that will have lower frequencies.

TABLE 9.2 PATTERN OF GRADIENT ACTIVATION DURING ONE TR INTERVAL IN A CONVENTIONAL SPIN-ECHO SEQUENCE

Time	SSG	PEG	FEG	Comment
0	On	Off	Off	90-degree RF pulse excites a slice (*z*-axis)
TE/4	Off	On	Off	Known amounts of phase added to spins (*y*-axis)
TE/2	On	Off	Off	180-degree RF refocusing pulse
TE	Off	Off	On	Echo created and signal detected (*x*-axis)

FEG, Frequency encode gradient; *PEG*, phase encode gradient; *RF*, radiofrequency; *SSG*, slice select gradient; *TE*, echo time.

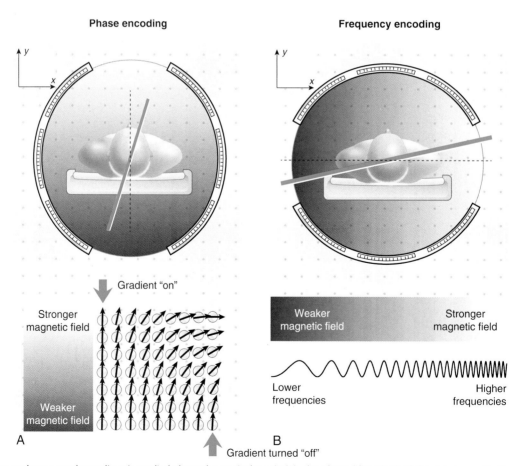

Phase encoding

Frequency encoding

Gradient "on"

Stronger magnetic field

Weaker magnetic field

A

Gradient turned "off"

Weaker magnetic field

Stronger magnetic field

Lower frequencies

Higher frequencies

B

Fig. 9.20 When a phase encode gradient is applied along the vertical *y*-axis (A), the phase (direction) of the net magnetization along this axis is changed from top to bottom, which provides information about signal location along the *y*-axis. When a frequency encode gradient is applied along the horizontal *x*-axis (B) during the echo formation, this provides information along the *x*-axis because lower frequencies are located where the magnetic field is lower, and vice versa.

Where the magnetic field is higher, the resulting signals will have higher frequencies in the detected signal. Fourier techniques can be used to extract the frequencies contained in echoes to determine the signal location. For a 256 × 256 image, each echo will need to be sampled 256 times, providing a row of data with 256 discrete values. Application of the FEG during echoes provides spatial information along the *x*-axis signals but does not provide any *y*-axis spatial data.

In between RF pulse and the resultant echoes, PEGs need to be applied along the *y*-axis. This phase encode direction must be perpendicular to both slice select (*z*-axis) and frequency encode (*x*-axis) directions (see Fig. 9.20). In the absence of any gradient, spins in a row of pixels along the *y*-axis will all rotate at the same speed and therefore point in same direction. A gradient along the *y*-axis will result in a magnetic field that varies along this direction. Spins in lower fields will rotate more slowly than spins in higher fields. Switching off this PEG will therefore result in the spins pointing in different directions, which is known as phase encoding. Different PEGs are applied when acquiring each echo. The number of PEGs (i.e., acquired echoes) determines the matrix size in this phase encode direction (*y*-axis). For a 256 × 256 MR image, 256 echoes would need to be acquired, and each echo would use a different value of the PEG along the *y*-axis.

References

Bushberg JT, Seibert JA, Leidholdt Jr EM, Boone JM. Magnetic resonance basics: magnetic fields, nuclear magnetic characteristics, tissue contrast, image acquisition. In: *The Essential Physics of Medical Imaging.* 3rd ed. Philadelphia: Wolters Kluwer; 2012:438–444.

Huda W: Magnetic resonance I. In: *Review of Radiological Physics.* 4th ed. Philadelphia: Wolters Kluwer; 2016:211–213.

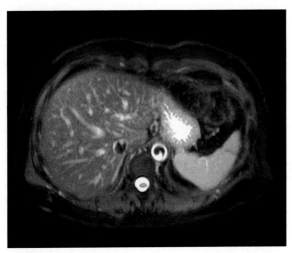

Fig. 9.21

1. An MR echo (signal) is analogous to which CT acquisition metric?
   A. Ray
   B. Projection
   C. Sinogram (acquired)
   D. Sinogram (filtered)

2. How does the size of the acquired *k*-space (MB) compare with the resultant MR image?
   A. *k*-space > MR image
   B. *k*-space = MR image
   C. *k*-space < MR image

3. What image quality aspect is controlled by the low spatial frequencies at the center of *k*-space?
   A. Contrast
   B. Resolution
   C. Mottle
   D. Artifacts

4. Which image quality aspect is controlled by the high spatial frequencies at the periphery of *k*-space?
   A. Contrast
   B. Resolution
   C. Mottle
   D. Artifacts

CASE 9.11

K-Space and MR Images

Fig. 9.21 Normal abdominal magnetic resonance image obtained using a half-Fourier technique, which reduces acquisition time and motion artifact.

1. **B.** A projection in CT is the transmitted intensities in linear detector array at one x-ray tube angle. In MR, each echo is a line of *k*-space, and analogous to a CT projection. In CT, we rotate the x-ray tube around the patient to get spatial information, whereas in MR we change the PEG for each echo.

2. **B.** When a 128 × 128 MR image is acquired, each echo is sampled 128 times, resulting in a row of 128 data points along the FEG direction, and 128 echoes are acquired with the data arranged in successive row (128 numbers) along the PEG direction. This acquired 128 × 128 data set (*k*-space) uses a two-dimensional Fourier transform to create a corresponding 128 × 128 MR image.

3. **A.** The low spatial frequencies are located at the center of *k*-space. If just the central region of *k*-space (i.e., low spatial frequencies) is used to reconstruct an image, the resultant image will just show the "large-scale structures" and lack any edges or detail. Accordingly, the overall image contrast will be apparent, but the spatial resolution will be very poor.

4. **B.** The higher spatial frequencies are located at the periphery of *k*-space. If just the outer region of *k*-space (i.e., high spatial frequencies) is used to reconstruct an image, the resultant image will just show the edges or detail that the image contains. When the large-scale structures are added to the edges and details, by including all the spatial frequencies, we will obtain a complete image.

Comment

In MRI, the acquired data set is called *k*-space, which is analogous to the sinogram data acquired in CT. Each echo produces one line of *k*-space, which is analogous to a projection in CT imaging. Stacking up all the acquired echoes creates *k*-space in MR, and stacking up all the acquired projections creates the sinogram in CT. The number of points along FEG is determined by digitization and provides one matrix dimension. The number of PEG is equal to the number of echoes, as each echo requires a different PEG and fills one line of *k*-space.

Consider an SE sequence, where each echo is sampled (digitized), resulting in a row of discrete numbers. If 256 echoes are acquired, and each sampled 256 times, we have a two-dimensional (2D) matrix that is 256 × 256 in size. The MR image is obtained by performing a 2D Fourier transform of *k*-space, which "extracts mathematically" the spatial information that was encoded by the application of the phase and FEGs.

A single pixel in *k*-space does not correspond to a single pixel in image space but instead makes up a portion of all the pixels in image space. Each data point in *k*-space refers to a spatial frequency, where low spatial frequencies are located at the center and high spatial frequencies are at the periphery. Low spatial frequencies at the center of *k*-space provide information for the large-scale structures, as well as the overall image contrast, but lacks edge information. Thus an image filling only the center of *k*-space will have the correct weighting (e.g., T1) but will appear blurry. Detail and edges are contained in the peripheral regions of *k*-space, which correspond to higher spatial frequencies. An image filling only the peripheral regions of *k*-space will appear as edges that lack proper contrast weighting (Fig. 9.22).

Several parameters can impact the resolution of an image. Consider body imaging where the matrix size for CT is 512 × 512, whereas the MR matrix size is typically 256 × 256. Pixel size is much larger for MR, and the resolution is correspondingly lower. A higher matrix size could be used but at the cost of longer scan times and worse SNR. Similarly, keeping the matrix size constant and decreasing the field of view will reduce the pixel size and correspondingly increase resolution but at the cost of decreased field of view and possible artifacts such as wrap around.

The spatial frequency halfway along the positive *x*-axis is the same as the spatial frequency halfway along the negative *x*-axis. It is thus possible to acquire half of *k*-space and achieve a reconstruction of the whole MR image. These symmetry properties of *k*-space permit the use of "half-Fourier" techniques (e.g., half-Fourier acquisition single shot turbo spin echo [HASTE], single shot fast spin echo [SS-FSE]) that are popular in body imaging where imaging time must be minimized to reduce motion artifacts. An MR image is generated by taking the 2D Fourier transform of *k*-space. The matrix size and data content of the *k*-space and MR image are generally identical.

References

Elster AD. Questions and answers in MRI. K-space (basic); 2017. http://mriquestions.com/k-space-basic.html. Accessed March 2, 2018.

Huda W. Magnetic resonance I. In: *Review of Radiological Physics*. 4th ed. Philadelphia: Wolters Kluwer; 2016:211–213.

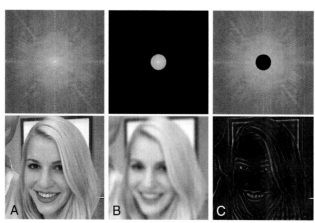

Fig. 9.22 Low spatial frequencies at the center of *k*-space provide information for large structures and overall image contrast (B). High spatial frequencies at the periphery of *k*-space provide the detail information and edges (C). Using all of *k*-space produces an image with both image contrast and detailed edges (A).

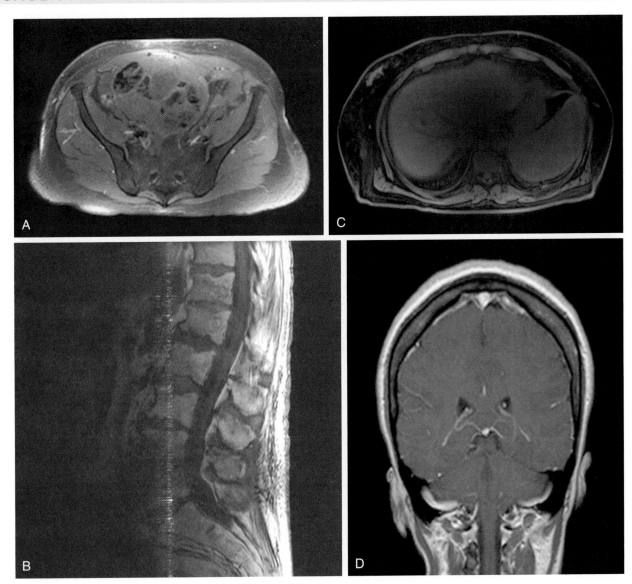

Fig. 9.23

1. How will doubling the magnetic field from 1.5 to 3 T increase the SNR in MRI (%)?
   A. <100
   B. 100
   C. >100

2. What coils would most likely be used to improve the uniformity of an MR scanner's main magnetic field?
   A. Shim
   B. RF
   C. Gradient
   D. Magnet

3. Which of these are properties of a magnetic field gradient?
   A. Peak gradient intensity
   B. Gradient rise time
   C. Gradient polarity (+ or −)
   D. A, B, and C

4. What material is used to shield an MR facility from external RF radiation?
   A. Lead
   B. Copper
   C. Concrete
   D. Nitrogen (liquid)

CASE 9.12

Instrumentation

Fig. 9.23 (A) Simulated fat sat failure due to inhomogeneities. (B) Zipper artifact due to radiofrequency (RF) leak. (C) Simulated dielectric artifact due to inhomogeneous RF. (D) Simulated distortion due to a nonlinear gradient.

1. **B.** The signal coming from the patient is proportional to the excess number of spins aligned along the main magnetic field versus antialigned. As the field strength increases, the excess number of aligned spins increases proportionally and thus so does the signal. Because the noise is not impacted by this, the SNR is doubled if the main field strength is doubled.

2. **A.** Shim coils are available on all clinical MR systems and are used to improve the magnetic field uniformity. This is necessary because the local environment of an MR facility (lifts, beams, girders, etc.) will distort the magnetic field and induce artifacts in the image.

3. **D.** Gradients are characterized by a strength (mT/m) as well as the time it takes to go from "no gradient" to the "full gradient." Gradients also have polarity (+ or −) depending whether the magnetic field increases or decreases as one moves along the gradient.

4. **B.** It is necessary to stop external RF signals (i.e., broadcast radio waves) from being picked up by the MR coils. This is usually achieved using copper shielding, which is placed around the room and is called a Faraday cage. The RF shielding also prevents the MRI RF coils from interfering with sensitive equipment outside the MR room.

Comment

Most clinical MR systems operate at 1.5 or 3 T. The SNR is generally taken to be directly proportional to the magnetic field strength so that 3 T systems have twice the SNR of those operating at 1.5 T. For this reason, research systems have been developed that operate at 7 and 11 T, offering very high SNR. There are several drawbacks to imaging at higher fields. First, T1 times increase and result in longer imaging times. Higher Larmor frequencies mean higher field imaging requires higher RF frequencies and an increase in RF absorption within the patient. Transient physiologic effects such as vertigo have been reported at high magnetic field strengths. Finally, several artifacts such as chemical shift are worse for high field imaging.

MRI relies on the idea that spins precess at very specific frequencies and that the machine can control them by controlling the magnetic field strength using gradients (Fig. 9.24). If the field strength varies unintentionally, signals can be mislocated and cause distortion (see Fig. 9.23D). Shimming the magnet means having the scanner automatically adjust small peripheral magnets (shims) to make the field more uniform. When a patient enters the magnet, they disrupt the uniform field (see Fig. 9.23A). The user may then actively shim the magnet, meaning that they indicate anatomy to be viewed prior to scanning and the scanner adjusts power to small magnets surrounding the bore to make the field uniform while the patient is inside.

Gradients are typically approximately 30 mT/m so that the magnetic field increases 30 mT when moving 1 m. Nonuniform gradients will result in distortions in the resultant MR image. When gradients are switched on and off, the changes in magnetic field will generate eddy currents in other coils or metal structures nearby. It is recognized that eddy currents impair scanner performance, which are associated with reduced image SNR and image artifacts. These artifacts are especially prominent in fast switching sequences such as echo planar imaging. Actively shielded gradient coils are used to reduce problems associated with eddy currents.

RF coils transmit and/or receive a signal. RF transmit bandwidth partially determines slice thickness. Thicker slices contain more spins and thus have higher signal and SNR. In large patients there may be significant loss of RF energy, which results in dielectric artifact. This is visualized as dark regions within the patient (see Fig. 9.23C). Dielectric artifact typically occurs in 3 T magnets because the higher magnetic field uses higher frequencies, which cause RF loss and has a wavelength about the size of an adult abdomen. RF from external sources will result in lines within the image called zipper artifact (see Fig. 9.23B). One line will be created for each RF frequency received. MR scanners use copper shielding in the walls to limit the transmission of radio broadcasts reaching the receive coil. If the MR room door is left open, the images will show zipper artifacts, with each line pertaining to a particular radio station broadcast.

References

Huda W. Magnetic resonance I. In: *Review of Radiological Physics*. 4th ed. Philadelphia:Wolters Kluwer; 2016:213–215.
Radiologic Society of North America. RSNA/AAPM physics modules. MRI: instrumentation. https://www.rsna.org/RSNA/AAPM_Online_Physics_Modules_.aspx. Accessed March 2, 2018.

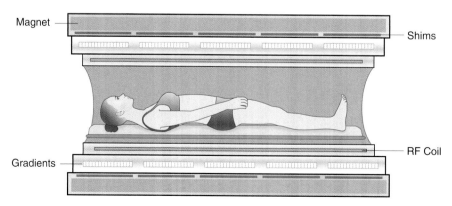

Fig. 9.24 A superconducting magnet carries a large current to create the magnetic field along the horizontal axis. Shim coils improve the magnetic field uniformity. Gradient coils change the magnetic field along any direction. Radiofrequency *(RF)* coils are used to send out RF pulses and receive echoes from the body. Note that the RF coil is closest to the body to give (and receive) the best possible signal.

CASE 10.1

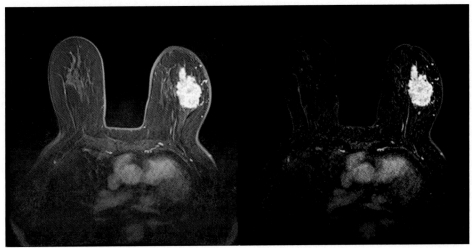

Fig. 10.1

1. In a conventional spin echo (SE) sequence, how many radiofrequency (RF) pulses are used in each repetition time (TR) interval?
   A. 1
   B. 2
   C. 3
   D. >3

2. In a conventional SE sequence, how many gradients are applied in each TR interval?
   A. 1
   B. 2
   C. 3
   D. 4

3. How many echoes are acquired when obtaining data for a 128 × 128 MRI image?
   A. 128/2
   B. 128
   C. 128 × 2
   D. 128^2

4. How long does it take to acquire data for a 128 × 128 MRI image using a fast spin echo (FSE) sequence (echo train length [ETL] of 4) with a TR of 1 second?
   A. 16
   B. 32
   C. 64
   D. 128

CASE 10.1

Spin Echo Imaging

Fig. 10.1 Breast MRI demonstrating a large left breast mass.

1. **B.** A conventional SE sequence starts with a 90-degree RF pulse, which produces the free induction decay (FID) signal. When the FID rapidly decays (T2* effects), a 180-degree refocusing RF pulse is applied at time echo time (TE)/2 to generate an SE at time TE. Accordingly, there are two RF pulses (at time zero and at time TE/2) in each TR interval.

2. **D.** A slice select gradient (SSG) is applied when the 90-degree RF pulse occurs and also when the 180-degree refocusing RF pulse occurs. A frequency encode gradient (FEG) is applied when the echo is generated, and a phase encode gradient (PEG) is applied prior to the FEG. Accordingly, four gradients are applied in each TR interval in conventional SE imaging.

3. **B.** In creating a 128 × 128 image, each echo will be sampled 128 times to create 128 pixels in the FEG direction. To create 128 pixels in the PEG direction requires 128 echoes.

4. **B.** In an SE sequence with TR = 1 second, it would take 128 seconds to acquire 128 echoes. In FSE with an ETL of four, there are four echoes acquired in each TR interval, so that the required 128 echoes can be obtained in a quarter of the time needed for standard SE sequences (32 seconds).

Comment

In SE imaging, a 90-degree RF pulse rotates longitudinal magnetization (i.e., M_z) into the transverse plane (i.e., M_{xy}), which rapidly dephases (T2*). Applying a refocusing 180-degree RF pulse at a time TE/2 will generate an echo at time TE (Fig. 10.2). SEs eliminate T2* dephasing and are weaker than the initial FID signal because of T2 losses caused by spin-spin interactions. Compared with the initial FID signal, a tissue with a T2 time of 50 ms has an echo intensity of 0.6 if the echo is produced at a TE of

25 ms and an echo intensity of 0.2 if the echo is produced at TE of 80 ms. Each echo is digitized to produce one line of k-space.

SE sequences consist of a 90-degree RF pulse followed by a refocusing 180-degree RF pulse. These repeat once for each TR interval. To generate a 128 × 128 MR image, an SE sequence must acquire 128 echoes, which results in an acquisition time of 128 × TR seconds. The "long time interval" between TE and TR can be used to acquire images corresponding to different "slices." For example, with TR of 500 ms and TE of 50 ms, up to ten 128 × 128 images could be acquired in 128 s, which is the time required to acquire just one slice (i.e., 128 × TR). The echo at 50 ms is used to produce the k-space for slice 1, an echo can be generated at 100 ms to produce a line of k-space for slice 2, another echo can be generated at 150 ms to produce a line of k-space for slice 3, and so on. A second echo pertaining to slice 1 would be obtained at 550 ms, a third echo pertaining to this slice would be obtained at 1050 ms, and so on.

For elliptical body shapes, more pixels are needed for the longer dimension (e.g., 192) in body images than for the shorter dimension (e.g., 128). The longer dimension is usually selected to be the frequency encode direction, with which each echo is sampled (i.e., digitized) 192 times. The basic pulse sequence is repeated 128 times to produce 128 different echoes and where the PEG is most likely to be applied along the shorter dimension. In body MR, the FEG is most likely applied along the (longer) lateral orientation and in head MR along the (longer) AP direction. It is possible to make the longer dimension the PEG direction, which could move motion artifacts into a different location because these occur along the PEG direction. However, choosing the longer direction to be the phase encode direction would require 192 echoes in time of 192 × TR, which is much longer than the standard imaging time of 128 × TR.

In FSE imaging, multiple echoes are acquired in each TR interval. Four echoes acquired in each TR interval corresponds to an ETL of four, with a different PEG applied to each echo. Because each echo generates a line of k-space, an ETL of four will reduce imaging time fourfold. Relative intensities of later echoes always get progressively weaker due to T2 dephasing,

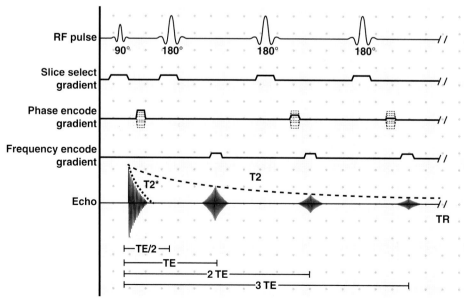

Fig. 10.2 In this fast spin echo sequence, three refocusing 180-degree RF pulses result in three echoes, with each one generating a line of k-space. The echo train length is three, and therefore imaging time is three times faster than a conventional spin echo sequence. The same slice select gradient is activated for every RF pulse and the same frequency encode gradient is activated for each echo, but each echo is acquired with a different value of the phase encode gradient. *TE,* Echo time; *TR,* repetition time.

and the FSE images can have lower signal-to-noise ratios (SNRs) than an SE sequence. Typical ETL values vary between 2 and 25, depending on the specific sequence used. Because many echoes are acquired in a single TR, each with its own TE, the "effective TE" that determines contrast weighting is the TE of the echoes that fill the center of k-space.

References

Huda W. Magnetic resonance II. In: *Review of Radiological Physics*. 4th ed. Philadelphia: Wolters Kluwer; 2016:221–224.

Radiological Society of North America. RSNA/AAPM physics modules. MRI: pulse sequences. https://www.rsna.org/RSNA/AAPM_Online_Physics_Modules_.aspx. Accessed March 2, 2018.

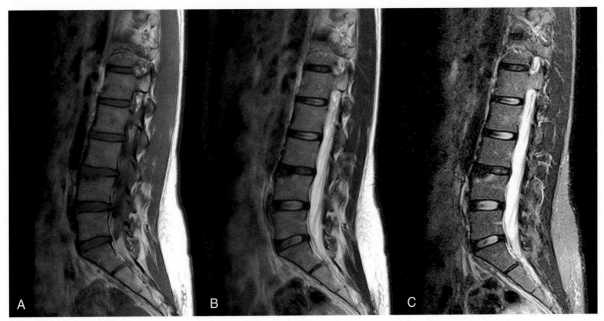

Fig. 10.3

1. Which tissue characteristic most likely affects the detected signal in SE imaging when TR is adjusted?
   A. T1
   B. T2
   C. T2*
   D. Spin density (ρ)

2. Which tissue characteristic most likely affects the detected signal in SE imaging when TE is adjusted?
   A. T1
   B. T2
   C. T2*
   D. Spin density (ρ)

3. Which of the following would most likely result in an image that exhibits T2 weighting?
   A. Long TR; long TE
   B. Long TR; short TE
   C. Short TR; long TE
   D. Short TR; short TE

4. Which of the following would most likely result in an image that exhibits only spin density (ρ) weighting (i.e., eliminates T1 and T2 weighting)?
   A. TR 2000 ms; TE 80 ms
   B. TR 2000 ms; TE 10 ms
   C. TR 400 ms; TE 80 ms
   D. TR 400 ms; TE 10 ms

CASE 10.2

T1 and T2 Weighting

Fig. 10.3 Acute L4 intravertebral disc herniation (Schmorl node). Marrow edema appears dark on T1-weighted images (A) and bright on T2-weighted images (B) with associated enhancement (C).

1. **A.** A short TR does not permit full recovery of longitudinal magnetization of tissues with long T1, so their signals will be weaker than tissues with short T1 that can fully recover even when TR is short. When TR is made very long, all tissues have time to fully recover, so the differences in signal between short and long T1 tissues disappear.

2. **B.** When TE is short, signals are not affected by T2 effects because tissues simply have not had sufficient time for transverse magnetization to dephase. When TE is long, tissues with short T2 will have dephased completely and give rise to no signal (void), whereas a tissue with a long T2 (e.g., fluid) will give a strong signal and appear bright.

3. **A.** T2 weighting is determined by the choice of the TE. A short TE will eliminate T2 weighting, but a long TE will emphasize differences in T2 properties of tissues. When TE is long, tissues with short T2 will be dark, whereas those with long T2 will appear bright.

4. **B.** A long TR eliminates T1 weighting because all tissues have sufficient time to fully recover. A short TE eliminates T2 weighting because all tissues will not have time to dephase. SE sequences with long TR and short TE only have spin density weighting.

Comment

Relaxation times T1 and T2 are tissue properties that are exploited using specific pulse sequences (Fig. 10.4). It is solely the *operator's choices* of TR and TE values in any pulse sequence that affect T1 and T2 tissue weighting. In SE imaging, the selected value of TR will affect the amount of contrast between tissues that have differences in T1. The choice of TR can thus be modified to either introduce or eliminate T1 weighting. Choice of spin TE will influence contrast between tissues that differ in their T2 values. The choice of TE can thus be modified to either introduce or eliminate T2 weighting. When both T1 and T2 effects are eliminated, regions with higher spin densities (ϱ) will appear brighter, whereas those with lower spin densities (ϱ) will appear darker.

TR choice affects T1 weighting because tissues with short T1 will recover longitudinal magnetization, whereas those with long T1 will not have recovered as much. Tissues with a short T1 should have full longitudinal magnetization (M_z) on the second, third, and all subsequent acquisitions, with all acquired echoes being "strong." Tissues with a long T1 have will have very little longitudinal magnetization (M_z) on the second, third, and all subsequent acquisitions, with the echoes being "weak." For short TRs, short T1 materials (e.g., fat) appear bright, and long T1 materials (e.g., cerebrospinal fluid [CSF]) appear dark, as depicted in Fig 10.4. A long TR value permits the magnetization in all tissues to fully recover and therefore generates no T1 weighting.

The reason why TEs influence T2 weighting is that long TEs reduce transverse magnetization for tissues with short T2 more than for tissues with long T2. T2-weighted images are obtained with a long TE. Reducing TEs will reduce the amount of T2 weighting in the image. Short TE values will result in very little T2 decay, thus showing minimal differences (i.e., contrast) between tissues that differ in T2 times. This eliminates T2 weighting. The longer the TE time, the higher the potential contrast between tissues that have different T2 values. T2-weighted SE images show long T2 tissues as bright, whereas short T2 tissues will appear dark. The long TE in T2-weighted images results in them having the lowest signal, although contrast between tissues is very good.

A long TR will eliminate T1 weighting because all tissues, irrespective of their T1 properties, will have sufficient time to fully recover. A short TE will eliminate T2 weighting because all tissues, irrespective of their T2 properties, will not have time to dephase. Thus the image contrast from a long TR and short TE sequence will not be dependent on T1 and T2 times. Instead

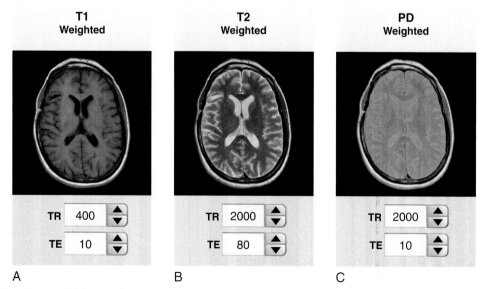

	T1 Weighted	T2 Weighted	PD Weighted
TR	400	2000	2000
TE	10	80	10
	A	B	C

Fig. 10.4 T1-weighted images (A) have a short repetition time *(TR)* and short echo time *(TE)*, whereas T2-weighted images (B) have long TR and long TE. A long TR eliminates T1 weighting, and a short TE eliminates T2 weighting, resulting in a proton density *(PD)* image (simulated in C).

the different signal strengths from each tissue comes from the different number of free spins that align along the magnetic field. The more free spins a material has, the brighter the signal in a proton (spin) density–weighted image will be. Because it has a long TR and short TE, proton-weighted images have the strongest signal, although not necessarily the highest contrast between tissues.

References

Bushberg JT, Seibert JA, Leidholdt Jr EM, Boone JM. Magnetic resonance basics: magnetic fields, nuclear magnetic characteristics, tissue contrast, image acquisition. In: *The Essential Physics of Medical Imaging*. 3rd ed. Philadelphia: Wolters Kluwer; 2012:423-427.

Huda W. Magnetic resonance II. In: *Review of Radiological Physics*. 4th ed. Philadelphia: Wolters Kluwer; 2016:221-224.

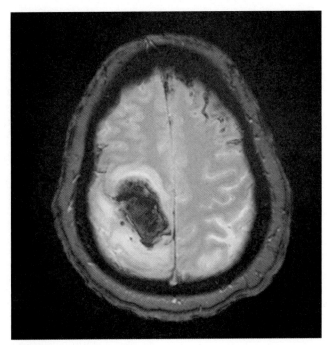

Fig. 10.5

1. Following a 15-degree RF pulse, which direction/axis will have the most amount of magnetization?
   A. M_z
   B. $-M_z$
   C. M_{xy}
   D. $M_z = M_{xy}$

2. How will replacing a conventional SE sequence with a gradient recalled echo (GRE) sequence likely affect the specific absorption rate (SAR)?
   A. Increase
   B. No change
   C. Reduce

3. How long will it likely take to acquire a 256×256 GRE image using a TR of 4 ms (seconds)?
   A. 0.1
   B. 1
   C. 10
   D. 100

4. Which type of weighting is likely introduced into MR images when an SE sequence is replaced by a GRE sequence?
   A. T1
   B. T2
   C. T2*
   D. ϱ

CASE 10.3

Gradient Recalled Echoes

Fig. 10.5 Susceptibility weighted sequence demonstrating signal dropout in the area of an intracerebral hematoma.

1. **A.** Following a 15-degree RF pulse, there will be more of magnetization along the positive longitudinal axis (M_z) than in the transverse plane (M_{xy}). Flip angles between 0 and 45 degrees will have more along the positive longitudinal magnetization, and those between 45 and 90 degrees will have more magnetization in the transverse plane.

2. **C.** Replacing a conventional SE sequence with a GRE sequence eliminates a 180-degree refocusing RF pulse. Because the application of RF pulses deposits energy in the patient, elimination of an RF pulse reduces the SAR.

3. **B.** A 256 × 256 image needs to acquire 256 echoes, and in each TR interval there will be one echo. The total imaging time will thus be 256 × 4 ms, or 1 second in total. Imaging times in GRE are reduced because TR times are much shorter than with conventional SE imaging.

4. **C.** T2* weighting is introduced into MR images when an SE sequence is replaced by a GRE sequence. An SE sequence can normally have T1, T2, and ϱ weighting but no T2* because magnetic field inhomogeneities are eliminated by the application of a 180-degree refocusing RF pulse.

Comment

A GRE pulse sequence is initiated using a flip angle that is less than 90 degrees. Fig. 10.6 shows three examples of GRE sequences that may commence with flip angles of 15, 45, and 75 degrees. For flip angles between 0 and 45 degrees, there will be more longitudinal magnetization than transverse, with the opposite being true for flip angles between 45 and 90 degrees. The amount of magnetization in the longitudinal direction following a 45-degree RF pulse is approximately 70%, and the transverse plane magnetization following this 45-degree RF pulse will also be approximately 70%. After any flip angle that is less than 90 degrees, the transverse component of magnetization (M_{xy}) will always guarantee that a signal in the form of an echo can be obtained. Because longitudinal magnetization (M_z) is also present, another RF pulse will always be able to result in transverse magnetization.

GRE sequences reverse the polarity of applied magnetic field gradients to generate echoes. An initial gradient is used to deliberately dephase the transverse spins (this speeds up the acquisition). Reversing the gradient will rephase the spins and produce an echo (one line of k-space). Dephasing and rephasing gradients are applied along the frequency encode direction. Leaving the second gradient "on" will also serve to provide frequency encoding during the echo creation. Because GRE does not need 180-degree refocusing RF pulses to generate echoes, these sequences usually have a lower SAR. The absence of a 180-degree RF refocusing pulse implies that less RF energy will be deposited in the patient.

The low flip angles used in GRE (some as low as 10 degrees) provide both signal (M_{xy}) and longitudinal magnetization (M_z), permitting the use of very short TRs. Because it is not necessary to wait for recovery of longitudinal magnetization, the use of short TRs means that imaging times can be markedly reduced. GRE imaging is therefore likely to be much faster than is generally possible with SE or even FSE. TR times in GRE sequences are tens of milliseconds, which are two orders of magnitude lower than typical TRs in SE sequences. This can be illustrated by 256 × 256 matrix–sized GRE sequences with a TR of only 10 ms. Generating this image requires 256 echoes (one for each line of space) that can be acquired in 2.6 seconds (256 × 0.01 s). By comparison, a T2-weighted SE sequence (TR 2000 ms) with the same matrix size would require more than 8 minutes of acquisition time (256 × 2 s).

A GRE sequence may generate T2* weighting, whereas a conventional SE has no T2* weighting but can exhibit T2 weighting depending on the choice of TE. In clinical practice, T2* relaxation is often the main determinant of GRE image contrast because T2* is not eliminated via a 180-degrees refocusing pulse. Areas of T2* dephasing, including blood clots and the presence of hemosiderin, will generally appear dark on GRE images. This is markedly different to SE imaging, where T2* effects generally cancel out when echoes are generated and will thus appear brighter. Both GRE and SE imaging have T2 weighting that can never be eliminated in MRI imaging. Because T2* is so much smaller than T2, GRE imaging is dominated by T2* effects, especially for relatively short TEs.

References

Elster AD. Questions and answers in MRI. Net magnetization (M): what is a gradient echo, and how does it differ from an FID? 2017. http://mriquestions.com/gradient-echo.html. Accessed March 2, 2018.

Huda W. Magnetic resonance II. In: *Review of Radiological Physics*. 4th ed. Philadelphia: Wolters Kluwer; 2016:224–225.

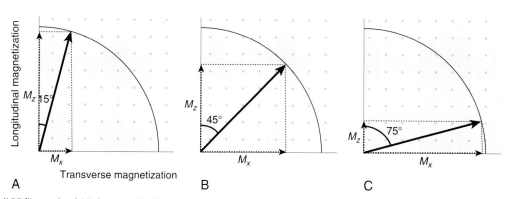

Fig. 10.6 A small RF flip angle of 15 degrees (A) will result in a large longitudinal magnetization (M_z) and a small transverse magnetization (M_{xy}), whereas the exact opposite is achieved using a large flip angle of 75 degrees (C). For a 45-degree RF flip angle, the longitudinal and transverse magnetizations will be equal (B).

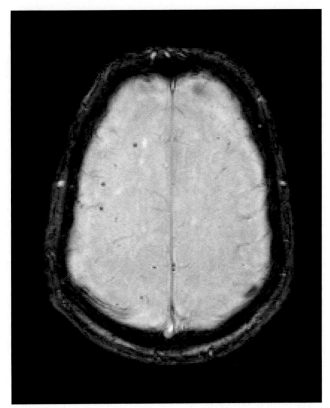

Fig. 10.7

1. On a GRE image, what is the most likely pixel intensity that represents tissue hemosiderin?
   A. Black
   B. Gray
   C. White
   D. A, B, and C possible

2. Which would most likely result in the highest T2* contrast in GRE imaging?
   A. Large flip angle and long TE
   B. Large flip angle and short TE
   C. Small flip angle and long TE
   D. Small flip angle and short TE

3. What likely happens to the amount of T1 weighting in a GRE image when the flip angle increases?
   A. Increases
   B. Unchanged
   C. Decreases

4. Which is a GRE pulse sequence?
   A. Fast low-angle shot (FLASH)
   B. Fast imaging with steady-state precession (FISP)
   C. Gradient recall acquisition in the steady state (GRASS)
   D. A, B, and C

CASE 10.4

GRE Imaging

Fig. 10.7 Susceptibility weighted image demonstrating foci of cerebral microhemorrhage.

1. **A.** Hemosiderin contains magnetic materials (iron) that will rapidly dephase any transverse magnetization in their vicinity. Accordingly, on GRE images any hemosiderin will always appear as dark areas with no signal.

2. **C.** A small flip angle reduces T1 weighting and a long TE will allow for T2* dephasing and thus more T2* weighting. GRE imaging do not have T2 weighting because T2* << T2.

3. **A.** If the RF flip angle is 90 degrees and the TR is short, the image is T1 weighted. Conversely, when the flip angle is very small, there will be negligible T1 weighting because there is always almost the full longitudinal magnetization present. Consequently, as the flip angle increases, we will obtain progressively more T1 weighting.

4. **D.** FLASH, FISP, and GRASS are all examples of GRE sequences.

Comment

Examples of GRE sequences include FLASH, GRASS, and FISP. In GRE sequences, any change in tissue susceptibility will result in signal loss because of these well-known T2* effects. The low flip angles mean lower transverse magnetization (M_{xy}) than would have been achieved with a 90-degree RF pulse, where all the longitudinal magnetization is flipped into the transverse plane (Fig. 10.8). As a result, sequences that use small flip angles (i.e., <90 degrees) will generate much weaker signals than SEs that commence with a 90-degree RF pulse. Because GRE images will have lower signals than SE sequences, they will generally appear of poorer image quality because the SNR is relatively low. Nonetheless, GRE images may still have excellent contrast between tissues and enable high-contrast lesions to be readily detected despite the high levels of noise.

In GRE, the choice of the initial flip angle is a major variable that affects the type of tissue contrast. A very small flip angle (e.g., 20 degrees) means that magnetization realigns quickly with the main magnetic field and that very little magnetization is ever flipped into the transverse plane. As a result, there is little time for tissue signal to differentiate because of T1 relaxation differences. For this reason, reducing the flip angle is generally associated with reducing the amount of T1 weighting. Conversely, a large flip angle allows for more differentiation from varying tissue T1 times and thus higher T1 weighting. In general, increasing the flip angle in GRE sequences is associated with increasing the amount of T1 weighting in the resultant image (Table 10.1).

GRE sequences also require a selection of TE and TR, both of which will affect the type of contrast in the resultant images. As the TR increases, magnetization for all tissues has more time to re-realign to the main magnetic field and reduce T1 weighting. However, if TR is kept short, then only short T1 tissues will appreciably realign, whereas long T1 tissues will not. This increases T1 weighting. When TE is kept short, there will be little time for tissues to dephase regardless of T2* time, which reduces T2* weighting. When TE is long, short T2* tissues will dephase and be dark, whereas long T2* tissues will not dephase

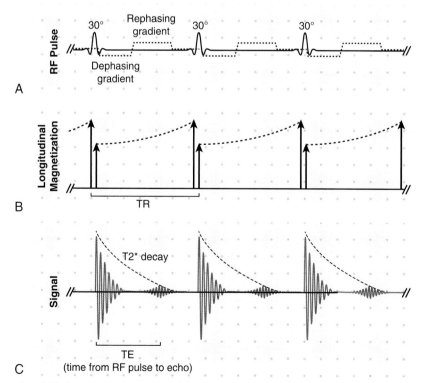

Fig. 10.8 (A) The middle series of 30-degree RF pulses in a gradient recalled echo sequence, which will reduce the longitudinal magnetization (B) and then regrows to an equilibrium value after a repetition time (TR). (C) After each 30-degree RF pulse, there is a rapid decay of the free induction decay signal under the influence of a dephasing gradient. There is a subsequent weak echo at echo time (TE), whose intensity is determined by T2*.

TABLE 10.1 SUMMARY OF HOW DIFFERENT TYPES OF
WEIGHTING CAN BE *INCREASED* IN GRADIENT
RECALLED ECHO PULSE SEQUENCES

T1 Weighted	T2* Weighted	Proton Density
Large flip angle	Small flip angle	Small flip angle
Short TR	Long TR	Long TR
Short TE	Long TE	Short TE

TE, Echo time; *TR*, repetition time.

much and will be bright so that T2* weighting increases. Increasing TR thus reduces T1 weighting, and increasing TE will most likely increase T2* weighting (see Table 10.1).

References

Hargreaves BA. Rapid gradient-echo imaging. *J Magn Reson Imaging*. 2012;36(6):1300-1313.

Huda W. Magnetic resonance II. In: *Review of Radiological Physics*. 4th ed. Philadelphia: Wolters Kluwer; 2016:224-225.

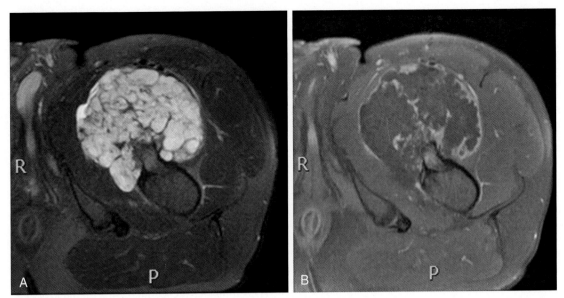

Fig. 10.9

1. If the initial tissue magnetization is along $+M_z$, where will this tissue magnetization be "pointing" following a 180-degree RF pulse?
   A. $+M_z$
   B. $+M_{xy}$
   C. $-M_z$
   D. $-M_{xy}$

2. Following a 180-degree RF pulse, at what time would a 90-degree RF pulse generate the largest FID signal?
   A. T1
   B. $2 \times$ T1
   C. $3 \times$ T1
   D. $4 \times$ T1

3. What should be done to increase the amount of T2 weighting in an inversion recovery (IR) pulse sequence (i.e., 180-degree RF pulse followed by a 90-degree RF after time to inversion (TI), and finishing with a 180-degree RF at time TE/2 later than TI)?
   A. Increase TI
   B. Reduce TI
   C. Increase TE
   D. Reduce TE

4. Which pulse sequence would most likely be used in musculoskeletal imaging of the toe to suppress the signal from fat?
   A. Short tau inversion recovery (STIR)
   B. Fluid-attenuated inversion recovery (FLAIR)
   C. FSE
   D. GRE

CASE 10.5

Inversion Recovery

Fig. 10.9 T2-weighted (A) and postcontrast T1-weighted (B) fat-saturated MRI in a patient with chondrosarcoma of the left hip.

1. **C.** In equilibrium, the tissue magnetization is aligned along the same direction as the external magnetic field, which by convention is denoted as the $+z$ axis. Following a 180-degree RF pulse, all the magnetization would be pointing in the $-z$ direction, opposite to the external magnetic field, and with none in the x-ray plane (i.e., no FID signal).

2. **D.** After $4 \times T1$, the magnetization will be fully recovered and once more point in the $+z$ direction, after passing through "zero" at a (known) time determined by the tissue T1. The longitudinal magnetization is fully recovered after $4 \times T1$ times, and this is when we would produce the highest FID signal by applying a 90-degree RF readout pulse.

3. **C.** At TI, longitudinal magnetization is "read out" by applying a 90-degree RF pulse to generate an FID. An SE is created by applying a subsequent 180-degree refocusing RF pulse at time TE/2, producing an echo at time TE. A short TE will introduce little T2 weighting, whereas a long TE will introduce a lot of T2 weighting.

4. **A.** In STIR, signal is suppressed from fat. FLAIR is a pulse sequence that suppresses signal from fluids (e.g., CSF). FSE and GRE sequences do not suppress the signal from any specific tissues.

Comment

IR pulse sequences make use of an initial inversion 180-degree RF pulse. Immediately following a 180-degree IR pulse, the longitudinal magnetization will be pointing in the $-z$ direction opposite the magnetic field direction. Recovery of the longitudinal magnetization occurs because of spin-lattice interactions and recovers to the $+z$ direction after passing through zero after four tissue T1 values. It is important to note that fat has zero longitudinal magnetization at a (known) short time interval because it recovers quickly (short T1). By contrast, fluids have zero longitudinal magnetization at much later time, because recovery takes a much longer time (long T1).

IR sequences follow the initial 180-degree pulse by a 90-degree pulse after time TI, which is referred to as the inversion time. RF pulses at TI are called "readout pulses" because these will rotate any longitudinal magnetization (M_z) into the transverse plane (M_{xy}). Applying the readout TI 90-degree pulse when the longitudinal magnetization of a tissue is zero will thus suppress the signal of the tissue (Fig. 10.10). When at TE/2 after TI a refocusing 180-degree pulse is applied, this will produce an echo at time TE (echo). Increasing the value of TE will increase the amount of T2 weighting in an IR sequence, and vice versa.

Fat suppression can be achieved using an IR, known as STIR sequence. At 1.5 T, STIR will have a TI value of approximately 180 ms, which serves to null the signal from fat. STIR is widely applied when fat saturation is necessary but the simpler chemical fat saturation methods fail (e.g., near implants). Because STIR is not dependent on the Larmor frequency, it provides more uniform fat suppression in the presence of magnetic field inhomogeneities. Postcontrast fat-saturated images are implemented

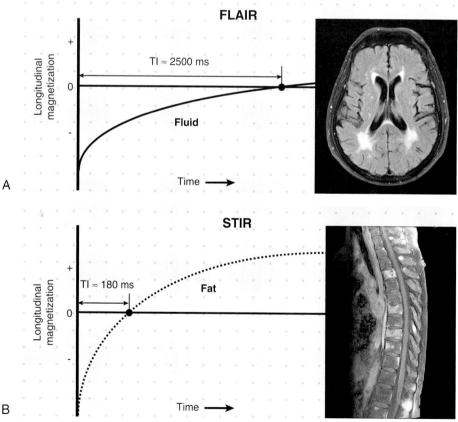

Fig. 10.10 At 1.5 T, (A) shows that following an initial 180-degree RF pulse, selection of a long inversion time (TI ~ 2500 ms) will suppress the signal from fluids. (B) Selection of a short inversion time (TI ~ 180 ms) will suppress the signal from fat. Transependymal edema is clearly seen as bright signal (A, *upper right*) next to dark *(nulled)* intraventricular cerebrospinal fluid. Similarly, hyperintense spinal metastases can be seen in (B) *(bottom right)* contrasted with the nulled fatty bone marrow. *FLAIR,* Fluid-attenuated inversion recovery; *STIR,* short tau inversion recovery.

in a variety of MR studies to demonstrate enhancing pathology and differentiate it from fat.

Signals from fluids are suppressed using FLAIR sequences. At 1.5 T, FLAIR would use a TI value of approximately 2500 ms to null the signal from CSF. Selective nulling of fluid is typically used to null the fluid signal in T2-weighted sequences (where fluid is bright) so underlying pathology can be detected. It is most widely used in brain imaging to null the very bright CSF signal. A variant of FLAIR is also used to null the signal from blood in cardiac imaging (black blood imaging). Here the IR sequence is typically called dual IR, or if a STIR sequence is also added, it is called triple IR.

References

Huda W. Magnetic resonance II. In: *Review of Radiological Physics*. 4th ed. Philadelphia:Wolters Kluwer; 2016:225–226.

Sensakovic WF. Regarding fat suppression in MRI, when are spectral techniques preferred over STIR, and vice versa? *AJR Am J Roentgenol.* 2015;205(3):W231–W232.

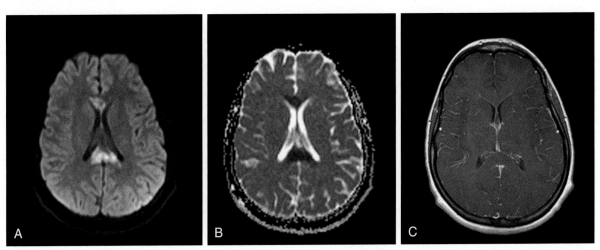

Fig. 10.11

1. Which factors affect the value of the b coefficient in diffusion-weighted imaging (DWI)?
   A. Gradient strength
   B. Duration
   C. Intervening interval
   D. A, B, and C

2. When the amount of diffusion increases in DWI imaging, what likely happens to DWI pixels and the value of the apparent diffusion coefficient (ADC)?
   A. Brighter and increase
   B. Brighter and decrease
   C. Darker and increase
   D. Darker and decrease

3. If a pixel intensity on a T2-weighted image is approximately the same as for a DWI image, what is the most likely value of the ADC (%)?
   A. ~100
   B. ~50
   C. ~10
   D. ~0

4. What is the most appropriate interpretation of a bright pixel on a DWI image as well as a high value on the ADC map?
   A. Low diffusion and long T2
   B. Low diffusion and short T2
   C. High diffusion and long T2
   D. High diffusion and short T2

CASE 10.6

Diffusion-Weighted Imaging

Fig. 10.11 Diffusion-weighted imaging (A), apparent diffusion coefficient map (B), and postcontrast T1-weighted image (C) in a patient with viral meningoencephalitis demonstrating diffusion restriction of the corpus callosum.

1. **D.** The "b" coefficient measures the overall effectiveness of a dephasing gradient. Dephasing will be increased by increasing the gradient strength, the gradient duration, and the interval between the two gradients.

2. **C.** If diffusion in a tissue increases, the pixel values in a diffusion-weighted image will get progressively weaker (i.e., darker) because of the increasing loss of signal. When the amount of diffusion is high, the calculated value of the ADC will be high.

3. **D.** If the pixel values on a T2-weighted and DWI image are essentially the same, then this implies that the protons have not moved, so the diffusion is negligible. Accordingly, the ADC value should be approximately zero and will appear dark (low) on an ADC map.

4. **C.** One reason for a bright (intense) pixel on a DWI image is the absence of any diffusion. However, if a tissue has an intrinsically very long T2 value, this can also result in a bright pixel on a DWI image. A high value of the ADC coefficient means a high diffusion and that this region has a high intrinsic T2 (AKA T2 shine-through).

Comment

Diffusion refers to the random motion of water molecules in tissues. Diffusion can be visualized by imagining placing a drop of dye in a glass of water. Over time the dye diffuses through the water, distributing itself uniformly. In a similar way, water diffuses through the body unless a boundary (e.g., interface between tissues) or disease process prevents it. It is this increase or decrease in diffusion that is imaged by diffusion sequences. Diffusion will be high in fluids like CSF and simple fluid but much lower in tissues. Diffusion in tissues reflect interactions with obstacles and barriers such as tissue planes and cell membranes. Molecular diffusion patterns can reveal tissue architecture in both normal and pathological tissue. Mapping white matter tracts in the human brain is an example of this.

DWI is used to identify and quantify the amount of diffusion in any given pixel. DWI sequences have two additional gradients added to a standard SE pulse sequence. DWI gradients are applied on either side of a 180-degree refocusing pulse in SE sequences where one serves to dephase spins and the second rephases spins by the same amount. The effects of the diffusion gradient depend on gradient strength (mT per meter) and gradient duration, as well as the time interval between these gradients, and are characterized by a "b" parameter. A T2-weighted image would have no diffusion gradients ($b = 0$ s/mm^2), with typical values of the b parameter being 500 and 1000 s/mm^2.

A minimum of two images are obtained in DWI sequences. The first image has no diffusion gradients (i.e., $b = 0$ s/mm^2) and is therefore a standard T2-weighted SE image. The second image has the diffusion gradients applied (e.g., $b = 1000$ s/mm^2), and a comparison of the two images provides quantitative information regarding the amount of diffusion present in each pixel. When there is no tissue diffusion, or very little, the gradient effects cancel out and the diffusion gradient image ($b = 1000$ s/mm^2) will be identical to the image with no diffusion gradient ($b = 0$ s/mm^2). However, when spins can diffuse into adjacent areas, echoes that create the diffusion images will become progressively weaker with increasing diffusion. As a result, pixels will appear darker where there is diffusion on the diffusion-weighted image (Fig. 10.12).

For each pixel, the values in the two images (i.e., $b = 0$ s/mm^2 and $b = 1000$ s/mm^2) can be used to estimate the amount of diffusion. This numerical value is known as the apparent diffusion coefficient (ADC) and is obtained at each pixel. If three images are acquired, using two different b values (500 and 1000 s/mm^2), the accuracy of the ADC parameter is improved. Bright areas on DWI images together with a low ADC generally indicate little diffusion. Bright areas on DWI images ($b = 1000$ s/mm^2) combined with a corresponding high ADC values indicate T2 shine-through. Tissues with very long T2 values are responsible for T2 shine-through, and a bright area on DWI should always be checked for the corresponding ADC value before any diagnosis is attempted.

References

Hagmann P, Jonasson L, Maeder P, et al. Understanding diffusion MR imaging techniques: from scalar diffusion-weighted imaging to diffusion tensor imaging and beyond. *Radiographics*. 2006;26 (suppl 1):S205–S223.

Huda W. Magnetic resonance II. In: *Review of Radiological Physics*, 4th ed. Philadelphia: Wolters Kluwer; 2016:226–227.

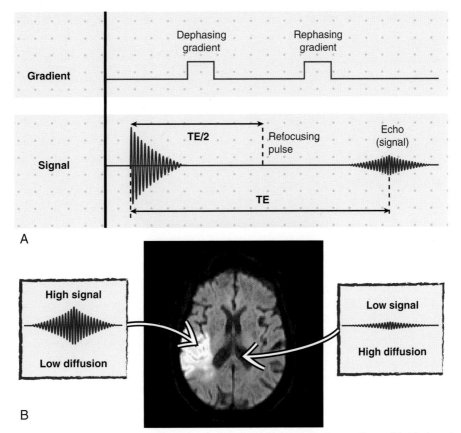

Fig. 10.12 The pulse sequence for a diffusion-weighted imaging sequence (A) has a dephasing gradient added before the 180-degree refocusing RF pulse in a spin echo sequence and an opposite rephrasing gradient added after the refocusing 180-degree RF pulse. (B) In the absence of diffusion, there will be a high signal, whereas increasing diffusion results in progressively weaker signals. *TE,* Echo time.

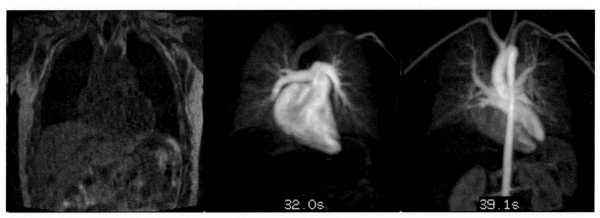

Fig. 10.13

1. What happens to spin-lattice interactions, and the corresponding T1 relaxation times, when gadolinium is added to blood?
   A. Increase and increase
   B. Increase and decrease
   C. Decrease and increase
   D. Decrease and decrease

2. What acquisition parameter change would most likely increase blood contrast in gadolinium-enhanced MR angiography (MRA) using conventional SE image sequences?
   A. Shorten TR
   B. Lengthen TR
   C. Shorten TE
   D. Lengthen TE

3. What is most likely enhanced when blood pool agents labeled with gadolinium are used to generate MR angiograms?
   A. Open veins
   B. Open arteries
   C. Thrombosed vessels
   D. A and B

4. Which type of image display is most likely used in gadolinium contrast–enhanced MRA?
   A. Magnitude
   B. Phase
   C. Maximum intensity projection (MIP)
   D. A and B

CASE 10.7

Angiography (Gadolinium Contrast)

Fig. 10.13 MR angiogram of the chest in a pediatric patient with pseudocoarctation of the aorta.

1. **B.** Gadolinium has many unpaired electrons with strong magnetic fields, so spin-lattice interactions will increase. When spin-lattice interactions increase, this will cause the magnetization to recover more quickly so that T1 is decreased.

2. **A.** The presence of gadolinium is identified by the fact that this will shorten T1 and thereby increase signal intensities on T1-weighted SE images (short TR). The choice of TE time will affect only any T2 weighting but not the contrast of blood containing gadolinium contrast.

3. **D.** Blood pool agents will increase the signal of the blood into which it is injected. As a result, the signal of blood inside of veins and arteries is increased. If a vessel is thrombosed, the vessel will not demonstrate an increased signal and thus will appear as darkness adjacent to bright blood.

4. **C.** Contrast-enhanced gadolinium studies use MIP displays, where the maximum intensity along a line is identified and used. A phase contrast (PC) image has flow in one direction shown in "white" and "black" in the other direction. A magnitude image displays the amount of tissue magnetization detected (i.e., ignores phase).

Comment

Most of MRA is performed with gadolinium contrast media. When patients are administered gadolinium contrast agents, this will shorten T1 of blood and increase signal intensities on T1-weighted images. Because T1 in the tissues of interest are substantially shortened, the TR can be reduced, so acquisition time decreases as well. Contrast is injected intravenously, and images are obtained both at the precontrast phase and during a first pass through the arteries. Subtraction of these two acquisitions will generate an image of arterial enhancement alone, as the difference signal from the surrounding tissues cancels out in the subtraction. This is because tissues with no gadolinium uptake will produce identical signal strength in the two images.

Correct timing is one of the most important considerations when obtaining high-quality MRA images. If the image is acquired too soon, then the contrast will not have reached the anatomy to be imaged and there will be no contrast enhancement or only edge enhancement. Blood pool agents will enhance both arteries and veins. If the image is acquired too late, then there may be venous contamination or no contrast left in the artery. It is also imperative that the patient not move between precontrast and postcontrast imaging, to avoid artifacts. Breath-hold imaging is generally performed in a short time, which will help to minimize motion artifacts.

Contrast-enhanced MRA images are generally viewed by looking at a stack of sections projected onto a single two-dimensional image, which is known as MIP (as seen in Fig. 10.14). A coronal MIP is created by displaying the pixel at each position with the highest signal. Contrast-enhanced MRA is acquired with T1 weighting because T1 is most impacted (shortened) by contrast agent. Artifacts can occasionally be seen with bright T1 tissues appearing on the MRA image (called shine-through). MRA typically includes fat suppression to minimize the bright fat signal that exists in T1 images.

MRA may be performed as an alternative study in patients with a contraindication to iodinated contrast media. MRA is also less susceptible to artifact produced by calcified vessels that is often encountered in CT angiography and also has the advantage of not exposing patients to ionizing radiation. Some of the drawbacks of MRA include increased exam time, higher cost, and inability to administer gadolinium in patients with renal failure. MRA is widely implemented in evaluating the carotid arteries and intracranial circulation, with superior image quality achieved using an MR contrast agent compared with time-of-flight (TOF) techniques. In addition, contrast-enhanced MRA offers increased SNR with fewer flow-related artifacts compared with TOF.

References

Chavhan GB, Babyn PS, John P, et al. Pediatric body MR angiography: principles, techniques, and current status in body imaging. *AJR Am J Roentgenol.* 2015;205:173-184.

Huda W. Magnetic resonance II. In: *Review of Radiological Physics,* 4th ed. Philadelphia: Wolters Kluwer; 2016:227-229.

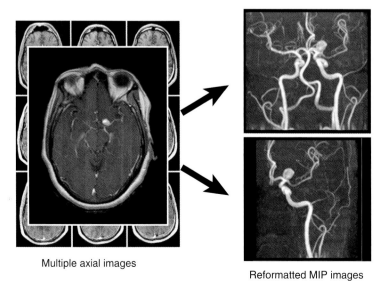

Multiple axial images

Reformatted MIP images

Fig. 10.14 Individual images will contain bright regions where there is flow in blood vessels *(left)* that are readily visualized using a maximum intensity projection *(MIP)* display *(right).*

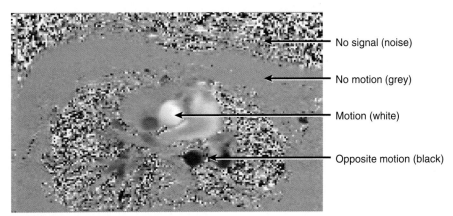

No signal (noise)

No motion (grey)

Motion (white)

Opposite motion (black)

Fig. 10.15

1. What should be adjusted to increase the amount of tissue saturation in TOF angiography?
   A. Increase TR
   B. Reduce TR
   C. Increase TE
   D. Reduced TE

2. Which type of flow will likely yield the highest signals in TOF angiography?
   A. In plane
   B. Perpendicular to plane
   C. 45 degrees to imaging plane
   D. A, B, and C equivalent

3. In a PC display, what is the appearance of stationary tissues?
   A. White
   B. Gray
   C. Black
   D. Any value possible

4. How many images are required to fully characterize three-dimensional (3D) flow using PC angiography?
   A. 2
   B. 4
   C. 6
   D. 8

CASE 10.8

Time-of-Flight and Phase Contrast Angiography

Fig. 10.15 Phase contrast image of the chest at the level of the ascending thoracic aorta.

1. **B.** Tissue saturation is affected by the choice of TR, where reducing TR increases tissue saturation, and vice versa. The choice of TE affects only the magnitude of the echo signal (T2 effect) and has no effect on tissue saturation, which is a T1-related effect.

2. **B.** In TOF angiography, a short TR is used to saturate stationary tissues so that when fresh (i.e., unsaturated) blood flows into the image plane, a 90-degree RF pulse will result in a large (bright) signal. The largest signals will occur for flow that is perpendicular to the image plane.

3. **B.** In a PC display, there will be three primary colors. Gray indicates the absence of flow, with "white" indicating flow in one direction and "black" indicating flow in the opposite direction.

4. **C.** In PC angiography, two images are required with added bipolar gradients to provide flow information along one axis, with the order in which bipolar gradients are applied reversed. To fully characterize flow along all three coordinates (i.e., x, y, and z) will require the generation of a total of six images, two for each of the three coordinates.

Comment

Information regarding blood flow can also be obtained using TOF, which relies on blood being tagged in one region and subsequently being detected in another region. Longitudinal magnetization of the flowing blood needs to be different from the longitudinal magnetization of the corresponding stationary tissues. When generating any image, stationary tissues will generally become saturated when short TR times are used, because longitudinal magnetization does not have sufficient time to recover. By contrast, fresh blood entering a slice is not "saturated" and will thus result in a higher signal than the saturated stationary tissue. Because flowing blood will be continuously refreshed during image acquisition, this fresh blood never becomes saturated (Fig. 10.16).

TOF needs to have images (slices) that are perpendicular to the blood vessels that are imaged. Venous and arterial flow can be differentiated by use of saturation pulses, which can be applied above or below the slice of interest. The most frequently used MRA techniques to perform TOF include GRASS and FISP GRE sequences. TOF is often used in the head and neck to evaluate the circle of Willis and carotid arteries. This technique is also useful in patients with renal failure who cannot receive iodinated contrast or gadolinium. One important limitation of TOF is the loss of signal both arising from turbulence and from very slowly flowing blood. This often results in overestimation of stenosis.

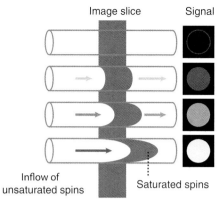

Image slice   Signal

Inflow of unsaturated spins   Saturated spins

Fig. 10.16 After a 90-degree RF pulse rotates the longitudinal magnetization into the transverse plane, this will be repeated after a repetition time (TR). When TR is short, there is insufficient time for recovery of the longitudinal magnetization, which becomes saturated and thus gives rise to low signals. When fresh blood (i.e., unsaturated spins) flows into the image plane, these result in a bright signal.

In PC, information regarding blood velocities is encoded by applying a bipolar gradient between a standard excitation pulse and the corresponding readout. A bipolar gradient consisting of a negative gradient followed by positive gradient is generally applied along a selected direction (axis). The total phase that is accrued during the application of these gradients will be zero for stationary spins but non-zero for spins that move (e.g., flowing blood). In a second image acquisition, the sequence of the bipolar gradients is reversed with a positive gradient followed by negative gradient. From the difference of the two images, the amount and direction of flow in each pixel can be readily calculated. Static tissues (e.g., muscle) will subtract out, giving essentially the same signal in each of the two images.

PC can compute the amount of flow in one direction at a time (e.g., x-axis). In a PC image, black is used to represent the maximum flow in one direction, whereas white will corresponds to the maximum flow in the opposite direction. In these images, gray values indicate stationary tissues. PC images also show areas such as lung that give little signal, which appears as a random mottled pattern and simply represents image noise (i.e., little signal). PC angiography offers "quantitative" measurements of blood flow that is not available with alternative techniques (i.e., TOF and gadolinium contrast). Three separate acquisitions along the x, y, and z directions, each consisting of two images, are needed to provide a complete quantitative representation of the total flow.

References

Huda W. Magnetic resonance II. In: *Review of Radiological Physics.* 4th ed. Philadelphia: Wolters Kluwer; 2016:227–229.

Kim SE, Parker DL. Time-of-flight angiography. In: Carr JC, Carroll TJ, eds. *Magnetic Resonance Angiography Principles and Applications.* New York: Springer; 2012:39–49.

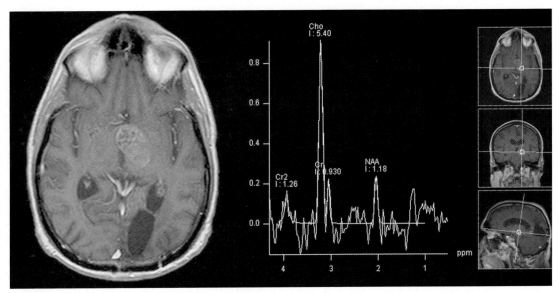

Fig. 10.17

1. How many echoes would be acquired to generate a 32 × 32 × 64 3D MR image?
   A. 32
   B. 96
   C. 32 × 32
   D. 32 × 32 × 64

2. How are echoes produced in echo planar imaging (EPI)?
   A. 90-degree RF pulses
   B. 180-degree RF pulses
   C. Switched gradients
   D. A and B

3. What happens to T2* and blood oxygenation level–dependent (BOLD) signal intensities when hemoglobin becomes oxygenated?
   A. Longer; increases
   B. Longer; decreases
   C. Shorter; increases
   D. Shorter; decreases

4. What is the most essential requirement for performing MR spectroscopy (MRS)?
   A. High magnetic fields
   B. Uniform magnetic fields
   C. Strong magnetic gradients
   D. Fast switching gradients

CASE 10.9

Advanced MRI

Fig. 10.17 MR spectroscopy performed in a patient with recurrent glioblastoma multiforme shows increased choline and decreased N-acetylaspartate *(NAA)*.

1. **C.** We can consider a 3D image ($32 \times 32 \times 64$) as consisting of 32 images, each of which is 32×64. A single 32×64 image would be generated by acquiring 32 echoes and sampling each of these 64 times to generate k-space with 32×64 data points. To generate 32 such 32×64 images would therefore require obtaining 32×32 echoes.

2. **C.** EPI commences with a 90-degree RF pulse, and a gradient (x-axis) rapidly dephases the transverse magnetization, and when this gradient is reversed, a gradient echo is produced. By repeatedly switching and reversing this gradient, a series of echoes are obtained where each echo corresponds to a line of k-space.

3. **A.** When oxygen is added to hemoglobin, the iron in this protein is effectively shielded from the surrounding fluid, and so the dephasing effect of Fe will be reduced. Since the dephasing effect of Fe is T2*, this means that T2* will get longer when hemoglobin becomes oxygenated and detected signal intensities increase.

4. **B.** Protons or other nuclei such as phosphorus-31 in different compounds have slightly different Larmor frequencies, and the detection of such differences is the essence of spectroscopy. Because the detected frequency is used to identify a specific chemical compound, it is *essential* that the magnetic field is extremely uniform.

Comment

In 3D imaging, a nonselective RF pulse produces transverse magnetization in a volume. In addition to the frequency-encoding gradient applied along the x-axis when echoes are obtained, two sets of orthogonal phase-encoding gradients are applied along z and y directions (there is no SSG). A $32 \times 32 \times 64$ image requires nearly 1000 echoes (32×32), each of which is sampled 64 times. Acquired data consist of 32 contiguous k-space slices, with each slice having 32 pixels along one axis (phase encode) and 64 along the other (frequency encode). Imaging time is long, being $32 \times 32 \times$ TR in this example. The extra phase encode direction means artifacts that propagate along the phase encode direction will appear twice in 3D sequences.

In EPI, a 90-degree RF pulse rotates the longitudinal magnetization (i.e., M_z) into the transverse plane (i.e., M_{xy}). Rapidly switched gradients are used in the frequency encode direction to produce many echoes, where one echo will produce one line of k-space (Fig. 10.18). A different PEG is applied for each echo, and the image data can be acquired within a single acquisition. In single-shot sequences, all phase encoding steps are obtained in a single TR interval, which can be as short as 100 ms. In such sequences, there is no TR parameter because all lines of k-space are acquired after the first RF pulse. In multishot sequences, the phase steps are divided into several "shots" so that an image with 256 phase steps could be divided into 4 shots of 64 steps each.

Functional MRI (fMRI) makes use of the amount of blood oxygenation, blood volume, or blood flow changes in the brain that is associated with neuronal activity. Oxygenated hemoglobin will reduce T2* because the added oxygen will "shield" the hemoglobin iron atoms, reducing dephasing effects. Brain activity will increase local venous blood oxygenation, which subsequently increases the intensity of the signal intensity from this region. This is called BOLD imaging. EPI sequences (T2*) weighting are commonly used, but the intensity changes are very small, typically a few percent. Mental activity such as visual, motor, or auditory brain function can be superimposed on high-resolution MR images in the form of color overlays.

MRS uses the slight difference in resonance frequency of protons or other nuclei such as phosphorus 31 found in metabolites. MRS requires strong and uniform static magnetic fields, with the water and fat peaks suppressed using techniques such as STIR. The acquired signal of a localized volume is obtained using spectroscopy sequences such as stimulated echo acquisition mode (STEAM) and point resolved spectroscopy (PRESS). Signals are collected without gradients to generate a localized chemical spectrum. Proton spectroscopy analyzes the presence of metabolites such as NAA, creatine, and choline, whereas phosphorus spectroscopy studies cellular metabolism from concentrations of inorganic phosphate, phosphocreatine, and adenosine triphosphate.

References

Elster AD. Questions and answers in MRI. MR spectroscopy I—basic concepts; 2017. http://www.mriquestions.com/mrs-i—basics.html. Accessed March 2, 2018.

Huda W. Magnetic resonance II. In: *Review of Radiological Physics*, 4th ed. Philadelphia: Wolters Kluwer; 2016:229–231.

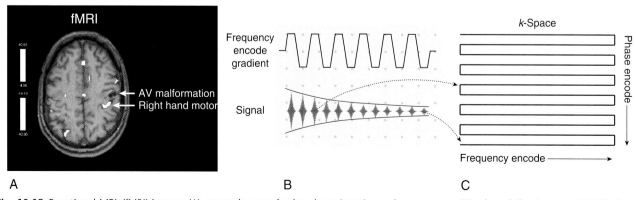

Fig. 10.18 Functional MRI *(fMRI)* images (A) can make use of echo planar imaging pulse sequences (B) where following an initial 90-degree RF pulse, gradients along the frequency encode direction are rapidly switched to create multiple echoes in a single TR interval. Each echo corresponds to a line of k-space, and a different phase encode gradient is applied between echoes to obtain the whole of k-space (C). *AV*, Arteriovenous.

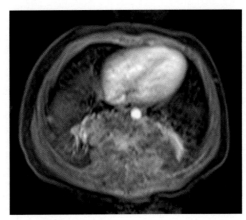

Fig. 10.19

1. If the limiting resolution of CT is taken to be 0.7 line pairs (lp) per mm, what is the most likely limiting spatial resolution of clinical MRI (lp/mm)?
   A. <0.7
   B. 0.7
   C. >0.7

2. Which characteristic is most likely to exhibit the smallest differences between two types of soft tissue (e.g., kidney and liver)?
   A. T1
   B. T2
   C. T2*
   D. Spin density (ϱ)

3. Which image acquisition parameter is likely to affect lesion contrast in GRE imaging?
   A. Flip angle (°)
   B. TE time (ms)
   C. TR time (ms)
   D. A, B, and C

4. Which type of "tissue weighting" pulse sequence is most likely to be used with gadolinium contrast agents?
   A. T1
   B. T2
   C. T2*
   D. Spin density (ϱ)

CASE 10.10

Resolution and Contrast

Fig. 10.19 Large enhancing posterior mediastinal mass in a child. Note the positive contrast of gadolinium in the heart, aorta, and mass in comparison with the negative contrast of air in the lungs.

1. **A.** The visibility of detail in CT (0.7 lp/mm) is approximately twice that of MRI. In practice, CT images use a 512×512 matrix size, which is typically at least twice as big as a good-quality MRI (256×256).

2. **D.** The difference in spin density for most tissues is very modest (<10%). Most tissues can be expected to have much larger intrinsic differences in their T1, T2, and T2* properties, with these differences being routinely exploited in clinical MRI.

3. **D.** Contrast in GRE images is always affected by flip angle, TE, and the TR. Flip angle and TR control the amount of T1 weighting. TE controls the amount of T2 (SE) or T2* (GRE) weighting in the image.

4. **A.** Adding gadolinium will reduce T1 times (increased spin-lattice interactions) that gives an increase in signal intensity on T1-weighted SE images (i.e., positive contrast agent). T2 and T2* effects are generally related to spin TEs, whereas spin density is not affected by contrast agents.

Comment

MR resolution refers to the amount of detail seen in MR images. MR of the brain uses a 25-cm field of view (FOV) and 128×128 matrix, making the pixels size approximately 2 mm per pixel (150 mm/128 pixels) and approximately 1 mm per pixel for a 256×256 matrix size. In clinical MR, limiting spatial resolution is approximately 0.3 lp/mm, based on a 256×256 matrix size. MR resolution in clinical practice is thus approximately 10 times worse than that of planar imaging, as chest x-rays have resolution of 3 lp/mm. Improving resolution requires smaller voxels, which will reduce the acquired signal and SNR. This is an important trade-off because SNR is generally of more importance in clinical work and most MRI scientist strive to increase SNR while ensuring that imaging times are not too long to be implemented into clinical practice.

The number of free protons to create signal is always part of contrast, although it is minimal in T1- and T2-weighted images. Nevertheless, bone (although not bone marrow) is black in all sequences partly because of lower density of free proton spins. The small number of free protons does not produce a measurable signal because of the extremely short T2 times exhibited by all solids, which are generally a few microseconds. Contrast agents that impact PD are used in gastrointestinal imaging and usually to fill the colon or similar structure. Water or mineral oil have many free protons and will cause a neutral to bright signal, whereas CO_2 has a low number of protons and will give no signal.

Tissue contrast in T1-weighted sequences depends on the differences in the amount of time it takes for tissue spins to relax back to alignment with the magnetic field. A long TR will allow all tissue to have all spins align to the field and thus no real difference in signal due to T1. A short TR means the tissues with short T1 (e.g., fat) will have more signal relaxed and ready to be flipped into the transverse plane for imaging; thus they will have a higher signal. Gadolinium contrast agents will shorten the T1 time of surrounding water protons and thus increase the signal on T1-weighted images. Abnormal enhancement can be seen in blood vessel malformations, tumors, and areas of hyperemia due to inflammation or infection.

Tissue contrast in T2-weighted images depends on the differences in the amount of time it takes for spins to dephase. A short TE means spins do not have much time to dephase and thus signal between tissues will not really depend on T2. A long TE allows more time for spins to dephase and thus tissue with a short T2 will be dark because they will dephase a lot. Tissues with a long T2 (e.g., CSF) will not dephase as much and thus will look bright (Fig. 10.20). Superparamagnetic iron oxide and other contrast agents purposely decrease T2 and T2* because of markedly increased dephasing. These contrast agents result in darker areas on T2-weighted images so that normal tissues (e.g., liver) darken, allowing bright underlying pathology to be more easily detected.

References

Hashemi RH, Lisanti CJ, Bradley Jr WG. Scan parameters and image optimization. In: *MRI: The Basics.* 4th ed. Philadelphia: Wolters Kluwer; 2018:187–196.

Huda W. Magnetic resonance II. In: *Review of Radiological Physics,* 4th ed. Philadelphia: Wolters Kluwer; 2016:231–232.

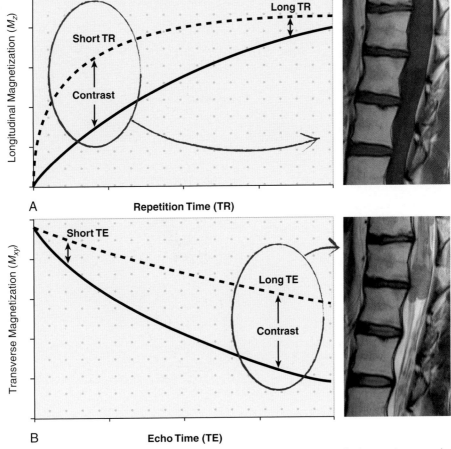

Fig. 10.20 *A* shows how a short repetition time *(TR)* provides T1 contrast in T1-weighted spin echo images between the short T1 tissue *(dotted line)* and the long T1 tissue *(solid line)*. *B* shows how a long echo time *(TE)* provides contrast in T2-weighted spin echo images between the long T2 tissue *(dotted line)* and the short T2 tissue *(solid line)*. Dropped metastases in a patient with medulloblastoma are clearly seen in the bottom T2-weighted image due to improved contrast between cerebrospinal fluid and tissue.

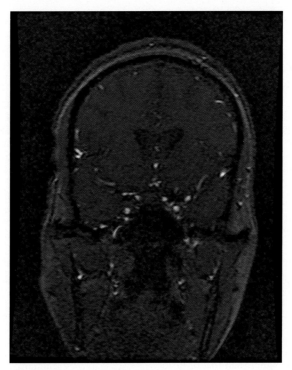

Fig. 10.21

1. What is the effect on the image signal and the image noise when four MR images are acquired and averaged together?
   A. Quadrupled and quadrupled
   B. Quadrupled and doubled
   C. Doubled and quadrupled
   D. Doubled and doubled

2. How will increasing the magnetic field from 1.5 to 3 T most likely affect the acquired SNR (%)?
   A. <100
   B. 100
   C. >100

3. When the receiver bandwidth is increased, what likely happens to the noise in the resultant images?
   A. Increases
   B. Unchanged
   C. Reduced

4. Which type of coil most likely has the lowest level of noise, and thereby the highest SNR?
   A. Body
   B. Head
   C. Surface
   D. A, B, and C similar

CASE 10.11

Signal-to-Noise Ratio

Fig. 10.21 Coronal MR venogram in a patient with venous sinus thrombosis. No signal is seen in the superior sagittal sinus.

1. **B.** If four images are added together, the signal (deterministic) must also be four times bigger. However, noise is a random phenomenon and will therefore not add in a linear manner. Random noise in four images when added together will only be twice as big so that four images added together will double the resultant SNR.

2. **B.** The SNR in MR is generally found to be directly proportional to the strength of the magnetic field. Thus doubling the field from 1.5 to 3 T also doubles the signal produced by tissue (and the SNR).

3. **A.** When the receiver bandwidth is increased because of the application of stronger magnetic field gradients, differences in Larmor frequencies will go up. It turns out that a high receiver bandwidth increases the detected noise.

4. **C.** Body coils have the highest level of noise, and surface coils have the lowest level of noise. In addition, surface coils are closer to tissues and result in the highest signals. As a result, the SNR of a surface coil is higher than that of head and body coils.

Comment

The most important determinant of image quality in MRI is the SNR (Fig. 10.22). Signal strength is directly proportional to the number of spins. Thus signal strength per pixel improves as pixel volume increases. Doubling each voxel dimension increases voxel volume and signal by a factor of eight (i.e., 2^3). Similarly, increasing the FOV increases the signal because doubling the FOV quadruples the pixel area and thereby quadruples the signal. Unlike in CT where the matrix size is always 512×512, MRI matrix size can vary and impact image quality. Doubling the matrix size (e.g., 128^2 to 256^2) while keeping the FOV the same will drop the voxel volume and SNR one-quarter but will improve resolution due to the smaller pixel size.

The tissue signal is directly proportional to how many spins align with the field versus how many align against it. In general, only a few spins per million protons align with the field, and it is these that contribute to the MR signal. The number of excess spins aligning with the field (and thus the signal) is proportional to the field strength. This proportionality has been found to be valid over a large range of magnetic field strengths (0.2 to 11 T). Although overall image quality is improved by increasing field strength, there are disadvantages to high field MRI. In general, patient heating and many of the artifacts found in MRI are worse as magnetic field strength increases. In addition, it is increasingly difficult to ensure field uniformity as the field strength increases.

The receiver bandwidth can be thought of as what frequencies the machine is listening to. When the receiver bandwidth is increased it will listen to more frequencies, but this has the drawback that it will also hear more noise. As a result, increasing receiver bandwidth will decrease SNR. Higher receiver bandwidth is also associated with larger read gradient strength. Accordingly, when strong gradients are used, which can help to reduce imaging artifacts, the receiver bandwidth will need to be increased. As a result, the higher receiver bandwidth will be associated with an increase in image noise. Increasing the FEG "trades" reduced image artifacts against an increase in noise.

Larger RF receiver coils create images with more noise than smaller ones and thus lower the SNR. RF receive-only coils often allow for parallel imaging, which decreases acquisition time but also decreases SNR. Setting the parallel acceleration factor to 4 will quarter acquisition time but also reduce SNR by 2. Receiver coils are often characterized by their number of channels. A channel is a complete electronics chain needed to turn the body signal into a digital signal for making images. A higher number of channels generally allows for faster and/or higher quality imaging, with the drawback of being more expensive.

References

Bushberg JT, Seibert JA, Leidholdt Jr EM, Boone JM. Magnetic resonance imaging: advanced image acquisition methods, artifacts, spectroscopy, quality control, siting, bioeffects, and safety. In: *The Essential Physics of Medical Imaging*. 3rd ed. Philadelphia: Wolters Kluwer; 2012:460–464.

Huda W. Magnetic resonance II. In: *Review of Radiological Physics*, 4th ed. Philadelphia: Wolters Kluwer; 2016:231–232.

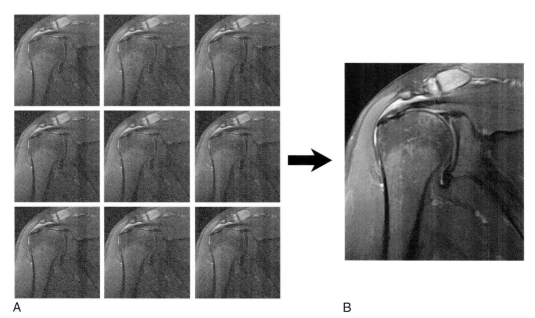

Fig. 10.22 Adding nine MRI images together (A) increases the (deterministic) signal ninefold but increases the (random) noise threefold ($9^{0.5}$). As a result, nine images taking nine times longer to generate will triple the resultant MR signal-to-noise ratio (B).

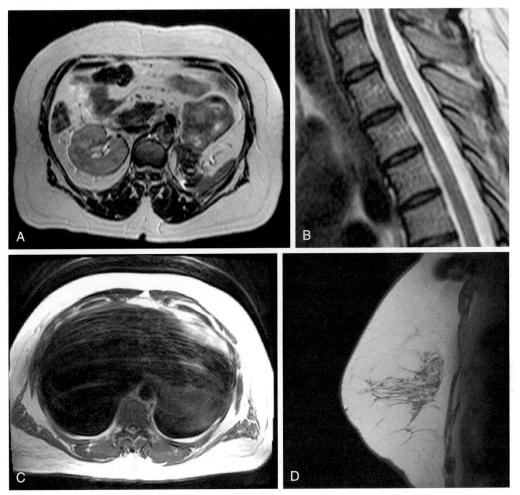

Fig. 10.23

1. Chemical shift artifact occurs along what direction?
   A. Phase encode
   B. Frequency encode
   C. Slice selection
   D. Oblique direction

2. What artifact is pictured in Fig. 10.23B?
   A. Motion
   B. Wraparound
   C. Truncation
   D. Chemical shift

3. What should be increased to help mitigate the motion artifact seen in Fig. 10.23C?
   A. ETL
   B. Pulse TR
   C. FOV
   D. Image matrix size (N^2)

4. How do you reduce susceptibility artifact as seen in Fig. 10.23D?
   A. Use gradient echo imaging
   B. Increase TE
   C. Increase receiver bandwidth
   D. Use a stronger magnetic field

CASE 10.12

Artifacts

Fig. 10.23 Chemical shift artifact around the kidney in a patient with a history of renal cell carcinoma and partial nephrectomy. Note the light rim on the right side of the kidney (where fat was artifactually shifted to) and the dark rim on the left (where signal void exists due to the shifted fat). (B) Truncation (Gibbs ringing) artifact is visualized as repeated edges inside the spinal cord. This artifact can mimic cord lesions or other pathology. (C) Abdominal MR of a patient experiencing shortness of breath. Inability to hold his or her breath led to chest wall motion during breathing, which is "ghosted" both inside and outside the patient in the anteroposterior direction. (D) Susceptibility artifact from a clip that had previously been inserted into the upper breast at the top of the image.

1. **B.** Chemical shift occurs because of mismapping due to the different Larmor frequencies of protons in fat versus that of protons in water. Because frequency differences cause the mismapping, the artifact occurs along the frequency encode direction.

2. **C.** When there are insufficient data because of a reduced matrix size, an artifact will be created that has the appearance of "ringing" in the form of multiple lines near edges. This is also sometimes called truncation, ringing, or Gibbs phenomena.

3. **A.** Motion artifact is typically mitigated by reducing the patient's motion. Increase the ETL in FSE imaging will cut acquisition time and thus reduce motion artifact. All the other options will increase acquisition time and thus make motion artifact worse.

4. **C.** Susceptibility artifact appears as a signal void and occurs due to dephasing of spins because of the properties of the tissues or objects in the local area. This can be mitigated by using SE imaging rather than gradient echo imaging, decreasing TE, increasing receiver bandwidth, and imaging at lower magnetic fields.

Inhomogeneity caused by the presence of foreign objects or differences in susceptibility between two adjacent tissues will alter the local magnetic field (see Fig. 10.23D). This increases local dephasing and causes signal voids (susceptibility artifact) and may alter the local Larmor frequency causing geometric distortion (especially in GRE sequences). These artifacts are minimized with short TE, lower B_0, use of SE or FSE, and large receiver bandwidth. Chemical fat saturation may also fail near such objects/interfaces and STIR sequences can be used for improved fat suppression because they are not dependent on the specific Larmor frequency of protons in fat versus those in water.

If there are insufficient data points (frequencies) in the image, a ringing artifact can occur, which appears as repeated lines near interfaces, also known as a truncation or Gibbs artifact (see Fig. 10.23B). Increasing matrix size or decreasing FOV can help with this artifact. If patient anatomy exists outside the FOV along the phase-encode direction, the scanner can become confused and wrap the anatomy into the image. This is called wraparound artifact, which is a result of aliasing and which can be mitigated by saturating tissue outside the FOV, using a larger FOV, or using the "no phase wrap" imaging option. If the patient moves during the scan, the scanner cannot localize the tissue correctly. This results in "ghosts" of the moving object across the entire image in the phase-encode direction. This can be mitigated by saturating the moving tissue or decreasing acquisition time to reduce the chance of patient motion (see Fig. 10.23C).

Chemical shift artifacts are caused by the slight difference in resonance frequency of protons in water and in fat, resulting in misregistration of fat and water (see Fig. 10.23A). Chemical shift artifacts can produce light and dark bands along the frequency encode direction at the edges of the kidney or the margins of vertebral bodies. Increasing the receiver bandwidth will reduce this artifact. Chemical shift artifacts can also occur for GRE images where the fat and protons within a pixel are alternately in and out of phase. This artifact will occur in all directions (not just the frequency encode) and is sometimes called India ink artifact. Changing GRE TE can result in bands that are alternately bright or dark depending on whether the fat/water is in or out of phase (T2* effect).

Flowing blood and CSF can result in ghosting along the phase-encoding direction. This is a specific type of motion often called pulsation artifact. It may be reduced by nulling the blood moving through the vessel. However, more often it is simply redirected by selecting the phase-encode direction so the artifact does not pass through tissue of interest. Another type of artifact, referred to as the "magic angle effect," occurs in tendons and nerves aligned at 55 degrees to the main field (Fig. 10.24). This results in longer T2 times (e.g., 5 ms) for those tissues and will appear bright on T1 and PD SE images (i.e., sequences with short TEs). This is especially important because this artificial signal increase often mimics disease in musculoskeletal imaging. Magic angle artifacts disappear on T2-weighted images (i.e., TE 100 ms).

References

Huda W. Magnetic resonance II. In: *Review of Radiological Physics*, 4th ed. Philadelphia: Wolters Kluwer; 2016:232.

Morelli JN, Runge VM, Ai F, et al. An image-based approach to understanding the physics of MR artifacts. *Radiographics*. 2011;31(3):849–866.

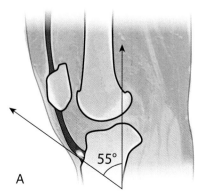
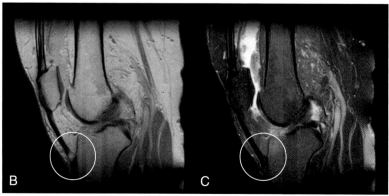

Fig. 10.24 Magic angle artifact schematic and simulation. When a tissue with a stranded texture such as a tendon is oriented at a specific angle in the magnetic field, the spin-spin interactions are limited (A). This means longer T2 times and thus a stronger signal on scans with relatively short echo time (TE) (e.g., T1 weighted) (B). Scans with longer TEs (e.g., T2 weighted) will not show a pronounced increase, because the increased signal will decay away with long TEs (C).

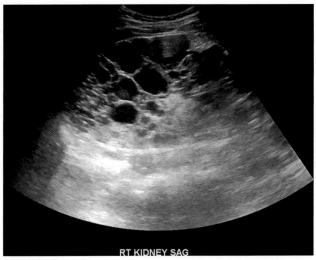

RT KIDNEY SAG

Fig. 11.1

1. What is varying along a sound wave?
   A. Pressure
   B. Temperature
   C. Charge
   D. Voltage

2. How is ultrasound velocity affected as frequency increases?
   A. Increased
   B. Unchanged
   C. Reduced

3. What is the approximate wavelength of a 1.5-MHz transducer (mm)?
   A. 0.01
   B. 0.1
   C. 1
   D. 10

4. Which has the highest ultrasound velocity?
   A. Lung
   B. Fat
   C. Tissue
   D. Bone

CASE 11.1

Sound

Fig. 11.1 Autosomal dominant polycystic kidney disease.

1. **A.** Sound is a pattern of pressure variations that propagates through a medium. There are no variations of temperature, charge, or voltage associated with any type of sound wave.

2. **B.** Ultrasound velocity is independent of frequency in any given medium. All the notes from a piano will travel to the listener at the same speed.

3. **C.** The wavelength is the velocity of sound divided by the frequency. In soft tissue, sound travels at 1500 m/s, so the wavelength is 1 mm (i.e., 1,500,000 mm/s divided by 1,500,000 Hz).

4. **D.** Bone has the highest sound velocity because it is the least compressible, and air has the lowest sound velocity because it is the most compressible. Fat and tissue have intermediate sound velocities.

Comment

Sound is pressure disturbance that travels away from its source (Fig. 11.2). Ultrasound is transmitted through tissue as patterns of alternating compression and rarefaction. At any point, pressure varies as a function of time in a cyclical manner (i.e., increases and decreases). If sound were frozen in time, the pressure along the wave would increase and decrease as a function of distance. Ultrasound waves propagate at a velocity, expressed in meters per second, which is the wavelength multiplied by the corresponding frequency. Frequency, expressed as cycles per second, is the number of oscillations in each second. The wavelength, expressed as meters, is the distance between two successive crests along the wave.

Frequencies are measured in hertz, with 1 Hz being 1 oscillation per second. Middle C on a piano has a frequency of 256 Hz, and harmonic frequencies are integral multiples of this value. Accordingly, 512 and 1024 Hz are harmonics of middle C. Audible sound ranges between 15 and 20 kHz. Ultrasound frequencies are those that are greater than 20 kHz. A small instrument such as a violin produces high frequencies, whereas a large instrument such as a double bass produces low frequencies. Diagnostic ultrasound uses transducers ranging between 1 and 20 MHz.

Ultrasound wavelengths are determined by the compressibility of a given material. At 1.5 MHz for soft tissue, the wavelength is 1 mm. At the same frequency of 1.5 MHz, wavelength would be 0.2 mm in air and approximately 3 mm in bone. The ultrasound wavelength always decreases with increasing frequency, and vice versa. Wavelengths are 0.1 mm at 15 MHz in soft tissue, which is 10 times lower than at 1.5 MHz. Ultrasound pulses are approximately two wavelengths long so that a 1.5-MHz transducer will generate 2 mm long pulses. The length of the ultrasound pulse determines axial resolution, which is half the pulse length (i.e., wavelength).

For a specified material, sound velocity is always independent of frequency. Consider different instruments in an orchestra that produce different frequencies but travel through air to the concert audience at the same speed. When velocity in each tissue is fixed, frequency and wavelength are inversely related so that higher frequencies have shorter wavelengths, and vice versa. Materials with low compressibility such as bone have high sound velocities. Materials that are easy to compress, such as air, have low sound velocities. Sound travels in soft tissue at 1540 m/s. Sound travels much more slowly in air and much faster in bone and is approximately 5% slower in fat.

References

Huda W. Ultrasound I. In: *Review of Radiologic Physics*. 4th ed. Philadelphia: Wolters Kluwer; 2016:237–238.

Bushberg JT, Seibert JA, Leidholdt Jr EM, Boone JM. Ultrasound. In: *The Essential Physics of Medical Imaging*. 3rd ed. Philadelphia: Wolters Kluwer; 2012:502–506.

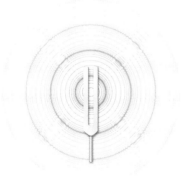

Fig. 11.2 Sound is a pattern of varying pressure characterized by a wavelength (distance between two successive peaks) and a frequency (number of oscillations at one point in 1 s). The tuning fork on the left has a shorter wavelength (and higher frequency) than the one on the right, but sound waves from both travel at the same velocity (wavelength × frequency).

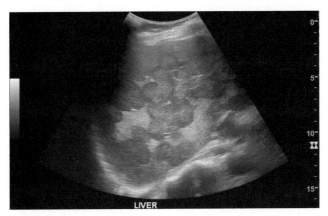

Fig. 11.3

1. Which of the following are ways in which ultrasound can interact with tissues?
   A. Reflections
   B. Scattering
   C. Refraction
   D. A, B, and C

2. Which interactions would be responsible for identifying the outline of the liver in ultrasound imaging?
   A. Specular reflections
   B. Nonspecular reflections
   C. Scattering
   D. Refraction

3. Which organ contents likely result in the least amount of scatter (i.e., anechoic)?
   A. Liver
   B. Kidney
   C. Prostate
   D. Bladder

4. What property is closely associated with the refraction of ultrasound beams?
   A. Resolution
   B. Contrast
   C. Mottle
   D. Artifacts

CASE 11.2

Interactions

Fig. 11.3 Multiple hepatic masses in a patient with metastatic breast cancer.

1. **D.** Ultrasound waves can interact with tissues by reflecting (specular and nonspecular), scattering, or refracting.

2. **A.** The outline of an organ such as the liver is obtained from specular reflections. These specular reflections occur from "smooth" surfaces.

3. **D.** Scattering occurs when sound interacts with structures smaller than the wavelength and is responsible for the inner contents of most organs. Because the bladder contents have no structure, this region will appear anechoic (black).

4. **D.** Refraction effects occur when sound travels from one type of tissue to another and there is a change in sound velocity. Refraction causes artifacts including shadowing behind a lesion (e.g., at lesion edge), as well as location errors (e.g., false aorta).

Comment

In ultrasound imaging, specular reflections occur from large smooth surfaces analogous to light being reflected from a smooth mirror (Fig. 11.4). A smooth surface is one that has variations that are larger than the wavelength of the ultrasound beam. For example, variations that are greater than 1 mm at 1.5 MHz would be considered smooth. Specular reflections result in echoes that travel back to the transducer and which are used to generate ultrasound images. Hyperechoic means higher acoustic intensity relative to the background signal, and hypoechoic means that there is lower acoustic intensity. Organs with fluids (e.g., bladder and cysts) have no internal structure and almost no echoes and will therefore appear to be black (i.e., anechoic).

Nonspecular (or diffuse) reflections occur with rough surfaces analogous to light diffusely scattered from a steamed-up mirror. A rough surface is one that has variations that are smaller than the wavelength of the ultrasound beam. For example, variations that are less than 0.1 mm at 15 MHz would be considered rough. Nonspecular reflections are generally weaker and are omnidirectional.

Most organs (e.g., kidney and liver) comprise of complex structures with many scattering sites. When the beam is scattered by diffuse reflections, a wave is generated that travels outward in all directions from the scatter source. The interference pattern of scattered waves from many scattering sites generates "speckle," which is a pattern of bright and dark spots that are characteristic of the tissue structure (unlike random noise). Despite their nonrandom characteristics, speckle still typically hinders diagnosis and vendors often provide image processing algorithms to smooth out the speckle in the image.

Refraction refers to a change in direction of an ultrasound beam when passing from one tissue to another. The physical basis of refraction is a difference in the speed of sound between two tissues. Ultrasound passing from one tissue to another with a different speed of sound keeps the same frequency, but the wavelength will change. Ultrasound imaging systems assume straight-line propagation, so refraction effects produce image artifacts in the form of spatial distortions. This is analogous to straight sticks that are partially immersed in water appearing bent because of refraction of light that has different velocities in air and water. Refraction will also generally result in dark shadows in ultrasound images when an ultrasound beam is "lost" after it has been refracted.

References

Huda W. Ultrasound I. In: *Review of Radiologic Physics.* 4th ed. Philadelphia: Wolters Kluwer; 2016:239–241.

Radiological Society of North America. RSNA/AAPM radiology physics educational modules. *Ultrasound—Concepts and Transducers.* https://www.rsna.org/RSNA/AAPM_Online_Physics_Modules_.aspx.

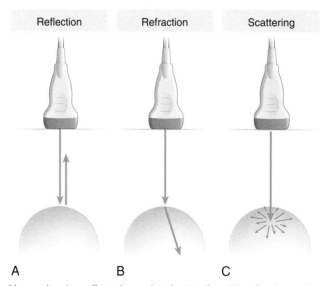

Fig. 11.4 Schematic of an ultrasound beam showing reflected sound at the interface (A), refraction as the sound passes from one type of tissue to another (B), and scattering from the internal structures of an organ (C).

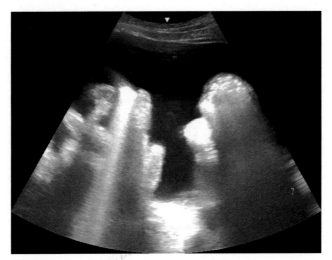

Fig. 11.5

1. Which material will likely have the highest acoustic impedance?
   A. High density and low sound velocity
   B. High density and high sound velocity
   C. Low density and low sound velocity
   D. Low density and high sound velocity

2. Which interface has the highest mismatch in acoustic impedance values?
   A. Lung:bone
   B. Lung:tissue
   C. Tissue:fat
   D. Fat:bone

3. What affects the amount of attenuation of an ultrasound beam?
   A. Tissue characteristics
   B. Ultrasound frequency
   C. Distance traveled
   D. A, B, and C

4. Which ultrasound interaction will most likely result in tissue heating?
   A. Reflection
   B. Refraction
   C. Scattering
   D. Attenuation

CASE 11.3

Impedance and Attenuation

Fig. 11.5 Large amount of ascites.

1. **B.** The acoustic impedance is the product of the material density and sound velocity. Air, which has a low density and low sound velocity, has very low acoustic impedance. Bone, which has a high density and a high sound velocity, has very high acoustic impedance.

2. **A.** Lung has very low acoustic impedance, and bone has very high acoustic impedance, so this pair has the highest impedance mismatch. Tissue and fat have acoustic impedances that are intermediate between the low of lung and the high of bone.

3. **D.** Ultrasound attenuation is dependent on tissue characteristics and is high in lung but low in fluids. Ultrasound attenuation is directly proportional to the operating transducer frequency as well as the total distance traveled by the sound wave.

4. **D.** Attenuation of an ultrasound beam will always result in heating because energy is being deposited. By contrast, interactions including reflections, scattering, and refraction carry the sound energy away from the interaction site, which therefore does not result in local heating effects.

Comment

Tissue acoustic impedance, generally referred to as Z, is the product of the density (ρ) and the sound velocity (v) and is expressed in Rayls. Air and lung have low acoustic impedances because they have low densities as well as low sound velocities. Bone and piezoelectric crystals have high acoustic impedance because they both have high densities as well as high sound velocity. Soft tissue has an acoustic impedance intermediate between these two extremes. Fat has an acoustic impedance that is slightly lower than that of tissue because its density and sound velocity are both a little less than that of soft tissue.

The acoustic impedance of the two tissues on either side of any interface determines the fraction of ultrasound reflected (Fig. 11.6). When there is a large difference in acoustic impedance at an interface (i.e., mismatch), most of the ultrasound energy is reflected. Strong echoes will always be produced from gas bubbles in abdominal imaging. A bone and tissue interface also reflects a substantial amount of the incident intensity so that imaging through bone is generally not possible. When acoustic impedances are similar, most of the ultrasound energy will be transmitted and the echoes produced at this interface will be weak. Fat and tissue interfaces have similar acoustic impedances so that a fat/tissue interface will result in weak echoes and most of the incident ultrasound energy is transmitted.

Decibels (dBs) are used to express relative intensities in ultrasound, which are calculated using a logarithmic scale. A negative decibel value refers to signal attenuation, and a positive decibel value refers to signal amplification. When the ultrasound beam intensity is reduced to 10%, this is expressed as -10 dB. A reduction to 1% is expressed as -20 dB and to 0.1% is expressed as -30 dB, and so on. Halving the sound intensity would be expressed as -3 dB. When the sound intensity increases 10-fold, this is expressed as $+10$ dB; increasing

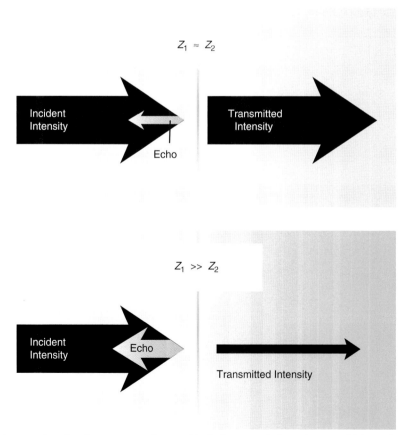

Fig. 11.6 The upper figure shows an interface between two tissues with similar acoustic impedances, which produces very weak echoes. The lower figure shows a large mismatch in acoustic impedance ($Z_1 \gg Z_2$), which results in large echoes and very little of the incident energy being transmitted through the interface.

100-fold is expressed as +20 dB, and increasing 1000-fold is expressed as +30 dB, and so on. A doubling of any intensity would be expressed as +3 dB.

Attenuation loss of ultrasound energy in tissues occurs as a result of scattering and absorption. Sound wave energy that is absorbed will be converted to heat. Ultrasound attenuation in tissue is exponential, expressed in terms of decibels, as described earlier, and is directly proportional to the ultrasound frequency. High-frequency transducers attenuate much more than those operating at low frequency. Attenuation is directly proportional to the distance traveled, where a doubling of the distance will double the amount of attenuation. A rule of thumb is that attenuation in tissue is 0.5 dB/cm per MHz. Fluids have very little attenuation and therefore transmit most of the incident ultrasound. By contrast, lung and bone have high attenuation coefficients, and any ultrasound transmitted into these organs will be rapidly attenuated.

References

Hangiandreou NJ. AAPM/RSNA physics tutorial for residents. Topics in US: B-mode US: basic concepts and new technology. *Radiographics*. 2003;23(4):1019-1033.

Huda W. Ultrasound I. In: *Review of Radiologic Physics*. 4th ed. Philadelphia: Wolters Kluwer; 2016:237-241.

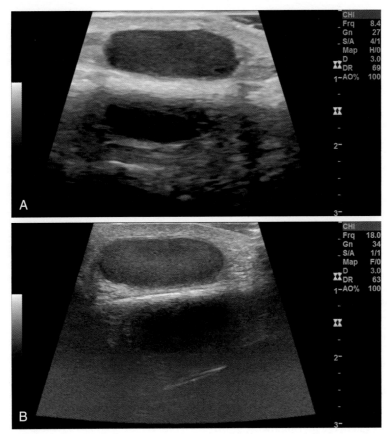

Fig. 11.7

1. What does an ultrasound transducer convert?
   A. Electrical energy (voltage spikes) to sound
   B. Sound (echoes) into electric pulses
   C. Digitizes voltages spikes and echoes
   D. A and B

2. How will increasing transducer thickness affect the transducer operating frequency?
   A. Increase
   B. No change
   C. Reduce

3. What is reduced when a backing layer is added to an ultrasound transducer crystal?
   A. Wavelength
   B. Frequency
   C. Sound velocity
   D. Pulse length

4. What is likely used to improve the transmission of sound energy into patients?
   A. Matching layer
   B. Backing layer
   C. Time gain compensation (TGC)
   D. Increased display gain

CASE 11.4

Piezoelectric Transducers

Fig. 11.7 Epidermal inclusion cyst visualized using a relatively low-frequency 8.4 MHz probe (A) and higher-frequency 18 MHz "hockey stick" probe (B).

1. **D.** A *transducer* is a generic term meaning a device that converts one type of energy into another type of energy. An ultrasound transducer converts electrical energy (voltage spikes) to sound and converts sound (echoes) into weak electric pulses.

2. **C.** A thick transducer will generate low frequencies (long wavelengths), whereas a thin transducer will generate high frequencies (short wavelengths). A transducer operating at 7.5 MHz has a transducer thickness of 0.1 mm, with thinner transducers having higher frequencies and vice versa.

3. **D.** Without a back layer, the pulse produced by a transducer would be very long. The addition of a backing layer will markedly reduce the pulse length. Having a short pulse is important because this improves axial resolution.

4. **A.** A matching layer will improve the transmission of ultrasound into a patient. Without a matching layer, there is a large mismatch between the high impedance of transducers and the much lower impedance of tissue, so little ultrasound would be transmitted into the patient.

Comment

Transducers convert one form of energy into another. A piezoelectric transducer will convert electrical energy into sound energy, and vice versa. Oscillating high-frequency voltages are produced electronically and then sent to the ultrasound transducer. Transducer crystals change shape in response to voltages at these electrodes. The induced crystal shape changes serve to increase (and decrease) pressure in front of the transducer, producing ultrasound waves that are propagated into the patient. Returning ultrasound echoes subject the crystal to pressure changes, which are converted into electrical signals. The electrical voltages induced by returning echoes are transferred to a computer and used to create ultrasound images.

Current ultrasound systems use broad-bandwidth transducers that are designed to generate more than one frequency. Operators can select the examination frequency to meet clinical requirements. The frequency at which any piezoelectric transducer is most efficient at converting electrical energy to acoustic energy is termed *resonance*. The piezoelectric element thickness determines the resonance frequency. Transducer crystals with thickness (t) are equal to one-half of the wavelength at resonance so that high-frequency transducers are thin and low-frequency transducers are thick. For example, a 0.1-mm crystal would resonate efficiently at 7.5 MHz and a 0.5-mm crystal would resonate efficiently at 1.5 MHz.

Medical ultrasound systems generally use pulsed transducers, which are designed to produce extremely short pulses. Blocks of damping material are placed behind transducers to reduce vibration and shorten pulses (Fig. 11.8). Conceptually this is similar to ringing a bell and then placing your hand on it to stop the vibrations (and the sound). By contrast, transducers without damping generate very long pulses (many wavelengths long). A short pulse is important because this affects the axial resolution. A typical transducer has a pulse that is two wavelengths long, with a corresponding axial resolution equal to the wavelength (i.e., half the pulse length).

Lead-zirconate-titanate (PZT) is one of the most common materials used to make ultrasound transducers. A matching layer is placed on the front surface of the transducer to improve energy transmission into the patient. The impedance of this matching layer is intermediate between the transducer and tissue and has a thickness that is one-fourth the wavelength of sound in that material. Ultrasound gel is often also applied to help beam transmission into the body. The presence of gel ensures that there is no air trapped between the transducer and the body. This is especially important because air will reflect 99% of the energy back to the transducer (leaving very little sent into the body for imaging).

References

Bushberg JT, Seibert JA, Leidholdt Jr EM, Boone JM. Ultrasound. In: *The Essential Physics of Medical Imaging*. 3rd ed. Philadelphia: Wolters Kluwer; 2012:513–519.

Huda W. Ultrasound I. In: *Review of Radiologic Physics*. 4th ed. Philadelphia: Wolters Kluwer; 2016:242–244.

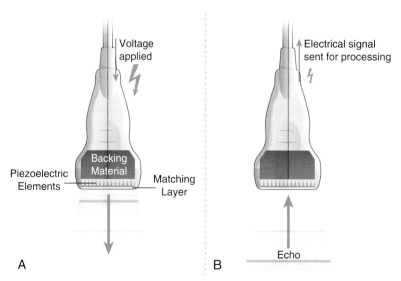

Fig. 11.8 An ultrasound transducer consists of an array of piezoelectric elements, a backing layer to reduce the pulse length, and a matching layer to improve transmission of sound into patients. When a voltage spike is incident on the transducer (A), a short ultrasound pulse is produced, and when an echo is incident on the transducer (B), an electrical signal is generated.

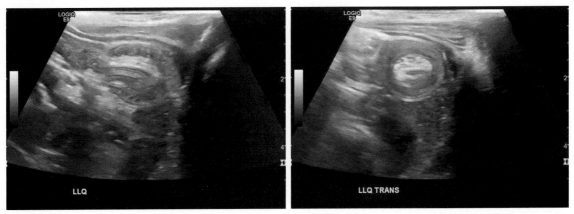

Fig. 11.9

1. Which "zones" can be used to characterize the ultrasound beam generated by a single transducer?
   A. Near
   B. Far
   C. Focal
   D. A, B, and C

2. How is the extent of the near field affected when the transducer size is increased?
   A. Increased
   B. Not changed
   C. Reduced

3. How can a single ultrasound transducer be focused?
   A. Acoustic lens
   B. Shaped transducer
   C. Time (phase) delays
   D. A, B, and C

4. Which type of spatial resolution could be improved by focusing a transducer?
   A. Lateral
   B. Elevational
   C. Axial
   D. A and B

CASE 11.5

Single Transducers

Fig. 11.9 Longitudinal and transverse ultrasound images in a child with colonic intussusception secondary to a pathologic lead point.

1. **D.** A single transducer will have a near zone where imaging is possible, as well as a far zone where the ultrasound beam diverges and imaging is not performed. Most clinical transducers will also be focused in one or two directions (lateral and/or elevational).

2. **A.** A doubling of the diameter of a cylindrical transducer will generally quadruple the extent of the near field. In multielement transducers with small single elements, several elements must be activated at once to increase the effective transducer size and achieve a satisfactory near field.

3. **D.** A single element can be focused to produce a narrow beam in both the lateral and elevational directions using time (phase) delays. This focusing can also be achieved using a shaped transducer or an acoustic lens, but the focal zone(s) will be fixed.

4. **D.** Focusing an ultrasound beam at a fixed distance in the image plane will improve lateral resolution performance. Additionally, focusing and reducing the slice thickness of the ultrasound beam at a fixed distance from the transducer will improve elevational resolution. Acoustic lens and shaped transducers can achieve either of these, or both, but only at a preselected distance from the transducer. Axial resolution is not affected by focusing.

Comment

The region used for ultrasound imaging is the near field of an ultrasound beam that is adjacent to the transducer. The far field starts where the near field ends and is the point at which the ultrasound beam diverges and the intensity falls off very rapidly. Ultrasound imaging is not possible in the far field. The near field is sometimes referred to as the Fresnel zone, and the far field is referred to as the Fraunhofer zone. Side lobes are emitted at angles to the primary beam and have reduced intensity. These will result in artifact in the resultant image because any detected echo is always assumed to have been obtained along the main beam.

The near field length increases with the transducer size (Fig. 11.10). Quantitatively, doubling the transducer size will quadruple the extent of the near field. When small elements are used in multielement probes, the firing of a single small element will result in a very short near field. To extend the length of the near field, it is customary to activate a small number of adjacent elements. This procedure is always adopted in linear arrays, and convex arrays (see later). The near field extent will also increase with increasing frequency, but this advantage is frequently offset by additional attenuation of the resultant ultrasound beam.

All current clinical ultrasound uses focusing of the ultrasound beam that causes convergence and narrowing of the beam. The point at which the beam is narrowest is referred to as the focal point. The focal zone is a region over which the beam is relatively narrow. Focusing an ultrasound beam will improve resolution performance. Focusing can occur both in the image plane (lateral) and in the image slice thickness (elevational), but not in the axial plane. For a single transducer, the amount of focusing is always fixed and determined by the characteristics of the ultrasound system.

For a single transducer, there are two ways by which the ultrasound beam can be focused, consisting of shaped transducers or using an acoustic lens. Ultrasonic focusing can be achieved using a transducer face that is concave and which will result in a beam that is narrowed at a fixed distance from the transducer. Alternatively, a concave acoustic lens can be positioned on the surface of the transducer, which can focus the beam at a fixed distance from the transducer. The performance of an acoustic lens is analogous to the use of optical lenses (spectacles) to correct deficiencies of the human eye.

References

Hangiandreou NJ. State-of-the-art ultrasound imaging technology. *J Am Coll Radiol.* 2004;1(9):691–693.

Huda W. Ultrasound I. In: *Review of Radiologic Physics.* 4th ed. Philadelphia: Wolters Kluwer; 2016:242–244.

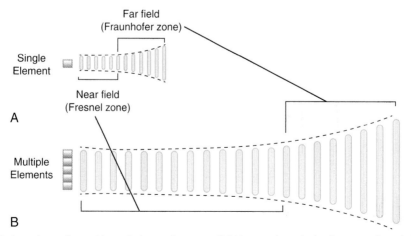

Fig. 11.10 A) When a single transducer element is used, the resultant near field is very short. Activating a number of elements (B), as is normally done with linear and convex arrays, results in a longer near field.

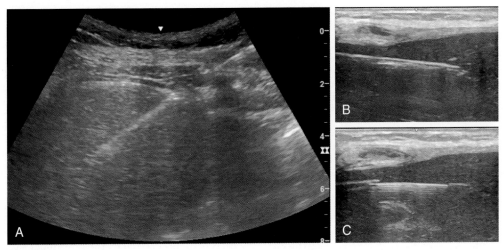

Fig. 11.11

1. How many individual elements are likely to be used in the construction of an ultrasound linear array transducer?
   A. 10
   B. 100
   C. 1000
   D. 10,000

2. What is the major benefit from using a linear array?
   A. Superficial visibility
   B. Large field of view
   C. Reduced thermal index
   D. Lateral resolution

3. Which type of resolution would most likely deteriorate the most with increasing depth on a convex (curvilinear) array?
   A. Axial
   B. Lateral
   C. Elevational
   D. A, B, and C

4. Which is most likely increased when spatial compounding is employed in ultrasound imaging?
   A. Lesion contrast
   B. Image noise
   C. Contrast-to-noise ratio
   D. A and B

CASE 11.6

Linear and Convex Arrays

Fig. 11.11 Percutaneous liver biopsy performed using a curved array in an adult (A) compared with liver biopsy in a pediatric patient using a linear array (B and C). Note the superior visualization of the biopsy needle. The biopsy tray can clearly be seen in *B* and *C*, while not visualized in *A*.

1. **B.** An ultrasound transducer most likely has of the order of 100 or so transducer elements. Importantly, 10 is (far) too few and would be inadequate to create an image. A total of 1000 is (far) too many because this would imply that the level of detail in ultrasound images would be superior to a CT image, which generally has "500 lines."

2. **A.** A linear array is characterized by a rectangular field of view and instantly recognized as such by any clinical image. Because the field of view is rectangular, superficial objects will be relatively easy to detect compared with alternative transducer types (i.e., convex, phased, annular).

3. **B.** With a convex array, the lines of sight diverge with increasing depth. As a result, the number of lines per unit distance is reduced, which must adversely impact the lateral resolution. Axial resolution is solely determined by the operating frequency, and the elevational resolution is often fixed or less affected.

4. **C.** Spatial compounding will not change the lesion contrast, but the addition of multiple images of the same object will reduce the amount of noise. As a result, the contrast-to-noise ratio improves, and lesions will therefore be easier to see.

Comment

Transducer arrays are multiple-element transducers, containing many small transducer elements. Vertical heights of each element are generally several millimeters, with the width less than a wavelength. An array can electronically scan the ultrasound beam without requiring any mechanical movement. To accomplish this, the transducer will activate a small group of elements to generate a line of sight and receive the incoming signal. It will then move on to the next group (shifting the beam in the process). Multielement arrays produce grating lobes that are similar to side lobes (i.e., ultrasound energy at oblique angles from the main beam). Grating lobes also give rise to the same types of artifacts as do side lobes (i.e., mispositioning of objects into the main image).

A linear sequenced array is constructed as a linear array of elements that are operated sequentially. A complete image frame is obtained by firing groups of elements from one end to the other end of the linear array. Linear arrays, which typically have 128 elements, will generate rectangular looking fields of view (Fig. 11.12). Linear arrays give superior superficial imaging without clutter or image distortion. Unfortunately, this advantage is at the cost of the field of view, which is narrow in comparison with other transducer geometries. Linear transducers are thus best for scanning relatively superficial anatomy such as the breast or extremities.

Convex arrays, also known as curvilinear, operate in a similar manner to linear array, with scan lines being produced perpendicular to the transducer face. On this type of transducer, beam steering is not needed because an image with an arc is naturally created by the transducer's geometry. Convex arrays produce a trapezoidal (sector) field of view, which is wider compared with a linear-array configuration and superior for imaging the abdomen (see Fig. 11.12). Convex sector systems can be designed to have a lateral resolution of less than 1 mm, but because of beam separation there will always be a decrease in lateral resolution. Unfortunately, the advantages of convex arrays come at the cost of poor imaging of superficial structures, which are often distorted by the sector image display.

A benefit of multielemental arrays is the possibility of steering the beam. This can be used to perform spatial compounding, sometimes called multibeam imaging, which uses multiple lines of sight to form one composite image. In spatial compound imaging, beams are steered from different angles, generally within 20 degrees from the perpendicular. Echoes from these directions are averaged together, referred to as being compounded, into one overall composite image. The contrast of the lesion will not change, but averaging multiple images with random noise will result in lower noise levels. Spatial compound imaging thereby provides an improved contrast-to-noise ratio, as well as superior margin definition. Compound imaging is used to image the breast, musculoskeletal injuries, and peripheral blood vessels.

References

Huda W. Ultrasound I. In: *Review of Radiologic Physics*. 4th ed. Philadelphia: Wolters Kluwer; 2016:244–246.

Kremkau FW. Transducers. In: *Sonography: Principles and Instruments*. 9th ed. St. Louis: Elsevier; 2016:47–59.

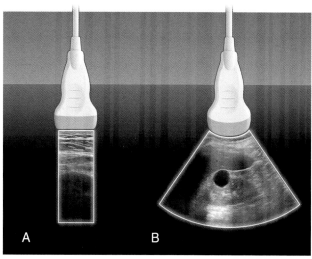

Fig. 11.12 The linear array (A) produces a rectangular field of view that is useful for visualizing superficial lesions in pediatric imaging. A convex array (B) produces a trapezoidal field of view that is important for imaging adult abdomens.

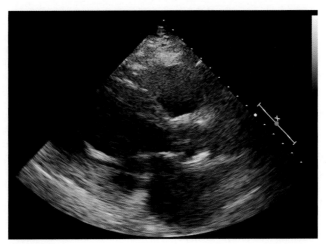

Fig. 11.13

1. Which type of transducer is most likely to produce a symmetrical ultrasound beam?
   A. Linear
   B. Convex
   C. Annular
   D. Phased

2. Which type of array would likely fire all the elements to generate a single line of sight?
   A. Linear
   B. Convex
   C. Phased
   D. A and B

3. What is a major benefit of a phased array transducer?
   A. Small footprint
   B. Electronic steering
   C. Electronic focusing
   D. A, B, and C

4. Which type of array can benefit from "electronic focusing" using variable time delays to activate individual elements?
   A. Linear
   B. Annular
   C. Phased
   D. A, B, and C

CASE 11.7

Annular and Phased Arrays

Fig. 11.13 Phased array echocardiogram in a patient with congestive heart failure.

1. **C.** An annular array will most likely be symmetrical because of the radial symmetry of this array. This type of symmetry is generally not expected for ultrasound beams generated by linear, convex, or phased arrays.

2. **C.** Phased arrays steer ultrasound beam direction (i.e., line of sight) by firing all their elements but with varying "time delays." Small adjustments are made to these "time delays" to steer and/or focus the resultant beams.

3. **D.** One important characteristic of a phased array is a small footprint, which can be used to introduce the beam into a narrow aperture (e.g., between ribs). Phased arrays use electronic steering to send the ultrasound beam in different directions, as well as (variable) electronic focusing to improve lateral resolution.

4. **D.** Multiple elements can be used to focus the beam to any desired depth, simply using electronically adjustable "time delays." In addition, such multiple elements can be triggered with varying "time delays" to steer the beam in any desired direction to achieve spatial compounding.

Comment

Annular arrays are created with rings that get progressively smaller, analogous to the cross section of an onion. Annular arrays are focused by applying an electrical pulse to each element in turn. Individual wave fronts from each ring will combine to create a composite pulse, focused to a specific depth. The focus depth depends on the time delay between the electrical pulses. Modification of the delay times between the pulses to each element enables the ultrasound beam to be focused to different depths. Annular array transducers have ultrasound pulses that are focused in two dimensions (vis-à-vis one in linear and curvilinear arrays). The symmetrical ultrasound beam produced by annular array can generally produce a thinner scan slice than other types of array.

Phased arrays have many small ultrasonic transducer elements, each of which can be fired in pulse mode independently. A phased array would typically have 96 elements, but these are fired collectively to generate a single beam at a particular angle that appears to emanate from the center of the transducer (line of sight). Varying the timing delays to individual transducer elements will results in a beam direction that can be changed (Fig. 11.14). In this manner the beam direction is steered electronically simply by adjusting the timing delays. Although the beam directions are continuously being changed, this is achieved electronically with no mechanical movement within the transducer.

A phased array sweeps out the ultrasound beam in a manner analogous to a searchlight, and the images are obtained from the detected echoes along each line of sight. Data from multiple beams can be assembled using sophisticated algorithms to generate image slices through the patient. Ultrasound images obtained with phased arrays appear to originate from a single point. However, on many systems, data from the most superficial region are "truncated" because of the limited use of these data. Phased-array transducers have a small footprint, so they can be used for cardiac imaging because they produce a beam that can image between the ribs.

In a phased array, a focused beam can be produced from the interference of the wavelets produced by each element. The time delays to successive elements will determine the depth of focus for the transmitted beam. These delay factors are applied to the elements during the receiving phase, thereby generating a dynamic focusing for the returning echoes. Arrays can electronically focus the transmitted beam at multiple depths, which can be "cherry picked" for focused regions of each lines along a given direction. The composite image will thereby have an increased effective focal zone. This type of electronic focusing can also be used by any array of elements (e.g., linear arrays and curvilinear arrays).

References

Huda W. Ultrasound I. In: *Review of Radiologic Physics.* 4th ed. Philadelphia: Wolters Kluwer; 2016:244–246.

Szabo TL, Lewin PA. Ultrasound transducer selection in clinical imaging practice. *J Ultrasound Med.* 2013;32(4):573–582.

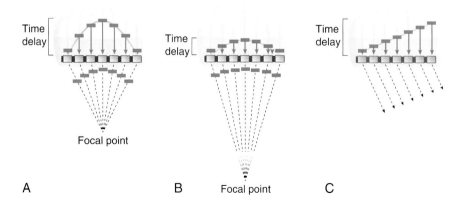

Fig. 11.14 Varying the time delays to multiple elements can be used to produce a short focal point (A) and in a subsequent pulse a longer focal point (B). Changing the time delays to multiple elements can also be used to steer the beam in any given direction (C).

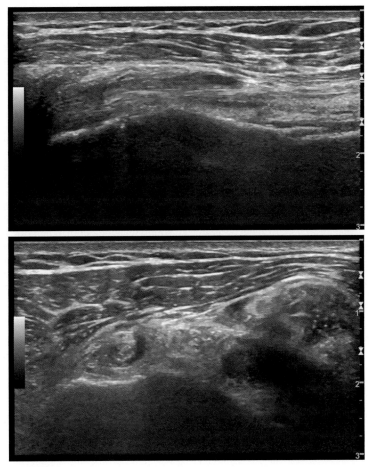

Fig. 11.15

1. What percentage of the 250-µs listening interval is most likely occupied by the ultrasound pulse (%)?
   A. <1
   B. 2
   C. 5
   D. >10

2. When the listening interval is 250 µs, what is the pulse repetition frequency (PRF) (kHz)?
   A. <4
   B. 4
   C. >4

3. What happens to the listening interval when PRF increases?
   A. Reduced
   B. Unchanged
   C. Increased

4. What is the nominal imaging depth for a 4-kHz PRF (250-µs listening interval) (cm)?
   A. 5
   B. 10
   C. 20
   D. 40

CASE 11.8

Lines of Sight

Fig. 11.15 Tenosynovitis of the biceps tendon.

1. **A.** Ultrasound pulses are extremely short and typically 1 μs or so. Consequently, less than 1% of the listening interval would be occupied by the pulse, with most of the time (>99%) taken up by the echo detection task.

2. **B.** If the listening interval is 250 μs, a pulse is sent out every 0.00025 s so that 4000 pulses will be emitted every second. Each pulse will generate a single line of sight, so 4000 image lines are created each second.

3. **A.** When the PRF increases, the listening interval will be reduced. What this means is that more lines of sight are created every second, but these will pertain to a reduced imaging depth.

4. **C.** An interface at a depth of 1 cm requires a round trip of 2 cm. Because ultrasound travels at a speed of 1540 m/s in tissue, the time for this round trip is 13 μs (i.e., distance of 0.02 m divided by velocity of 1540 m/s). When one listens for 250 μs, the depth is nearly 20 cm, ideal for imaging an adult abdomen.

Comment

An ultrasound sound transducer generates a very short pulse, typically two wavelengths long, which is sent out along a selected direction. This pulse is then followed by a much longer waiting interval, which is 250 μs for a 4-kHz PRF. Technically, the listening interval is known as the pulse repetition period (1/PRF), and during this time the transducer will be detecting echoes. Each pulse will be able to generate one line of data (Fig. 11.16). Images are created by generating many lines of sight that are sequentially directed to cover the region of interest.

Information about the location of reflectors and sources of scatter within the patient is obtained from the time at which echoes are detected. Consider an interface at a depth of just 1 cm so that for an echo to be detected requires a round trip of 2 cm. Ultrasound travels at a speed of 1540 m/s in tissue, so that the time for this round trip is 13 μs, which is the distance traveled (i.e., 0.02 m) divided by the ultrasound velocity in soft tissue (i.e., 1540 m/s). Long listening intervals permit the detection of "late" echoes, thereby increasing the effective ultrasound imaging depth. When one listens for 250 μs, the depth is nearly 20 cm, and halving the listening time (i.e., doubling the PRF) will also halve the imaging depth.

The PRF is the number of separate pulses, corresponding to the separate lines of sight that are sent in 1 s. The most common PRF value is 4 kHz, which corresponds to 4000 pulses per second. An ultrasound transducer that is operated at a PRF of 4 kHz will thus be able to produce 4000 lines of sight. In this example, if 100 lines of sight are used to create a complete ultrasound frame (image), there will be 40 such images created every second. If multiple focal zones are used, this requires increasing the number of pulses (two focal zones require two pulses) and thus will require more time (slower frame rate) or a doubled PRF (at the cost of a shallower maximum imaging depth).

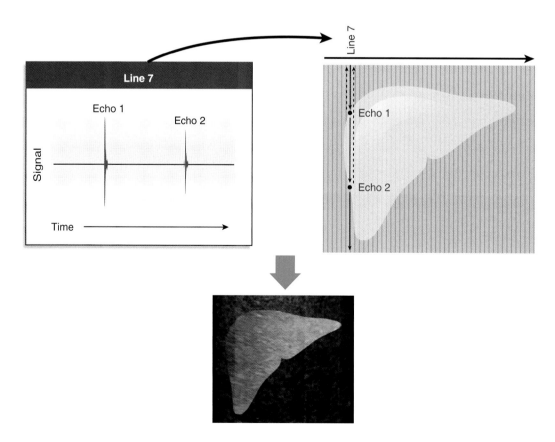

Fig. 11.16 The upper left image shows the seventh listening interval that is used to create the seventh line of an ultrasound image. This listening interval contains two echoes from the left edge of the liver. The resultant ultrasound image will likely have 100 or so such lines, and increasing the number of lines per image would likely improve lateral resolution performance.

Line density (LD) is the lines in an image divided by the corresponding (angular) field of view. When the LD increases, either by increasing the number of lines or by reducing the field of view, it will generally improve lateral resolution. Another way to increase LD is reducing the frame rate, but this will also reduce the temporal resolution. LD can also be increased by increasing the PRF, but because this requires the listening time to be reduced, the imaging depth will decrease.

Reducing the field of view is good for imaging small objects that require high temporal resolution (e.g., fetal heart).

References

Huda W. Ultrasound II. In: *Review of Radiologic Physics*. 4th ed. Philadelphia: Wolters Kluwer; 2016:251–252.

Kremkau FW. Transducers. In: *Sonography: Principles and Instruments*. 9th ed. St. Louis: Elsevier; 2016:59–66.

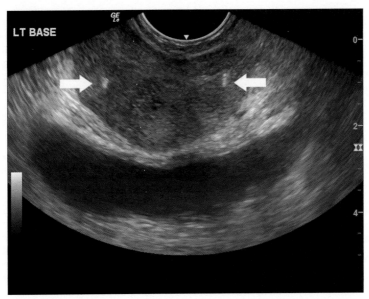

LT BASE

Fig. 11.17

1. What is the name given to correction of tissue attenuation in an ultrasound system?
   A. Depth gain compensation (DGC)
   B. TGC
   C. Swept gain
   D. A, B, and C

2. To maintain image brightness when imaging a fatty liver, how should the amount of DGC adjustment be modified?
   A. Increased
   B. Unchanged (increase power)
   C. Reduced

3. Which can be increased to increase the imaging depth of an ultrasound system?
   A. Image display gain
   B. PRF
   C. Transducer operating power
   D. A or B

4. What is the most likely frame rate in real time ultrasound imaging (frames per second)?
   A. <30
   B. 30
   C. >30

CASE 11.9

Depth Compensation and Frame Rates

Fig. 11.17 Transrectal ultrasound of the prostate showing two gold fiducial markers *(arrows)* within the base of the prostate. Note the anechoic bladder along the lower half of the image.

1. **D.** The common name for attenuation correction with depth in ultrasound is depth gain compensation (DGC) or time gain compensation (TGC). Other names include swept gain and time variable gain.

2. **A.** A fatty liver attenuates more than other soft tissue so that the amount of TGC that needs to be applied must be increased to generate an image with satisfactory intensities. In this way the amount of TGC that has been applied can provide valuable clues about any abnormal pathology.

3. **C.** Transmit power must be raised to increase the imaging depth. Increases in the display gain will result in a brighter-looking image, but the intrinsic image properties such as imaging depth are not affected by simply changing the appearance of an image. An increase in PRF reduces the listening interval, which always reduces imaging depth.

4. **B.** In real-time ultrasound imaging, images are typically created and displayed at a rate of 30 frames per second. At this frame rate, any motion in the body can be followed in real time with no evidence of "flicker."

Comment

An echo from any reflector at the surface travels no distance and thus undergoes no attenuation. Echoes from the same reflector located at a depth of 1 cm will be weaker because the ultrasound echo has traveled 2 cm (round trip) and will have been attenuated by this tissue thickness of 2 cm. Echoes from a depth of 10 cm will be much weaker because of attenuation in traveling 20 cm (round trip). Uncorrected echo data show distant echoes as being weaker than superficial echoes.

Ultrasound scanners can compensate for the increasing attenuation that occurs with depth. Compensation is achieved by increasing the electronic signal amplification (gain) with increasing echo return time. Correcting for echo attenuation in this manner is called depth gain compensation (DGC). DGC is also referred to as time gain compensation (TGC), time-varied gain (TVG), and swept gain. DGC is implemented so equal reflectors have the same brightness in the ultrasound images regardless of their depth in tissue. DGC controls may be adjusted by the operator, and the pattern of control settings that results in a satisfactory image may provide potential diagnostic information.

If increasing amplification by the DGC controls is required, this indicates additional attenuation in an organ such as a fatty liver. This will result in the liver having a coarsely hyperechoic echotexture. By comparing echogenicity of an organ to an "internal control" (e.g., comparing the liver and kidney), diagnostic information can be obtained. For example, if the renal cortex appears hyperechoic compared with the liver parenchyma, medical renal disease is suggested. DGC may be applied to the entire field of view or be applied at specified depths. There is a column of sliders on the ultrasound machine corresponding to different depths that can be altered to optimize visualization of specific regions or to even out image brightness.

Image frame rates for real-time ultrasound are 30 frames per second. This frame rate is sufficient to permit ultrasound imaging so that motion is followed in "real time." The frame rate in ultrasound is thus like that encountered in fluoroscopy, and both of these imaging modalities have a similar temporal resolution. Accordingly, the temporal resolution of real-time ultrasound is 1/30th second, the time it takes to create on ultrasound frame (image). Multiplication of the frame rate and the lines per image always will produce the PRF. When the frame rate changes, the operator may make adjustments to factors such as the line density, the field of view, and the imaging depth. Ultrasound thus offers "trade-offs" between these parameters, as depicted in Fig. 11.18.

References

Hangiandreou NJ. State-of-the-art ultrasound imaging technology. *J Am Coll Radiol.* 2004;1(9):691–693.

Huda W. Ultrasound II. In: *Review of Radiologic Physics.* 4th ed. Philadelphia: Wolters Kluwer; 2016:252–253.

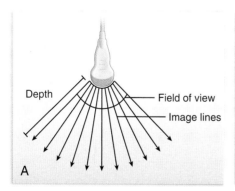

Fig. 11.18 (A) The number of lines in an image divided by the angular field of view defines the line density. One way of increasing the line density is to reduce the frame rate, which will reduce the temporal resolution performance. (B). Another way to increase the line density is by increasing the pulse repetition frequency *(PRF)* (C), but this will also reduce the imaging depth because of the reduction in the listening interval.

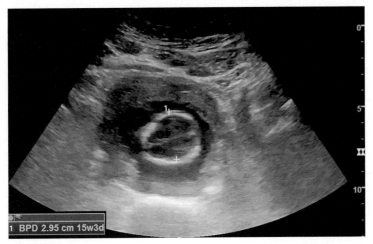

Fig. 11.19

1. Which display mode will provide the most accurate measurement of distance between two interfaces?
   A. A
   B. B
   C. M
   D. All equivalent

2. Which of the following encodes reflection strength as image brightness?
   A. A-mode
   B. Cardiac M-mode
   C. Two-dimensional (2D) gray scale
   D. B and C

3. Which display mode will likely be used to monitor the performance of the tricuspid valve?
   A. A
   B. B
   C. M
   D. All equivalent

4. Which imaging modality has an image matrix size that is most similar to that of ultrasound?
   A. Positron emission tomography (PET)
   B. MR
   C. CT
   D. Photospot

CASE 11.10

Image Display

Fig. 11.19 Biparietal diameter (BPD) measurement.

1. **A.** A-mode plots the echo intensity as a function of time (i.e., distance). A-mode is generally considered the modality that offers the highest accuracy for distances, especially when these are "small."

2. **D.** Both cardiac M-mode (motion) and 2D gray-scale ultrasound imaging would use B-mode, where the intensity of an echo is depicted as a pixel intensity. In A-mode, the amplitude of the echo is explicitly shown on the vertical axis where the horizontal axis is time (distance).

3. **C.** M-mode (time-motion mode) plots distance (echo depth) as a function of time. In M-mode, a stationary interface would produce a horizontal line. A valve that moved toward (or away from) the transducer would show a variable pattern and provide valuable data on valve function.

4. **C.** Ultrasound images use a 512 × 512 matrix, the same as is currently used for clinical CT. The matrix sizes of PET and MR are both less than 512 × 512, and that of photospot imaging is 1024 × 1024.

Comment

Transducer data are sent to a scan converter for processing, which enables format of image acquisition to be converted into that required for image display. Scan converters generate images from echo data acquired from distinct beam directions and which are then displayed on a monitor. To obtain a geometrically correct image, the acquired image data need to be interpolated into the display grid. Accordingly, scan conversion is essentially an interpolation process. To perform the scan conversion, the geometry of the transducer grid and the screen grid are required together with their location with respect to each other.

The magnitude of returning echoes along a line provides information about acoustic impedance differences between tissues along this line. In an A-mode (amplitude) display, depth is along the horizontal axis and echo intensity is displayed directly on the vertical axis. A-mode imaging is primarily used in ophthalmology. In M mode (or time-motion) display, time is shown on the horizontal axis and the corresponding depth is shown along on the vertical axis. The brightness of each pixel is proportional to the strength of the reflected ultrasound wave. M mode displays time-dependent motion, such as cardiac valve motion. Time resolution is especially important in M mode imaging because quantitative measurements (e.g., the time between valve openings) can provide the physician with important diagnostic information.

In B-mode, meaning brightness, the echo intensity is shown as a brightness value along each line of sight (Fig. 11.20). This will create a 2D cross section of the underlying structures consisting of numerous B-mode (brightness mode) lines. Using this gray-scale depiction, highly reflective surfaces, such as bone, would be white (hyperechoic), whereas low-reflectivity surfaces such as muscle would be gray (isoechoic). Where there is no reflection, such as the fluids in the urinary bladder, the appearance would be totally black (hypoechoic or anechoic). In this display mode, all the deeper structures would be displayed on the lower part of the screen, with superficial structures appearing on the upper part. Two-dimensional gray-scale imaging, as well as T-M displays, encodes reflection strength as image brightness.

Images in ultrasound are displayed in a 512 × 512 matrix, with 1 byte [8 bit] being coded for each pixel. Because ultrasound uses only 1 byte (8 bits) to code for each pixel, it permits the display of only 256 shades of gray (see Fig. 11.20). This is less than in CT, which uses 12 bits per pixel and can display 4096 different Hounsfield unit values. Each ultrasound image (frame) contains 0.25 MB of information (i.e., 512 × 512 × 1 Bytes). The data content of ultrasound images is thus intermediate between those of MR (e.g., 0.125 MB) and CT (0.5 MB). For real-time imaging, with 30 frames being generated every second, the rate at which data are being acquired would be 7.5 MB/s.

References

Huda W. Ultrasound II. In: *Review of Radiologic Physics.* 4th ed. Philadelphia: Wolters Kluwer; 2016:252–253.

Kremkau FW. Instruments. In: *Sonography: Principles and Instruments.* 9th ed. St. Louis: Elsevier; 2016:104–106.

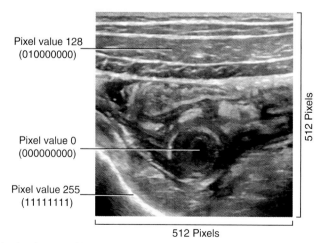

Fig. 11.20 A typical image matrix size in ultrasound imaging is 512 × 512, which corresponds to a total of a quarter of a million pixels. Each pixel is coded using 1 byte (8 bits), which can depict 256 shades of gray ranging from 0 (black) to 255 (white).

CASE 12.1

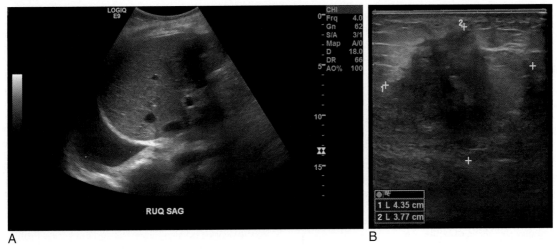

A

B

Fig. 12.1

1. Which type of array would most likely be used to image a
   1-month-old infant?
   A. Linear
   B. Curvilinear
   C. Annular
   D. Phased

2. Which type of array would most likely be used to image an
   adult abdomen?
   A. Linear
   B. Curvilinear
   C. Annular
   D. Phased

3. Which type of array would most likely be used to image the
   circle of Willis transcranially?
   A. Linear
   B. Curvilinear
   C. Annular
   D. Phased

4. What is the most likely frequency of a breast ultrasound
   transducer (MHz)?
   A. Low (2 MHz)
   B. Moderate (5 MHz)
   C. High (10 MHz)
   D. Very high (20 MHz)

CASE 12.1

Clinical Transducers

Fig. 12.1 Right pleural effusion seen during liver ultrasound using a curvilinear transducer (A) and BIRADS 5 breast mass seen using a linear transducer (B).

1. **A.** A linear array is characterized by a rectangular field of view (FOV) and instantly recognized as such by any clinical image. Because the FOV is rectangular, superficial objects will be relatively easy to detect compared with alternative transducer types (i.e., convex, phased, annular).

2. **B.** Convex (curvilinear) arrays, as opposed to linear arrays, would be used to image an adult abdomen because of the much larger FOV that these types of multielemental transducers offer. The FOV increases with increasing imaging depth, which is exactly what is required in adult abdominal ultrasound imaging.

3. **D.** There are only limited "windows" to image through the skull (orbits and temporal lobes), so an ultrasound imaging system with a small footprint is essential. A phased array has a very small footprint and is the instrument of choice when imaging the circle of Willis.

4. **C.** Breast ultrasound makes use of high-frequency transducers, generally between 10 and 15 MHz. The average thickness of a compressed breast is only 6 cm, so penetration is adequate, and the high frequency offers excellent axial spatial resolution performance (~0.1 mm).

Comment

Linear arrays (Fig. 12.2) are used when superficial structures are being evaluated and greater depth penetration is not needed, such as in pediatric imaging. Pediatric probes are likely to have a smaller footprint compared with those used for adults and operate at higher-frequency varieties, generally in excess of 7 MHz. Linear arrays are used to image superficial structures such as the carotid arteries, leg veins, thyroid, subcutaneous lymph nodes, and testicles. Linear arrays can also be used to screen small children for appendicitis and intussusception without use of radiation. In these applications, the limited FOV determined by the physical size of the linear array is clinically acceptable.

Convex arrays, also referred to as curvilinear arrays, are operated at moderate frequencies typically around 4 MHz. They are used to image the abdomen due to improved visualization of deeper structures in comparison with linear arrays. Convex arrays are commonly used to image the liver, gallbladder, spleen, kidneys, urinary bladder, aorta, inferior vena cava, and pancreas. Like linear arrays, convex arrays are also used during ultrasound-guided procedures where the probe type is selected based on the anatomic target.

Transcranial imaging of the brain and vasculature uses acoustic windows through the skull such as the temples or eyes. Transcranial phased array probes use low frequencies of approximately 2 MHz to image blood vessels. Because breasts are relatively thin, approximately 60 mm on average, high resolution is achieved by using high-frequency transducers. A 10-MHz breast transducer will generally have sufficient penetration in an average breast. Very-high-frequency transducers greater than 20 MHz do not adequately image deeper tissues in the breast and are used only for niche applications where imaging extremely superficial areas is required.

Specialized endo-arrays (see Fig. 12.2) are located at the end of the probe and also known as end-fire arrays. Probes with wide FOVs (90 to 150 degrees) for imaging the pelvic region may fit through small openings (transrectal). Miniature intracavitary (or intravascular) ultrasound probes have high-frequency transducers located on the tip of a catheter that permit intravascular imaging with excellent resolution and can image blood vessels in great detail. Real-time ultrasound data can be saved as a cine clip, be reconstructed into a two-dimensional (2D) map, or be used for procedural guidance while stenting or ballooning complex stenoses.

References

Huda W. Ultrasound II. In: *Review of Radiological Physics*. 4th ed. Philadelphia: Wolters Kluwer; 2016:253–254.

Szabo TL. Ultrasound transducer selection in clinical imaging practice. *J Ultrasound Med*. 2013;32:573–582.

Linear Probe

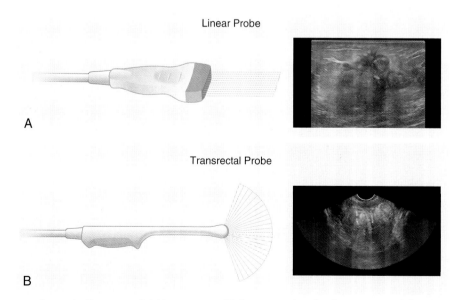

A

Transrectal Probe

B

Fig. 12.2 (A) A Linear array demonstrating a superficial breast mass. (B) A transrectal probe demonstrating fiducial markers in the base of the prostate. Specialized probes are useful for evaluating specific anatomic regions.

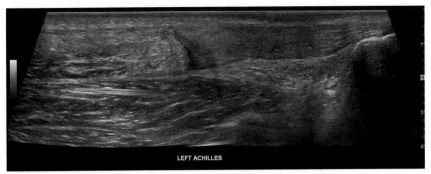

LEFT ACHILLES

Fig. 12. 3

1. Which resolution is improved the most when a linear array is replaced with a 1.5-dimensional (1.5D) array?
   A. Axial
   B. Lateral
   C. Elevational
   D. Temporal

2. Which body region is most likely to benefit from the use of extended FOV imaging?
   A. Musculoskeletal
   B. Pediatric
   C. Cardiac
   D. Abdomen

3. If the fundamental frequency is 3 MHz, which frequency would most likely be used to perform harmonic imaging (MHz)?
   A. 4.5
   B. 6
   C. 7.5
   D. 9

4. What kind of gases can be used in bubbles in contrast imaging in ultrasound?
   A. Air
   B. Nitrogen
   C. Perfluorocabons
   D. A, B, or C

CASE 12.2

Advanced Techniques

Fig. 12.3 Full-thickness Achilles tendon tear seen here on extended field of view imaging.

1. **C.** A 1.5D array would improve the elevational resolution (AKA slice thickness) by sending out several pulses where the elevation focusing occurs at different depths. A composite image can be stitched together by "cherry picking" the best parts of each line of sight along every given direction.

2. **A.** Extended FOV imaging is mainly used in musculoskeletal imaging, where the type of anatomy (limbs) makes this a powerful tool.

3. **B.** For a fundamental frequency of 3 MHz, harmonic imaging would be performed at 6 MHz. Although 9 MHz is also a harmonic frequency of 3 MHz, there would be too much attenuation at this high frequency to make use of it in ultrasound imaging.

4. **D.** Air, nitrogen, and perfluorocabons have all been used as encapsulated microbubbles in contrast ultrasound imaging.

Comment

A 1.5D array has a large number of elements in the scan plane (>100) and a much smaller number (~6) in the slice thickness (elevational) direction. Focusing in the elevational direction can be used to reduce the effective slice thickness and improve elevational resolution (Fig. 12.4). By producing several lines, each focused at a different depth, the thinnest part of each line be used to produce a composite image with improved overall elevational resolution. A 1.5D array is likely to have comparable lateral and elevational resolution. Volume imaging uses 2D arrays containing thousands of transducer elements (e.g., 64 × 64). With 2D arrays, sound waves are sent out at different angles, and the echoes are processed by a computer program to reconstruct three-dimensional volume image.

Static B-mode techniques can be used to generate extended FOV, which will enable a large subject area to be obtained in a single image. Extended FOV images are obtained when a probe is moved over the area of interest. Images are "stitched" together electronically to generate in a single slice image covering the area of interest (e.g., full-length view of the Achilles tendon). Extended FOV imaging is beneficial when large areas are being visualized and there is limited motion, such as in musculoskeletal. Abnormalities including fluid collections, muscle injuries, and tumors can be seen relative to landmarks such as joints or tendons.

Ultrasound harmonic imaging uses broadband transducers and uses a frequency that is twice the fundamental frequency of the emitted pulse. High frequencies are produced from nonlinear ultrasound interactions with tissues, and harmonic imaging selectively tunes the receiver to a higher harmonic frequency. The second harmonic (twice the fundamental frequency) is most frequently used because third harmonics and above have too much attenuation. Cardiac systems transmit at frequencies at approximately 2 MHz and receive signals at frequencies twice that range. Harmonic imaging eliminates fundamental frequency clutter (noise). Patients with thick and complicated body wall structures benefit from harmonic imaging.

Contrast agents for vascular and perfusion imaging are encapsulated microbubbles containing air or insoluble gases such as perfluorocarbons. Microbubbles are generally smaller than red blood cells, which facilitates tissue perfusion. Intense ultrasound echoes are generated from the large difference in acoustic impedance between gas and surrounding fluids and perfused tissues. Microbubbles produce harmonic frequencies, which permits use of pulse inversion harmonic imaging. In this mode, a standard one pulse and an additional inverted pulse with a reversed phase both travel along the same beam direction. Addition of the received echoes will produce zero signals for soft tissues but not for microbubbles. Pulse inversion harmonic imaging improves the sensitivity of ultrasound to contrast agents and reduces imaging artifacts due to reverberations.

References

Anvari A, Forsberg F, Samir AE. Primer on the physical principles of tissue harmonic imaging. *Radiographics*. 2015;35(7):1955–1964.

Huda W. Ultrasound II. In: *Review of Radiological Physics*. 4th ed. Philadelphia: Wolters Kluwer; 2016:254–255.

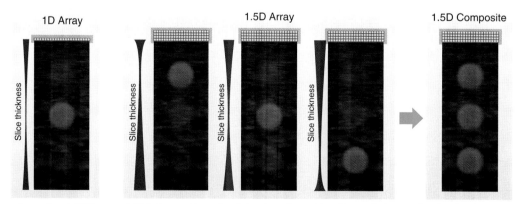

Fig. 12.4 A one-dimensional *(1D)* array *(left)* is focused to a fixed depth and can identify one lesion at the depth where the ultrasound slice thickness is minimized by the use of an acoustic lens (or shaped transducer). A 1.5-dimensional *(1.5D)* array creates three images where the slice thickness is minimized at three separate depths (second, third, and fourth images). A composite image *(right)* can be created using the best bits of the three acquired images to show all three lesions at three different depths.

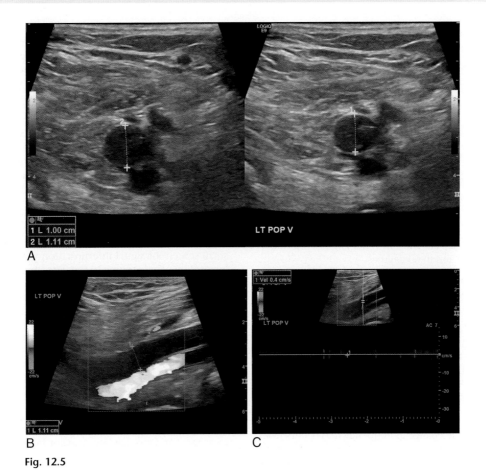

Fig. 12.5

1. Which factors affect the Doppler frequency shift?
   A. Operating frequency
   B. Doppler angle
   C. Reflector velocity
   D. A, B, and C

2. What is the ideal Doppler angle (degrees)?
   A. Low (0 to 30)
   B. Intermediate (30 to 60)
   C. High (60 to 90)

3. What is the most likely maximum frequency shift from high blood flow (100 cm/s) in clinical Doppler ultrasound (Hz)?
   A. 10
   B. 100
   C. 1000
   D. 10,000

4. How many transducers are normally used in continuous Doppler?
   A. 1
   B. 2
   C. 3
   D. >3

CASE 12.3

Doppler Physics

Fig. 12.5 Left popliteal vein deep venous thrombosis showing no compressibility (A), no color flow (B), and no spectral Doppler signal (C).

1. **D.** The ultrasound operating frequency and the reflector velocity both affect the magnitude of the frequency shift in Doppler ultrasound. By far the most important parameter is the Doppler angle (at a Doppler angle of 90 degrees there is no frequency shift, whereas at 0 degrees there is the maximum frequency shift).

2. **B.** At low angles, there are problems of refraction and reflection. At high angles, the frequency shift is very sensitive to the Doppler angle. The ideal Doppler angle is intermediate, or between 30 and to 60 degrees.

3. **C.** In Doppler ultrasound, most of the frequency shifts will likely be in the range of 300 to 3 kHz, so 1 kHz is the best available answer.

4. **B.** Continuous Doppler uses two transducers that are on all the time. One transducer sends out ultrasound while the second transducer detects echoes. The Doppler signal is obtained by comparing the frequencies of the emitted and detected sound, with the frequency shift proportional to the blood velocity.

Comment

The Doppler effect is the change in frequency that occurs from any moving "sound source," such as reflections from a flowing blood cell. Objects that move toward the ultrasound transducer reflect soundwaves at a higher frequency, and those that move away from the transducer will reflect sound waves at a lower frequency. The angle between the ultrasound beam and the moving object is known as the Doppler angle(θ). Doppler frequency shifts are directly proportional to $\cos(\theta)$. The maximum frequency shift will occur when the reflector is moving directly toward the transducer (dθ = 0 degree) or moving directly away (θ = 180 degree). There is no Doppler shift at a 90-degree Doppler angle, no matter how fast the reflectors are moving. It is important to always remember that Doppler measures only the shift in frequency, not the reflector velocity per se.

At any Doppler angle, frequency shifts are directly proportional to reflector velocity, so that a doubling in the velocity would also double the corresponding measured Doppler frequency shift. Doppler frequency shifts are also proportional to the ultrasound operating frequency and are typically in the audio range (hundreds of Hz). Increasing the operating ultrasound frequency will also increase the frequency shift. Doppler can evaluate blood flow in vessels based on the backscatter of ultrasound pulses from blood cells (Fig. 12.6). In addition to Doppler frequency shifts, pulsed Doppler always explicitly provides depth information.

Continuous-wave (CW) Doppler uses continuous transmission and reception of ultrasound waves to detect movement. This is achieved by using two dedicated transducers where one continuously sends out ultrasound waves at the transmit frequency while another receives ultrasound reflections at a different frequency. The magnitude of the flow is obtained by the difference in the emitted and received frequencies, which is generally in the audible range. CW Doppler probes are usually compact and handheld. There is a speaker and/or a small screen for signal interpretation with the option to wear headphones for listening to the detected frequency shifts. This device is useful when a patient's pulse is not palpable and it is necessary to confirm the presence of blood flow. This is very useful when assessing a patient with critical limb ischemia. The heart rate can also be obtained, which is useful when performing a bedside examination of a pregnant patient to detect and measure the fetal heart rate. The frequency shifts through the cardiac cycle will generate a characteristic "whoosh" sound.

Because pulses are not being emitted, CW Doppler does not permit identification of the specific region where the wave is reflected. The frequency shift originates from reflectors somewhere along the path of the ultrasound beam, which is referred to as the CW Doppler line. A major benefit of CW Doppler is its ability to measure very high velocities, typically more than 1 m/s. This high accuracy occurs because the beam is always transmitting, and thus undersampling can never lead to aliasing artifact. High blood flows are often seen in pathologies of the heart such as an aortic stenosis. Because CW Doppler cannot identify the exact position of flows, it is often used in tandem with pulsed wave Doppler–CW to determine max velocity and pulsed wave to pinpoint location.

References

Huda W. Ultrasound II. In: *Review of Radiological Physics*. 4th ed. Philadelphia: Wolters Kluwer; 2016:256–257.

Kremkau FW. Doppler principles. In: *Sonography: Principles and Instruments*. 9th ed. St. Louis: Elsevier; 2016:133–183.

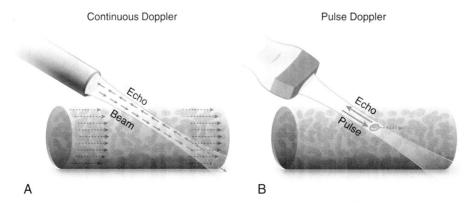

Continuous Doppler   Pulse Doppler

A   B

Fig. 12.6 In pulsed Doppler (B), the location of the source producing the echo is known exactly from the time it takes for the echo to arrive at the transducer. In continuous Doppler (A), the frequencies of the emitted and reflected sound are obtained using two transducers so that the origin of the echoes is only known in a general sense.

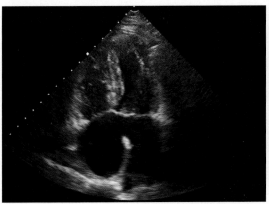 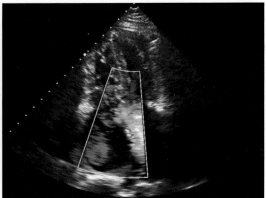

Fig. 12.7

1. What is the best descriptor of the pattern of blood flow in a normal blood vessel?
   A. Uniform
   B. Linear
   C. Laminar
   D. Turbulent

2. What type of image is used to identify a region of interest where flow is to be assessed using Doppler analysis?
   A. A mode
   B. B mode
   C. M mode
   D. A, B, or C

3. How do the pulse repetition frequencies (PRFs) in Doppler ultrasound compare with PRFs used in conventional B-mode imaging?
   A. Higher
   B. Comparable
   C. Lower

4. Which clinical conditions can be identified by the flow pattern exhibited using spectral Doppler?
   A. Portal hypertension
   B. Right-sided heart failure
   C. Tricuspid regurgitation
   D. A, B, and C

CASE 12.4

Flow and Spectral Doppler

Fig. 12.7 Primum-type atrial septal defect demonstrating color Doppler flow across the defect on echocardiography.

1. **C.** Most blood vessels exhibit laminar flow, with a parabolic flow profile that is highest at the vessel center and with reduced flow near the vessel walls.

2. **B.** A conventional B-mode image is generally obtained to identify the region of interest where the subsequent Doppler information is to be obtained.

3. **A.** PRFs used in Doppler ultrasound are generally higher than the PRFs used in conventional B-mode imaging. In B-mode imaging the most common PRF is 4 kHz, whereas in Doppler ultrasound this is likely to be increased to 8 kHz.

4. **D.** Portal hypertension, right-sided heart failure, and tricuspid regurgitation can all be identified from the shape and size of the spectral waveform.

Comment

Laminar flow is the term used to characterize the flow of blood throughout most of the human body. This term relates to concentric layers of blood moving in parallel down the length of blood vessel that are found in the human circulatory system. The center of the vessel has the highest blood velocity, which is reduced as one moves toward the blood vessel wall. The Bernoulli principle dictates that flow velocities will increase when a blood vessel narrows and will be reduced if the blood vessel gets larger. When the diameter of a vessel is halved, the cross-sectional area is reduced to a quarter, and the corresponding flow is therefore quadrupled. Turbulent flow will occur if the vessel shape is disrupted by objects such as plaque deposition or stenoses.

Doppler generally uses PRFs of approximately 8 kHz, which are higher than PRFs used in B-mode imaging. Only echoes from a region of interest contribute to the Doppler signal so that earlier and later echoes are not analyzed. Multiple pulses are directed along the same scan line to obtain many signals (samples) pertaining to each pixel. Duplex ultrasound combines real-time B-mode imaging with Doppler detection, where the B-mode images permit the selection of a region of interest and permit the Doppler angle to be estimated. B-mode images provide information on patient anatomy, and the corresponding Doppler shifts provide information on flow in a region of interest.

Spectral analysis will display frequency shift on the vertical axis as a function of time on the horizontal axis. The frequency shift information represents blood flow, providing that the Doppler angle is accounted for. At a given moment in time, the intensity of any given frequency shift is displayed as a brightness value. Positive velocities (i.e., positive frequency shifts) are placed above the horizontal axis, whereas those that are negative are placed direction below the horizontal axis. When flow is fairly uniform, the space between the spectral line and baseline is black. Turbulent flow containing a mixture of vectors will result in spectral broadening, which means that the spectral line is wider and the spectral window about the baseline is filled (Fig. 12.8).

Because flowing blood is pulsatile, the spectral characteristics vary with time. Right-sided heart failure, portal hypertension, and tricuspid regurgitation have characteristic spectral waveforms that permit these clinical conditions to be diagnosed with a high level of accuracy. A number of measures have been developed from the shape of the waveform detected in spectral analysis. These include resistance index (RI) (also called resistive index or Pourcelot index), systolic/diastolic (S/D) ratio (sometimes called the A/B ratio), and pulsatility index (PI).

References

Boote EJ. AAPM/RSNA physics tutorial for residents: topics in US: Doppler US techniques: concepts of blood flow detection and flow dynamics. *Radiographics.* 2003;23(5):1315–1327.

Huda W. Ultrasound II. In: *Review of Radiological Physics.* 4th ed. Philadelphia: Wolters Kluwer; 2016:256–258.

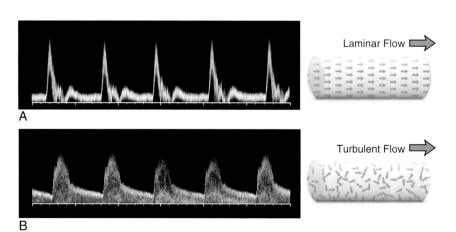

Fig. 12.8 Normal carotid waveform with laminar flow (A) demonstrating a clear area under the waveform contrasted with poststenotic turbulence causing spectral broadening (B).

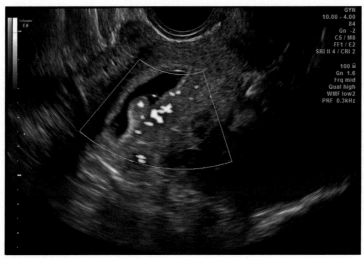

Fig. 12.9

1. What flow characteristic(s) does color Doppler normally display?
   A. Magnitude
   B. Direction
   C. Pressure
   D. A and B

2. In color Doppler, what does a light blue pixel most likely represent?
   A. Flow toward transducer
   B. Flow away from transducer
   C. Turbulent flow
   D. High flow

3. What is increased when acquiring data that is to be used to exhibit power Doppler (vs. color Doppler)?
   A. Intensity
   B. Frequency
   C. Sampling
   D. None (same input data)

4. Which characteristics are generally attributed to power Doppler images?
   A. No angular dependence
   B. No aliasing
   C. High sensitivity
   D. A, B, and C

CASE 12.5

Color and Power Doppler

Fig. 12.9 Saline-infused sonohysterogram demonstrating an endometrial polyp. Power Doppler images show internal vascularity of the lesion.

1. **D.** In color Doppler, we obtain information about the direction of flow from the color, where typically red indicates flow toward the transducer and blue is away from the transducer. The underlying B-mode image also provides information on the Doppler angle. The color intensity provides information regarding the magnitude of the average detected frequency shift from each pixel.

2. **B.** In color Doppler, a light blue pixel would indicate a modest negative frequency shift that likely results from modest flow away from the ultrasound transducer. It will not flow toward the transducer, which would be red, or turbulent/high flow.

3. **D.** Color and power Doppler use the same acquired data regarding the frequency shifts, but these are processed differently. Color Doppler determines the average frequency shift, whereas power Doppler simply records the total number of frequency shifts and ignores the magnitude of these shifts.

4. **D.** Key characteristics of power Doppler are the fact that there is no angular dependence and absence of aliasing artifacts. In addition, because power Doppler "just shows flow," it is excellent at detecting slow flow.

Comment

Color Doppler makes use of Doppler frequency shifts to provide a 2D visual display of moving blood, which is encoded in the form of colors (i.e., red and blue). Flow information is obtained from the average of a relatively large (e.g., 50) number of samples obtained from each pixel. Color Doppler provides data over a selected region of interest regarding both direction and magnitude of flow that is present. Current clinical practice typically encodes red pixels to indicate motion toward the transducer, whereas blue pixels would indicate motion that is moving away from the transducer.

Color Doppler data are superimposed on a standard (black and white) B-mode image. Color Doppler may show flow in vessels that are too small to see by imaging alone and allows complex blood flow to be visualized. Clinical examples of this include hyperemia associated with infection and blood flow inside of solid tumors. Using color Doppler, a nonvascular collection or hematoma can be clearly differentiated from a solid mass, which may have an identical gray scale appearance. Turbulent flow may appear as green or yellow colors due to aliasing and display of colors at the extremes of the color scale. This is most commonly seen when imaging turbulent arterial flow such as in arterial stenosis or when the color scale is improperly set.

Power Doppler makes use of the same data as color Doppler, and the ultrasound intensities (i.e., power used) are thus identical in power and color Doppler. The only difference between power and color Doppler is how these identical acquired data are processed and subsequently displayed. This can be illustrated by taking 100 pulses sent out at approximately 90 degrees to a blood vessel, where 45 show small negative frequency shifts, 45 small positive frequency shifts, and only 10 are exactly zero. Color Doppler interprets these as having as a mean frequency shift of zero and would therefore depict this as a dark region (i.e., no flow). By contrast, power Doppler analyzes these same data as 90 frequency shifts and displays this as an orange color with 90% of the maximum intensity (Fig. 12.10).

Power Doppler is more sensitive in detecting blood flow than is color Doppler, with the trade-off of losing directional

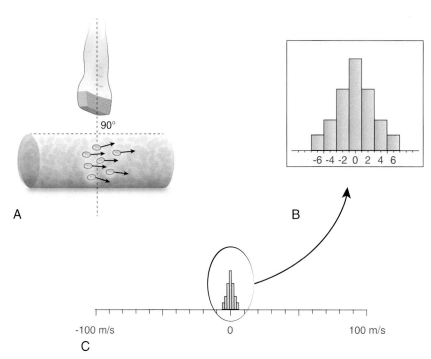

Fig. 12.10 The results of analyzing the frequency shifts detected in 100 consecutive pulses from one pixel is represented (B). When processing 45 small positive shifts, 45 small negative shifts, and 10 shifts that are exactly zero, color Doppler (C) shows an average of zero, whereas power Doppler (A) shows 90% of the echoes have nonzero frequency shifts.

information. Doppler angles have little effect on the displayed color intensity, where the intensity at 90 degrees would be slightly less than that at 0 degrees for a given level of flow. Power Doppler intensities show little variation with the direction of flow and have the same appearance for flow toward and away from a transducer. Of note is that power Doppler does not generally show aliasing artifacts. In clinical practice, power Doppler is more sensitive than color Doppler and is therefore ideal for detecting slow blood flow.

References

Huda W. Ultrasound II. In: *Review of Radiological Physics*. 4th ed. Philadelphia: Wolters Kluwer; 2016:257–258.

Kremkau FW. Doppler principles. In: *Sonography: Principles and Instruments*. 9th ed. St. Louis: Elsevier; 2016:133–183.

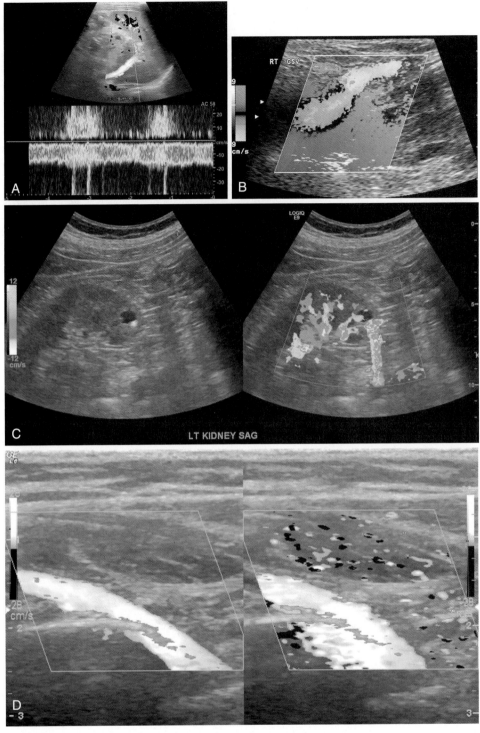

Fig. 12.11

1. What artifact is depicted in Fig. 12.11A?
   A. Aliasing
   B. Shadowing
   C. Enhancement
   D. Reverberation

2. What artifact is depicted in Fig. 12.11B
   A. Streak
   B. Ghost
   C. Mistuning
   D. Flash

3. What artifact is depicted in Fig. 12.11C?
   A. Refraction
   B. Side lobes
   C. Comet tail
   D. Twinkle

4. What artifact is depicted in Fig. 12.11D?
   A. Ringing
   B. Bleed
   C. Septal penetration
   D. Beam hardening

CASE 12.6

Doppler Artifacts

Fig. 12.11 (A) Aliasing. (B) Flash artifact. (C) Twinkle. (D) Bleed.

1. **A.** Aliasing occurs when we do not sample the moving blood quickly enough to accurately reproduce its direction and velocity. This is typically visualized as the blood in the center of a vessel appearing to suddenly be slower or in the opposite direction of the flow at the vessel walls.

2. **D.** Flash artifact occurs when movement (typically from a heartbeat) causes tissues to vibrate at the Doppler frequency. This is visualized as a burst of a single color through most of the tissues in the FOV.

3. **D.** Twinkle artifact occurs when the ultrasound beam interacts with a rough surface with impedance very different to soft tissue. The appearance of multiple colors is thought to be an interference pattern created by refraction of the beam by the object.

4. **B.** Color bleed occurs when the Doppler gain is set too high. The gain amplifies very small tissue vibrations and these appear as noisy color pixels overlaid on normal tissue.

Comment

The ultrasound machine samples flowing blood by repeatedly sending out pulses that reflect off the blood cells and are then read by the machine. The machine can determine blood velocity by comparing the original pulse frequency and phase to that of the reflected pulse. A user selects how often a pulse is sent out to sample the moving blood, the PRF. If the PRF is set too low, then we will not sample enough to accurately estimate the velocity (Fig. 12.12). In this case aliasing occurs. Aliasing appears as a sudden change in flow speed and direction. The artifact is typically reduced by increasing the PRF, and the PRF should be at least twice the Doppler frequency shift. For example, a 1-kHz Doppler shift requires a minimum PRF of 2 kHz to avoid aliasing.

If the Doppler image is filled with a sudden burst of color (typically low velocity), this is most likely a flash artifact. Flash artifact is caused by a sudden movement of the transducer or the patient and may be observed when performing color Doppler on a kicking fetus. Similarly, rhythmic pulsation of the heart and blood flow can cause tissue vibrations that create the flash artifact. In the case of the artifact caused by heart motion and vessel pulsation, the artifact can be mitigated by setting the wall filter parameter to filter out those frequencies and thus not show the color pixels in the image.

Twinkle artifact generally occurs behind a strong attenuator with a rough surface, such as stones or calcifications. It appears as a pattern of rapidly fluctuating reds and blues in the Doppler overlay in regions where there cannot be flow. It is speculated that this occurs due to interference patterns from refractions caused by the strong attenuator and is generally very dependent on machine settings. This artifact is especially useful for diagnosing small urinary calculi. Twinkle artifact is generally more sensitive for detecting small stones (i.e., urolithiasis) than by the detection of acoustic shadows.

Modern Doppler imaging is typically duplex imaging, meaning that there is the original B-mode gray scale image with a color overlay representing velocity. A color priority setting tells the machine when the B-mode gray scale is low (representing vessel lumen) and to show Doppler color instead of gray scale pixels. Color gain is a setting that amplifies the Doppler signal. If this is set too high, then Doppler color may "bleed" or "bloom" outside the vessel and into the surrounding tissue. Decreasing gain will reduce color bleed but will also remove color from slow flow (e.g., flow during diastole).

References

Hindi A, Cynthia P, Barr RG. Artifacts in diagnostic ultrasound. *Rep Med Imaging*. 2013;6:29–48.

Huda W. Ultrasound II. In: *Review of Radiological Physics*. 4th ed. Philadelphia: Wolters Kluwer; 2016:259–260.

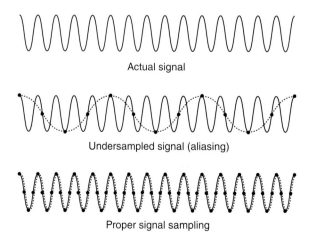

Actual signal

Undersampled signal (aliasing)

Proper signal sampling

Fig. 12.12 Top image is the actual signal we wish to sample and reproduce. If we sample too infrequently (*dots in middle image*), then the signal we reproduce with our machine (*dotted line*) will have a lower frequency than the actual frequency. This will result in a poor reconstruction (i.e., artifact). To correctly reproduce the frequency, we must sample at least twice per wavelength of the original signal, as shown by the dotted line in the lower image.

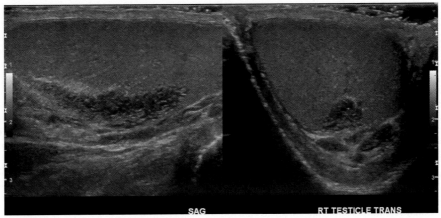

SAG   RT TESTICLE TRANS

Fig. 12.13

1. What is the axial resolution of a phased array transducer operating at 15 MHz (mm)?
   A. 0.01
   B. 0.1
   C. 1
   D. 10

2. If the axial resolution is of a clinical ultrasound transducer is 1 mm, what would be the most likely lateral resolution (mm)?
   A. 0.25
   B. 1
   C. 4
   D. Frequency dependent

3. Which image characteristic is most closely related to the elevational resolution?
   A. Pixel size
   B. Slice thickness
   C. Crystal width
   D. FOV

4. What is the most likely temporal resolution of real-time B-mode ultrasound (ms)?
   A. 0.3
   B. 3
   C. 30
   D. 300

CASE 12.7

Resolution

Fig. 12.13 Prominent rete testis, a benign incidental finding.

1. **B.** The axial resolution is simply the ultrasound wavelength. Because the wavelength of a 1.5-MHz transducer is 1 mm, that of a 15-MHz transducer is 10 times less, or 0.1 mm.

2. **C.** Empirical evidence has shown that the lateral resolution in ultrasound imaging is generally four times worse than the axial resolution. Although this is rather surprising, the same pattern is found at all frequencies currently used in clinical ultrasound.

3. **B.** The terms elevational resolution and slice thickness are interchangeable and refer to the same "dimension" in an ultrasound image. There is no direct relationship between elevational resolution and either axial or lateral resolution performance.

4. **C.** Real-time B-mode ultrasound normally uses a frame rate of 30 frames per second. To create a single frame requires 1/30th of a second, so the temporal resolution is approximately 30 ms, which is similar to fluoroscopy.

Comment

Axial resolution pertains to the ability of an ultrasound system to successfully separate (i.e., resolve) two objects lying along the axis that the beam travels. The pulse length is the primary

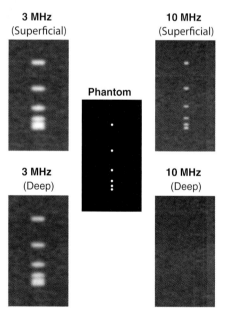

Fig. 12.14 Images obtained of the resolution phantom with a 3-MHz probe *(left)* and a 10-MHz probe *(right)*. The upper images obtained at a very shallow depth show that the lateral resolution is four times worse than the axial resolution and is also markedly better at 10 MHz. The lower images show that the low frequency has good penetration, whereas no data are seen at high frequencies at depth because of the increased attenuation of ultrasound at 10 MHz.

determinant of axial resolution. Axial resolution is equal to one-half of the ultrasound pulse length, or wavelength, and is generally independent of imaging depth. Because an increase in the transducer frequency will reduce the pulse length, using higher frequency invariably improves axial resolution (Fig. 12.14). At 1.5 MHz, axial resolution is approximately 1 mm, and at 15 MHz, the axial resolution is an order-of-magnitude better at 0.1 mm. Increasing transducer frequency will always improve axial resolution but will also increases attenuation that reduces beam penetration. Consequently, the choice of ultrasound transducer involves a trading of spatial resolution and the corresponding imaging depth.

Lateral resolution pertains to the ability to resolve, or separate, two adjacent objects in a patient in the resultant ultrasound image. The width of an ultrasound beam is the key determinant of lateral resolution performance. A secondary factor is the number of lines per frame, which improves lateral resolution when the line density is increased. Focusing the ultrasound beam will generally improve lateral resolution. With multiple elements, it is possible to have a "variable" focal length where "cherry picking" of multiple lines is used to improve lateral resolution. When multiple focal lengths are generated, this requires the use of more pulses to generate a single line of sight and is likely to result in a reduced frame rate as fewer lines of sight are generated in a single second. Lateral resolution usually deteriorates at increased distances from the transducer because of an increase in the beam width beyond the focal zone. Empirical data support the well-known rule of thumb that lateral resolution is approximately four times worse than axial resolution, and this will be true at most frequencies used in clinical ultrasound.

Resolution in the plane perpendicular to the image plane is known as elevational resolution. Effective or nominal slice thickness is another term for elevational resolution, and with focused beams this resolution will depend on depth. Elevational resolution generally depends on the distance (depth) from the transducer surface. Height of transducer elements, generally a few millimeters or so, will be the critical factor that determines elevational resolution. Elevational resolution is markedly improved using 1.5D arrays where multiple beams are transmitted along each line of sight, and each one is focused to a different depth. Cherry picking of these multiple beams permits an improved, and more uniform, elevational resolution to be achieved.

Temporal resolution refers to the ability to detect moving objects in their true sequence. The frame rate, which is the number of frames generated per second, determines temporal resolution. A real-time B-mode image normally operates at a frame rate of 30 frames per second so that each frame requires 1/30th second to be created, and this would be the normal temporal resolution. Temporal resolution can be affected by changing the lines used to create an image. If the line density or FOV is reduced, the temporal resolution can increase. Another way of increasing the temporal resolution is to use a higher PRF, which provides more "lines of sight" in a given time interval but at the price of a reduced imaging depth.

References

Huda W. Ultrasound II. In: *Review of Radiological Physics*. 4th ed. Philadelphia: Wolters Kluwer; 2016:258–259.

Radiological Society of North America. RSNA/AAPM physics modules. Ultrasound: image quality, artifacts and safety. https://www.rsna.org/RSNA/AAPM_Online_Physics_Modules_.aspx. Accessed February 27, 2018.

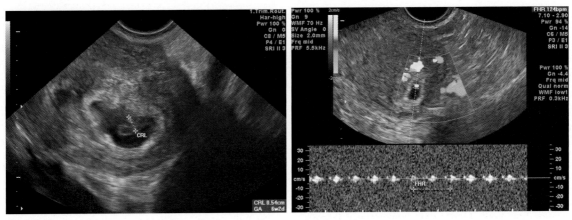

Fig. 12.15

1. What units are normally used to express intensities in diagnostic ultrasound?
   A. mW
   B. mW/cm
   C. mW/cm^2
   D. mW/cm^3

2. Which are used to express intensities in diagnostic ultrasound?
   A. Spatial peak and pulse average
   B. Spatial peak and temporal average
   C. Spatial average and pulse average
   D. Spatial average and temporal average

3. Which likely uses the highest intensities in diagnostic ultrasound?
   A. B-mode (adults)
   B. B-mode (infants)
   C. M-mode
   D. Doppler

4. Increasing which ultrasound factors are likely to increase the ultrasound intensity?
   A. PRF
   B. Beam focusing
   C. Penetration depth
   D. A and B

CASE 12.8

Intensities in Ultrasound

Fig. 12.15 Normal 6-week intrauterine pregnancy.

1. **C.** Intensities in ultrasound are expressed in mW/cm², which is the power (i.e., mW) per unit area (i.e. cm²).

2. **B.** Ultrasound intensities refer to the highest intensity across the transducer face (spatial peak). Ultrasound intensities also look at the value averaged over the pulse and the listening interval (temporal average).

3. **D.** Intensities in Doppler are approximately 50 times higher than in B-mode imaging. M-mode imaging is generally approximately four times higher than conventional B-mode real-time ultrasound imaging.

4. **D.** Increasing the PRF and the amount of focusing will increase ultrasound beam intensities. As the ultrasound penetrates into the patient, there will be attenuation of the beam which is proportional to the penetration depth. As the beam is attenuated, the intensities will be reduced.

Comment

Spatial peak intensity is the point of highest intensity in the beam. Spatial intensities are normally observed to be highest in the center of the beam over a very small area. The spatial average is averaged over the transducer area and is more indicative of the intensity to which most tissue is subjected. Similarly, the pulse average is the pulse intensity that is averaged over the duration of the pulse. The temporal average refers to an averaging over the (short) pulse duration, as well as over the much longer listening time. Clearly the pulse average will be much higher than the temporal average. The temporal average is more indicative of the average energy to which most tissue is subjected and inherently includes mechanisms, such as blood flow, that help to disperse the energy.

Ultrasound intensities are specified along the central beam axis (spatial peak) and averaged over time (temporal average). This spatial peak and temporal average intensity is used because it will best predict patient thermal effects (Table 12.1). Spatial peak relates the highest instantaneous energy given to tissue and temporal average accounts for blood flow and other energy dissipation (i.e., cooling) mechanisms. The pulse average intensity is markedly

higher than the temporal average intensity (Fig. 12.16) and will be an indicator of ultrasound bioeffects such as cavitation. Intensities in B-mode ultrasound are 10 mW/cm². It is notable that M-mode is four times higher than B-mode, and Doppler ultrasound can have intensities that are 50 times higher than B-mode.

Focusing in the lateral dimension will increase ultrasound intensities. When transducers are focused in the elevational plane, ultrasound intensities will be further increased. If the effective diameter of a beam is halved, then the area is reduced to 25% and the ultrasound beam intensity will be increased by a factor of four. Another factor that increases the ultrasound intensity is the PRF. When the PRF is doubled, there will be twice as many pulses emitted by an ultrasound transducer so that the beam intensity will also be doubled. It is important to remember the other trade-off for using a lower PRF is a lower framerate (all else being equal) and in Doppler the possibility of aliasing.

High-intensity focused ultrasound (HIFU) has been developed to treat a range of disorders, including malignant lesions. Focused ultrasound uses an acoustic lens or a shaped transducer to concentrate multiple intersecting beams of ultrasound on a target lesion. In HIFU, the sound wave must penetrate the patient as it is being focused onto a lesion. To achieve good penetration with minimal attenuation losses requires the use of low frequencies (e.g., <2 MHz). Where the beams converge at the focal point, the energy can have thermal and/or mechanical effects. HIFU is normally performed with real-time B-mode ultrasound imaging to enable treatment targeting and monitoring.

References

Huda W. Ultrasound II. In: *Review of Radiological Physics*. 4th ed. Philadelphia: Wolters Kluwer; 2016:259–260.

Miller DL. Safety assurance in obstetrical ultrasound. *Semin Ultrasound CT MR*. 2008;29(2):156–164.

TABLE 12.1   INTENSITY METRICS IN ULTRASOUND

Intensity Metric	Acronym	Comments
Spatial average temporal average	SATA	Lowest intensity
Spatial average pulse average	SAPA	Higher than SATA
Spatial peak temporal average[a]	SPTA	Correlates to thermal damage
Spatial peak pulse average	SPPA	Highest intensity and correlates with cavitation effects

[a]Intensity that is specified for clinical systems.

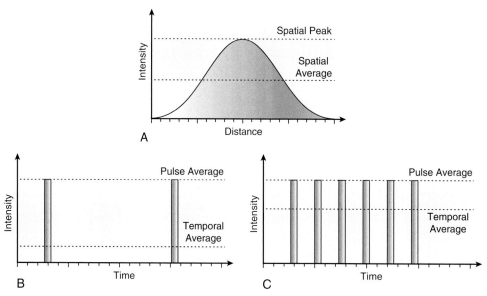

Fig. 12.16 (A) The spatial peak is higher than the spatial average and typically covers only a very small area in the center of the transducer. (B) The pulse average is much higher than the temporal average because the transducer transmits energy for only a fraction of the time it listens for echoes. (C) As the PRF increases (i.e., listening time decreases), the temporal average increases and comes closer to the pulse average.

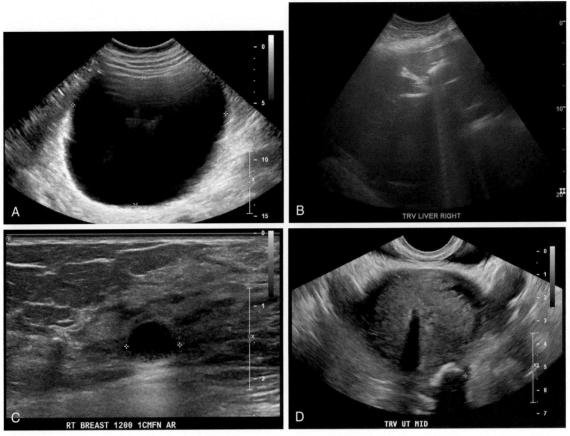

Fig. 12.17

1. What is the artifact in Fig. 12.17A?
   A. Truncation
   B. Motion
   C. Reverberation
   D. Ghosting

2. What is the artifact in Fig. 12.17B?
   A. Speed displacement
   B. Side lobes
   C. Ring-down
   D. Through transmission

3. What is the artifact in Fig. 12.17C?
   A. Through transmission
   B. Septal penetration
   C. Mach bands
   D. Partial volume

4. What is the artifact in Fig. 12.17D?
   A. Shadowing
   B. Refraction
   C. Grating
   D. Mirror

CASE 12.9

Image Artifacts

Fig. 12.17 (A) Reverberation artifact along the anterior bladder caused by layers of fascia. (B) Ring-down artifact in the liver in the setting of pneumobilia. (C) Through transmission seen along the posterior aspect of a simple cyst in the breast. (D) Shadowing of a calcified uterine fibroid.

1. **C.** Fig. 12.17A shows the reverberation artifact where there are multiple echoes of the same object. Truncation, motion, and ghosting artifacts are generally observed in MR imaging.

2. **C.** Fig. 12.17B shows ring-down artifact, which requires the presence of fluid and gas bubbles for its formation. Speed displacement, side lobes, and through transmission are ultrasound artifacts but bear no resemblance to ring-down artifacts.

3. **A.** Fig. 12.17C is an example of through transmission, AKA enhancement, which is a telltale sign of a fluid-filled cyst. Septal penetration occurs in nuclear medicine, Mach bands are seen on mammograms of well-defined lesions, and partial volume artifacts are seen in CT imaging.

4. **A.** Fig. 12.17D is an example of shadowing, which is very useful for the identification of strong attenuators and/or reflectors such as gallstones. Refraction, grating, and mirror artifacts occur in ultrasound imaging but bear no resemblance to shadowing artifacts.

Comment

Ultrasound artifacts occur because assumptions made in creating images are violated (Fig. 12.18). The machine assumes the received signal is from a pulse in the main beam. Side lobes and grating lobes are beams oblique to the main beam, and reflections from these are incorrectly interpreted as coming from the main beam. This causes objects oblique to the beam to appear under the beam in the image. The machine also assumes that the main beam travels in a straight line. Sound is a wave and thus will refract (i.e., bend) when passing between tissues. This results in objects being displaced (like how the bottom of a pool appears closer than it actually is) and can cause intensity variations as part of the beam is bent away leaving little or no energy underneath it (hypointense) and more energy laterally where it is bent (hyperintense).

The machine assumes that attenuation is uniform. When a tissue with much higher attenuation (e.g., calcification) is in the beam path, the signal will be very weak distal to the object and thus the image appears hypointense in that region (i.e., shadowing). Similarly, when a tissue has much lower attenuation (e.g., simple cyst), the beam distal to the cyst is very strong, and thus the image appears hyperintense in that region (i.e., enhancement). Although both shadowing and enhancement are artifacts, these are very helpful in identifying the presence of specified features. For example, kidney stones and gallstones can be easily identified by the presence of shadowing, whereas a fluid-filled cyst will always exhibit enhancement.

There is also an assumption that the pulse travels through tissue with a speed of 1540 m/s. If an appreciable amount of the path is bone, air, or other material with a different speed of sound, then the object will appear too deep or shallow depending on whether the speed of sound is slower or faster, respectively.

The machine assumes the signals are the result of one reflection from the object surface, but the beam may reflect off one or more surfaces during its travels. This results in a mirror artifact that displaces an object in the image relative to its actual position. This is typically seen when the beam traverses a large smooth surface such as the diaphragm. Similarly the beam may reflect between two surfaces repeatedly (each time it reflects it sends a little energy back to the machine to make image). This results in the surface appearing in the image multiple times along the path of the beam and getting weaker as it gets deeper. If the gap between the reflections is very small, then the series of surfaces are very close together on the image and form an artifact reminiscent of a comet tail.

Ring down is another type of reverberation artifact. This occurs due to the vibration of tiny amounts of liquid surrounded by gas pockets. The ring-down artifact is often used to support the diagnosis of emphysematous cholecystitis.

References

Feldman MK, Katyal S, Blackwood MS. US artifacts. *Radiographics.* 2009;29:1179–1189.

Huda W. Ultrasound II. In: *Review of Radiological Physics.* 4th ed. Philadelphia: Wolters Kluwer; 2016:241–260.

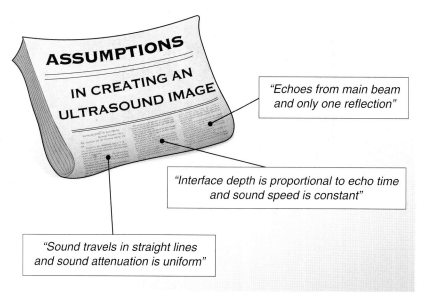

Fig. 12.18 Artifacts occur in ultrasound imaging when fundamental assumptions made by the system are violated.

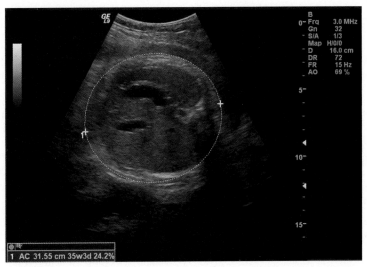

Fig. 12.19

1. Which can be accredited by the American College of Radiology (ACR)?
   A. Ultrasound
   B. Breast ultrasound
   C. Ultrasound therapy
   D. A and B

2. Which are likely to be found in an ultrasound quality control (QC) phantom?
   A. Tissue equivalent materials
   B. Resolution details
   C. Low-contrast reflectors
   D. A, B, and C

3. How often are ultrasound probes QC tested for uniformity, geometric accuracy, and display?
   A. Daily
   B. Weekly
   C. Monthly
   D. Semiannually

4. Which of the following can be identified using QC phantoms?
   A. Faulty transducer
   B. Nonuniformities
   C. Extraneous noise
   D. A, B, and C

CASE 12.10

Quality Control

Fig. 12.19 Fetal ultrasound showing an estimated gestational age of 35 weeks 3 days by abdominal circumference measurement *(white dotted line)*.

1. **D.** The ACR offers accreditation in both diagnostic and breast ultrasound. Ultrasound can also be used therapeutically (see Case 12.8), but this is not accredited by the ACR.

2. **D.** A QC phantom would likely contain tissue equivalent materials, resolution details (see Fig. 12.19), and low-contrast reflectors. Regular measurements are essential to permit operators to objectively assess ultrasound imaging performance.

3. **D.** The ACR ultrasound accreditation requires that facilities assess image uniformity, geometric accuracy, and ultrasound display (and more) at least every 6 months.

4. **D.** Problems that can be identified by performing routine QC measurements include detection of faulty transducer (elements), image nonuniformities, and extraneous noise.

Comment

Tissue mimicking phantoms can quantitatively and qualitatively assess image quality by analyzing echo intensities. Phantoms can measure all three resolutions (axial, lateral, and elevational) as a function of depth for each available transducer frequency. Uniform phantoms will detect faulty transducers, as well as problems of nonuniformity, artifacts, and extraneous sources of noise. The maximum depth with useable signal and the signal-to-noise ratio of one or more standard targets are also measured (Fig. 12.20).

Phantoms help to differentiate problems associated with image formation (echoes) from those associated with image display. Transducers should be checked for being intact and to ensure that there are no significant cracks and delamination. A common artifact is partial signal loss due to broken elements on the transducer. These appear as rays of signal loss emanating from the transducer. Electrical integrity should also be checked for frays and other problems.

ACR accreditation programs are available for diagnostic ultrasound and breast ultrasound. ACR accreditation is always dependent on the existence of a formal QC program. QC tests recommended by the ACR accreditation program need to be documented, as well as corrective action taken being available for review (Table 12.2). ACR phantom testing is identical for ultrasound and breast ultrasound. In practice, the most frequently encountered problems relate to maladjusted monitors, which affect the display of images, not the actual acquired images themselves.

Measurements taken in ultrasound are very important to diagnosis. As such, it is important to ensure that measurements made with digital calipers on the image are accurate. This is typically accomplished by measuring between two marks in the phantom with a known distance. Similarly, sufficient contrast to identify lesions are important. Phantoms will have several sizes of anechoic cylinders (similar to a simple cyst) and several slightly hyperechoic cylinders (similar to lesions) to test this.

References

Bushberg JT, Seibert JA, Leidholdt Jr EM, Boone JM. Ultrasound. In: *The Essential Physics of Medical Imaging*. 3rd ed. Philadelphia: Wolters Kluwer; 2012:568–575.

Huda W. Ultrasound II. In: *Review of Radiological Physics*. 4th ed. Philadelphia: Wolters Kluwer; 2016:259–260.

TABLE 12.2 SUMMARY OF TYPICAL ANNUAL TESTS PERFORMED ON ULTRASOUND IMAGING SYSTEMS AND THEIR PURPOSE.[a]

Annual Testing	Purpose
Physical and mechanical inspection	Make sure equipment has not physically degraded to the point of being unsafe.
Image uniformity and artifact survey	Ensures that pathology is due to the patient and not an artifact of the scanner.
Geometric accuracy	Make sure measured sizes are accurate. This is especially important in mechanical probes.
Electronic image display performance	Make sure the sonographer/radiologist at the machine sees an accurate image on the machine screen.
Primary interpretation display performance	Make sure the radiologist sees an accurate image on the read station.

[a]Note that the American College of Radiology requires quality control measurements on at least a 6-month basis to ensure that these ultrasound systems continue to function in a satisfactory manner. This routine quality control testing is less extensive and performed semiannually by the facility's technologist (or a service engineer), while annual surveys are much more detailed and are performed annually by the medical physicist.

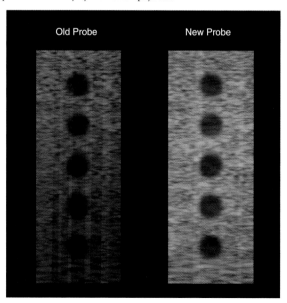

Fig. 12.20 By using a standard phantom, it is easy to see the degradation of image quality of an "old probe" *(simulated on the left)* in comparison with a "new probe" *(simulated on the right)*. Quality control measurements are objective and can accurately depict the loss of image quality, something that would be very difficult to achieve using clinical images alone.

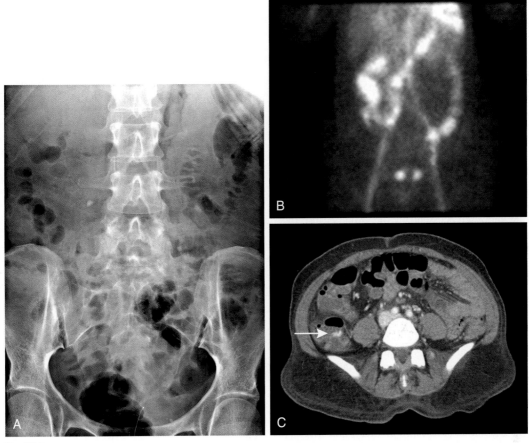

Fig. 13.1

1. How will increasing the x-ray beam quality (half-value layer) likely affect organ dose per unit kerma area product conversion factors (D_{organ}/KAP)?
   A. Increase
   B. No change
   C. Reduce

2. Compared with a full lung scan, how will scanning only half the lung length in a chest CT examination most likely impact on the lung dose (%)?
   A. <50
   B. 50
   C. >50

3. For a fixed administered activity, how will an increase in patient size most likely affect the resultant organ doses?
   A. Increase
   B. No change
   C. Reduce

4. Which types of risk can be estimated from a knowledge of organ doses?
   A. Carcinogenic
   B. Hereditary
   C. Deterministic
   D. A, B, and C

CASE 13.1

Organ Doses

Fig. 13.1 Imaging of a patient with active bleeding of the cecum, including abdominal x-ray (A), nuclear medicine bleeding scan (B), and CT angiogram of the abdomen (C). For the x-ray, the radiation incident on the patient is the kerma area product (Gy-cm2), and for the CT the radiation incident on the patient is the dose length product (mGy-cm). In the nuclear medicine scan, the organ dose is proportional to the administered activity (MBq).

1. **A.** When the x-ray beam quality increases, the beam becomes more penetrating and will therefore deliver higher doses to deeper-lying organs and tissues. In radiography and fluoroscopy, D_{organ}/KAP factors generally increase with increasing x-ray beam quality.
2. **B.** When only half of the lung is irradiated, the lung dose will be approximately one-half of the dose to this organ when it is fully irradiated. At constant CT x-ray tube output, the lung will absorb only half the radiation that it absorbs when fully irradiated.
3. **C.** When patient size increases for the same administered activity, patient organ doses will generally be reduced. The energy being deposited is essentially the same because the administered activity is fixed, but the organs have greater mass so there is "more dilution" of the absorbed energy.
4. **D.** Organ doses can be used to predict the likelihood of a deterministic effect (e.g., skin burn from the peak skin dose [PSD]). Organ doses can also be used to predict the risk of carcinogenesis in any organ and the risk of genetic effects from gonad doses.

Comment

Consider a radiograph where the intensity of the radiation is given by the air kerma (K_{air}, Gy) and the total radiation incident on the patient is K_{air} × Area (cm^2). The organ dose is the energy absorbed (mJ) from the x-ray beam divided by the corresponding organ mass (kg) and is expressed in mGy. Medical physicists can generate organ dose conversion factors to convert an incident K_{air} × Area (KAP) into an organ dose (D_{organ}). Conversion factors are expressed as D_{organ}/KAP and consider both the x-ray beam characteristics (quality, quantity, and area) and patient characteristics (age, size, body region, and projection). Superficial organs such as the skin will have doses that are comparable with the incident air kerma, exit doses in a normal-sized adult will be approximately 1% of the incident radiation, and the dose in the middle will be approximately 10%. Doses to organs that are not directly irradiated will always be very low and are generally of no clinical significance.

In CT, the radiation intensity incident on the patient is expressed as the CT dose index (CTDI$_{vol}$ (mGy), which controls the amount of mottle in any given patient examination. For a scan length of L (cm), the total amount of radiation is the dose length product (DLP), which is given by CTDI$_{vol}$ × L mGy-cm. Medical physicists can generate organ dose conversion factors to convert the total radiation used (i.e., DLP) into an organ dose (i.e., D_{organ}/DLP), which consider patient characteristics and the region that is scanned. For a head CT scan (CTDI$_{vol}$ = 60 mGy and DLP = 900 mGy-cm), the dose to the brain and eye lens will be close to 60 mGy. For an abdominal CT scan (CTDI$_{vol}$ = 15 mGy and DLP = 450 mGy), the liver dose will be close to 20 mGy.

In nuclear medicine, organ doses in each patient will be proportional to the administered activity (A). Medical physicists have generated conversion factors that enable dose to any organ or tissue in any sized patient to be determined using such D_{organ}/A conversion factors. Organ doses are generally directly proportional to the administered activity and depend on the specific uptake of a radiopharmaceutical by a given organ. The smaller the organ taking up the activity, the higher the organ doses, and extremely high doses could be delivered if the fetal thyroid were to take up radioiodine (Fig. 13.2). For most diagnostic nuclear medicine studies with technetium-99m (^{99m}Tc)-labeled pharmaceuticals, the maximum organ doses are generally between 20 and 50 mGy.

Organ doses can be used to assess the risk of a deterministic effect, which may occur with superficial organs (e.g., skin) that receive relatively high doses. Organ doses can also be used to assess the risk of stochastic effects, which predominantly refers to carcinogenesis for virtually all diagnostic radiologic examinations. When the gonads receive a high dose, the hereditary risk can become important. In most diagnostic examinations, there are many organs and tissues that are exposed, and any dose quantity needs to account for all the exposed organs and tissues. It is possible to combine the doses to all irradiated organs into an "effective dose" to provide a single metric that conveniently describes the amount of radiation received by a patient in any diagnostic radiologic examination, as discussed in Case 13.4.

References

Bushberg JT, Seibert JA, Leidholdt Jr EM, Boone JM. X-ray dosimetry in projection imaging and computed tomography. In: *The Essential Physics of Medical Imaging*. 3rd ed. Philadelphia: Wolters Kluwer; 2012:384–385.

Huda W. Patient dosimetry. In: *Review of Radiological Physics*. 4th ed. Philadelphia: Wolters Kluwer; 2016:49–50.

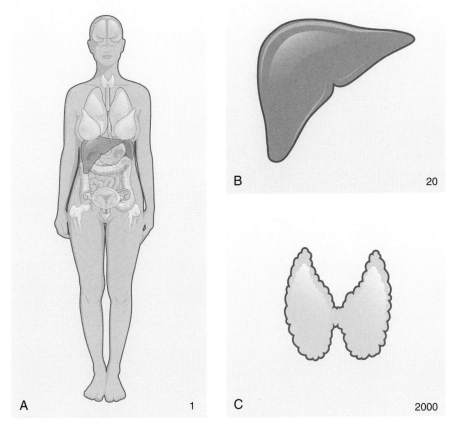

Fig. 13.2 When a given amount of activity (1 MBq) used in diagnostic nuclear medicine is uniformly distributed in an average sized patient, there will be a given whole-body dose (A). If this activity is deposited in the liver, the liver dose will be approximately 20 times higher (B), and when the same activity is deposited in the thyroid, the thyroid dose will be 2000 times higher (C).

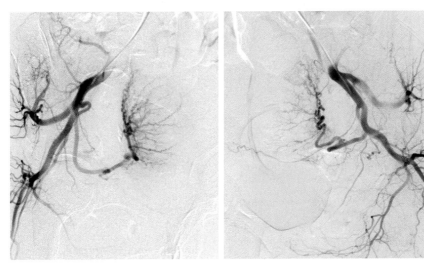

Fig. 13.3

1. What must be considered when converting the interventional reference point (IRP) K_{air} into a corresponding patient skin dose?
   A. Air/tissue differences
   B. Patient backscatter
   C. Table attenuation
   D. A, B, and C

2. Which is the most likely value of dose index (DI), defined as (PSD)/(IRP K_{air}), encountered in interventional radiology procedures performed as US Academic Medical Centers (AMCs)?
   A. <1
   B. 1
   C. >1

3. What percentage of patients undergoing interventional radiology procedures at AMCs is likely to exceed a PSD of 5 Gy (%)?
   A. <1
   B. 1
   C. >1

4. What PSD (Gy) is considered the threshold dose for a sentinel event by the Joint Commission (JC) accreditation body?
   A. 2
   B. 5
   C. 10
   D. 15

CASE 13.2

Skin Doses

Fig. 13.3 Uterine artery embolization performed at opposite obliquities, reducing skin dose when compared with the use of a single field. Collimation should also be implemented whenever possible.

1. **D.** When converting IRP K_{air} into a PSD, medical physicists must account for air/tissue differences, patient backscatter, and table attenuation. Additional factors include the location of the patient relative to the IRP and (any) beam overlap when multiple beams are used.

2. **A.** The most likely value of DI is close to 0.6, which means that the IRP K_{air} most likely overestimates the true patient PSD. However, there will be a relatively small number of procedures (<5%) for which IRP K_{air} would underestimate the PSD.

3. **A.** Less than 1% of patients undergoing interventional radiology procedures at AMCs would be expected to have a PSD more than 5 Gy. Note that at this value of PSD, radiation burns would likely occur 10 days post irradiation.

4. **D.** The JC considers a PSD more than 15 Gy to be a sentinel event that would trigger an inspection and require a root cause analysis to be performed by the medical institution.

Comment

To convert an IRP K_{air} value into the corresponding patient PSD requires the expertise of a medical physicist. Factors that would need to be considered are the physical absorption differences between air and tissue, x-ray beam attenuation by the table, and the presence of backscatter radiation when patients are exposed. Differences in the IRP location relative to the patient skin, as well as overlap from multiple projections, are also important geometric factors that have a significant impact on the PSD.

PSD can be measured or calculated and subsequently compared with the IRP K_{air} (Fig. 13.4). The DI can then be computed as the true PSD divided by the corresponding IRP K_{air}. Studies of clinical interventional radiology procedures performed at AMCs show that most DI values lie between 0.5 and 0.8.

This is a reassuring finding because it indicates that on average the IRP K_{air} presented to an interventional radiology (IR) operator overestimates the PSD. Accordingly, the actual tissue effects danger is slightly lower than what is displayed by the IR system for most patients. Nonetheless, there will be occasional procedures for which IRP K_{air} would underestimate the PSD. Current scientific data suggest that a DI of greater than 1 might occur in less than 5% of procedures, and a consultation with a medical physicist would help to identify such cases in clinical practice.

PSD in nephrostomy placements, inferior vena cava (IVC) filter placements, and pulmonary angiography are unlikely to exceed 2 Gy. Highest PSDs occur in complex cases such as transjugular intrahepatic portosystemic shunt (TIPS) and embolization procedures. It is estimated that PSDs exceed 5 Gy in up to 20% of spine neuroembolization procedures performed to treat arteriovenous malformations and tumors. Approximately one-quarter of interventional radiology procedures will exceed a PSD of 2 Gy at an AMC. It has been estimated that only 1 in 10,000 procedures result in a serious radiation burn, which is defined as one that requires some type of major clinical intervention (e.g., skin grafts).

The IRP K_{air} provided by the vendor is generally taken to be the PSD, which will likely be a conservative estimate of the true PSD. Radiation burns are never expected at less than 2 Gy, so no action would be required in such cases. It is prudent to inform the patient when the estimated dose is between 2 and 5 Gy, identifying where and when a burn might occur. At greater than 5 Gy, many departments would ask a medical physicist to assess the validity of the IRP K_{air} as the PSD. In these cases, it would likely be prudent for a healthcare provider to directly contact a patient 1 to 2 weeks after the procedure to inquire about possible radiation burns. The JC considers an unintended skin dose in excess of 15 Gy to be a sentinel event. This is set at the limit of necrosis, and facilities must perform a root cause analysis and develop an action plan in response.

References

Hricak H, Brenner DJ, Adelstein SJ, et al. Managing radiation use in medical imaging: a multifaceted challenge. *Radiology.* 2011;258(3):889–905.
Huda W. Radiation risks. In: *Review of Radiological Physics.* 4th ed. Philadelphia: Wolters Kluwer; 2016:63–64.
Miller DL, Balter S, Schueler BA, et al. Clinical radiation management for fluoroscopically guided interventional procedures. *Radiology.* 2010;257(2):321–332.

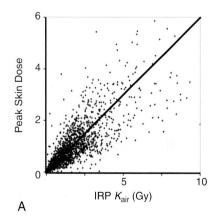

A

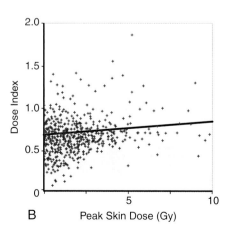

B

Fig. 13.4 (A) The typical correlation between the peak skin dose (PSD) and interventional reference point *(IRP)* K_{air} shows that the slope is approximately 0.6. (B) The dose index (i.e., IRP K_{air} divided by PSD) as a function of PSD shows that the average value is approximately 0.6 and that in most cases, but not all, IRP K_{air} overestimates the true PSD.

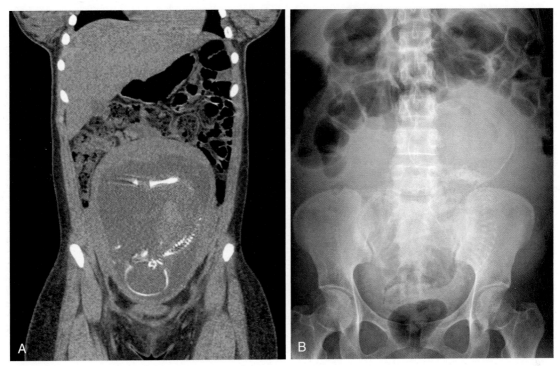

Fig. 13.5

1. What is the most likely embryo dose for a patient undergoing a chest CT examination (mGy)?
   A. <1
   B. 1
   C. >1

2. What is the most likely maximum embryo dose for a woman undergoing abdominal/pelvic fluoroscopy by the time the first 5-minute alarm sounds (mGy)?
   A. 0.1
   B. 1
   C. 10
   D. 100

3. What is the most likely embryo dose to a normal-sized patient who undergoes a routine abdominal pelvic CT scan using a $CTDI_{vol}$ of 15 mGy?
   A. <15
   B. 15
   C. >15

4. Which nuclear medicine scan most likely results in the lowest embryo dose?
   A. Technetium-99m–labeled macro aggregated albumin (MAA)
   B. Technetium-99m–labeled diethylenetriaminepentacetate (DTPA)
   C. Technetium-99m–labeled red blood cells (RBCs)
   D. A, B, and C are similar

CASE 13.3

Conceptus Doses

Fig. 13.5 CT (A) and radiograph (B) images obtained from two pregnant women.

1. **A.** The embryo is never directly irradiated during a chest CT examination. The embryo dose will be very low (<0.1 mGy).

2. **C.** The embryo dose rate during posteroanterior (PA) fluoroscopy is approximately 2 mGy/min so that after 5 minutes of fluoroscopy, the maximum embryo dose is approximately 10 mGy. This embryo dose is approximately 10 times higher than would be received during a single abdominal pelvic x-ray examination (anteroposterior [AP] projection).

3. **C.** For a (long) abdominal pelvic scan in an average-sized patient (i.e., one that wholly irradiates the embryo), the embryo dose at a $CTDI_{vol}$ of 15 mGy would be approximately 20 mGy (i.e., D_{uterus}/DLP is approximately 1.4).

4. **A.** The embryo dose for a technetium-99m–labeled MAA scan would both be low (<1 mGy) because the activity is concentrated in the lungs. The embryo dose for a technetium-99m–labeled DTPA scan would be higher (~5 mGy), as the activity is excreted through the urinary bladder. Although dose to the embryo is low with technetium-99m–labeled RBCs distributed throughout the body, the dose from an MAA scan would be lower.

Comment

For all x-ray–based examinations, if the x-ray beam does not directly irradiate the embryo, the corresponding dose will be low. Therefore examinations of the head, chest, and all the extremities have negligible embryo doses. Head and extremity CT imaging results in negligible embryo and fetal doses. In a chest CT embryo, doses are low and normally would be less than 0.1 mGy. Because internal scatter is the main source of embryo dose during chest CT, placing lead aprons on the patient's abdomen will result in negligible dose benefit. Lead aprons during head and chest CT provide little benefit, but these can help to reassure anxious patients and may be used for this purpose.

For an AP projection abdominal radiograph in a standard-sized adult patient, the embryo dose is approximately 1 mGy. Doses will generally be reduced for PA and lateral projections. In a normal-sized patient undergoing a PA abdominal/pelvic projection in fluoroscopic image, the dose rate at the embryo is approximately 2 mGy/min. When patient size increases, the amount of radiation used will generally also be increased for examinations performed using automatic exposure control (AEC). However, the larger patient size will also increase the attenuation within the patient, and it is reasonable to assume that the increased tissue thickness at least partially compensates for the increased entrance radiation exposure.

Embryo doses are directly proportional to the total amount of radiation used in a CT scan, namely the $CTDI_{vol}$ used to perform the scan. Embryo dose estimates should also take account of scan length and patient size (Fig. 13.6). In normal-sized females (23 cm AP), an abdominal/pelvic CT scan, performed using $CTDI_{vol}[L]$ of 15 mGy and DLP[L] of 450 mGy-cm, results in an embryo dose of approximately 20 mGy. At fixed radiation techniques ($CTDI_{vol}$), a smaller patient with an AP dimension of 18 cm would have an embryo dose 20% higher, and a larger patient with an AP dimension of 28 cm would have an embryo dose 20% lower. When the embryo is exposed in multiphase body CT scans, the cumulative embryo dose is the sum of the doses associated with each phase.

Embryo doses in nuclear medicine will depend on the biodistribution pattern of the administered radiopharmaceutical. The highest doses will normally be obtained when the administered activity is taken up by organs near the uterus and when a significant amount is excreted through the urinary bladder. For activity that is remote from the uterus, such as in lung scans, embryo doses will generally be low (<1 mGy). When uptake of activity is much closer to the uterus (e.g., a DTPA examination) the embryo doses are found to be in the range of 5 to 10 mGy.

References

Huda W. Patient dosimetry. In: *Review of Radiological Physics*. 4th ed. Philadelphia: Wolters Kluwer; 2016:49–50.

Huda W, Randazzo W, Tipnis S, et al. Embryo dose estimates in body CT. *AJR Am J Roentgenol*. 2010;194(4):874–880.

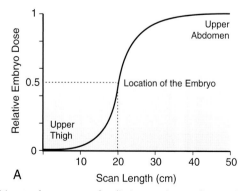
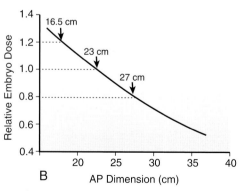

Fig. 13.6 In addition to the amount of radiation used to perform a CT scan (i.e., $CTDI_{vol}$), embryo doses also depend on the scan length and patient size. (A) The relative embryo dose increases with scan length from the upper thigh region to the upper abdomen. (B) Changing the patient anteroposterior *(AP)* dimension affects the resultant relative embryo dose.

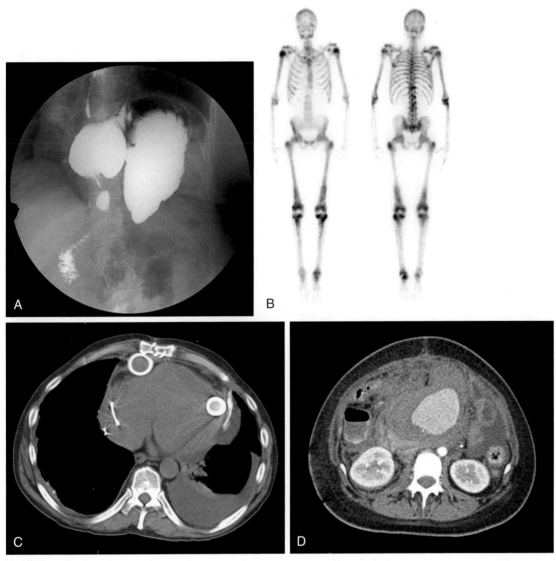

Fig. 13.7

1. Which of the following organs is the most radiosensitive for cancer induction?
   A. Adrenals
   B. Colon
   C. Kidney
   D. Heart

2. How many chest x-rays would likely deliver the same patient dose as a chest CT examination?
   A. <100
   B. 100
   C. >100

3. Which examination likely delivers a dose that is closest to that received by a patient from natural background in 1 year?
   A. Chest x-ray
   B. Abdominal x-ray
   C. Head CT
   D. Embolization (IR)

4. How does the effective dose (E) from a chest CT scan likely compare with the occupational radiation exposure received by an IR fellow in a year (5 mSv)?
   A. Greater than
   B. Comparable to
   C. Less than

CASE 13.4

Effective Dose

Fig. 13.7 Four examples of scans that have moderate effective doses, ranging between 1 and 10 mSv. Upper gastrointestinal examination (A) showing organoaxial gastric volvulus, bone scan in a patient with diffuse osseous metastasis (B), CT chest showing a left ventricular assist device (C), and CT abdomen showing a mesenteric pseudoaneurysm (D).

1. **B.** The colon is one of the five most radiosensitive organs for the induction of cancer. The other four are the breast, (females) red bone marrow, lungs, and stomach.

2. **B.** The effective dose of a typical chest CT examination (5 mSv) is most likely 100 times bigger than a typical chest x-ray examination (0.05 mSv).

3. **C.** The average American receives approximately 3 mSv each from natural background, which is close to that of a head CT scan (1 to 2 mSv). A chest x-ray would have an effective dose less than 0.1 mSv, an abdominal x-ray would likely be 0.3 mSv, and a procedure such as an embolization would be well over 10 mSv.

4. **B.** A chest CT scan would likely have an effective dose of 5 mSv, which is the effective dose from occupational exposure that an IR fellow would likely receive in 1 year's exposure. Note that the IR fellow would need to wear two badges (above/below lead apron) to obtain a reliable estimate of their effective dose.

Comment

The concept of total radiation detriment has been developed by the International Commission on Radiological Protection (ICRP). To obtain an estimate of the total harm (i.e., detriment) clearly requires use of judgment factors pertaining to the relative importance of fatal cancers, nonfatal cancers, and genetic effects. Tissue-weighting factors (W_T) are the contribution of each organ to the total detriment associated with a uniform whole-body exposure and are averaged over both age and sex. Table 13.1 shows the values for W_T currently recommended for radiation protection practice by the ICRP and how these compare with prior values. Remainder organs consist of the adrenals, gall bladder, heart, kidney,

pancreas, prostate, small intestine, spleen, thymus, and uterus/cervix. W_T are crude nominal indicators of the relative radiosensitivity of any organ that will undergo periodic revisions as our understanding of radiation risks improves with the passage of time.

For a given radiologic examination, a simple listing of such "organ doses" will not assist radiology practitioners to grasp "how much radiation patients receive." This goal can be achieved with the effective dose, which considers the equivalent dose to every organ together with each organ's relative radio sensitivity (W_T) (Fig. 13.8). This E metric is obtained from the product of the equivalent dose (H) to an organ by the corresponding organ weighting factor (W_T), and then by summing up for all irradiated organs. Effective doses are the uniform whole-body dose that would produce the same patient detriment as any radiologic exam. Effective doses are normally expressed in mSv, the unit for equivalent dose, whereas organ-absorbed doses are expressed in mGy. In radiology, an organ dose of 1 mGy from x-rays corresponds to an organ equivalent dose of 1 mSv. Patient detriment from a body CT with an effective dose of 5 mSv is comparable with this patient receiving a uniform whole-body dose of 5 mGy, where each organ receives exactly this amount of radiation.

Effective doses enable comparisons of the radiation pattern deposited in the same patient from various types of x-ray examination. Fluoroscopy patterns of energy deposition (dose distributions) can be directly compared to those in radiography, mammography, CT, and nuclear medicine. Effective doses less than 0.1 mSv, such as chest x-ray examinations, are classified as "very low." Effective doses between 0.1 and 1 mSv, such as most body radiographic examinations, are classified as "low." Effective doses between 1 and 10 mSv, such as gastrointestinal (GI) studies and CT scan of the major parts, are classified as "moderate." Effective doses more than 10 mSv, such as generally occur in complex interventional radiology procedures, are categorized as "high." Use of effective dose also enables direct comparisons of radiation exposures in radiology with natural background doses, as well as with regulatory dose limits.

References

Huda W. Patient dosimetry. In: *Review of Radiological Physics*. 4th ed. Philadelphia: Wolters Kluwer; 2016:54–57.

Radiological Society of North America. RSNA/AAPM physics modules. Fundamentals: radiation measurement and units. https://www.rsna.org/RSNA/AAPM_Online_Physics_Modules_.aspx. Accessed February 27, 2018.

TABLE 13.1 TISSUE-WEIGHTING FACTORS ASSIGNED BY THE INTERNATIONAL COMMISSION ON RADIOLOGICAL PROTECTION[a]

Tissue	TISSUE-WEIGHTING FACTOR, W_T		
	ICRP 26 (1977)	ICRP 60 (1990)	ICRP 103 (2007)
Bone marrow (red)	0.12	0.12	0.12
Breast	0.15	0.05	0.12
Lung	0.12	0.12	0.12
Stomach	—	0.12	0.12
Colon	—	0.12	0.12
Thyroid	0.03	0.05	0.04
Bladder	—	0.05	0.04
Esophagus	—	0.05	0.04
Liver	—	0.05	0.04
Bone (surface)	0.03	0.01	0.01
Skin	—	0.01	0.01
Salivary glands	—	—	0.01
Brain	—	—	0.01
Remainder	0.30	0.05	0.12
Gonads	0.25	0.20	0.08

[a]In Publication 26 (1977), Publication 60 (1990), and Publication 103 (2007).
ICRP, International Commission on Radiological Protection.
All tissues refer to carcinogenesis except for the gonad, which refers to the induction of hereditary effects that occur in the offspring of an exposed individual.

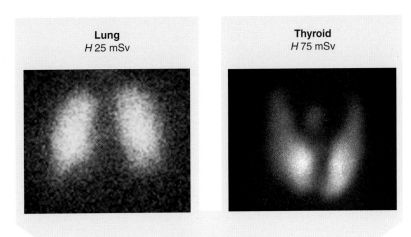

Lung
H 25 mSv

Thyroid
H 75 mSv

Effective Dose 3 mSv

Fig. 13.8 A lung scan *(left)* with lung dose of 25 mSv has an effective dose of 3 mSv (25 mGy/mSv lung dose/dose equivalent multiplied by the lung tissue-weighting factor W_T of 0.12). A thyroid scan *(right)* with a thyroid organ dose of 75 mSv also has an effective dose of 3 mSv (75 mGy/mSv organ dose/dose equivalent multiplied by the thyroid tissue-weighting factor W_T of 0.04). For a given patient, the total detriment from these two examinations may be taken to be approximately comparable.

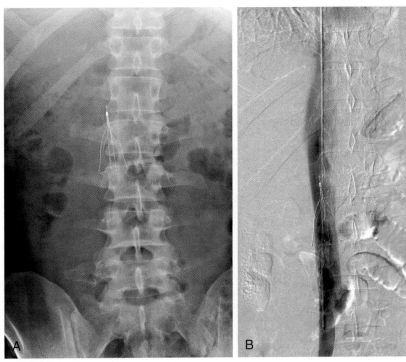

Fig. 13.9

1. What must be considered when generating factors to convert incident radiation (KAP) into patient organ and effective doses?
   A. Beam quality
   B. Body region
   C. Projection used
   D. A, B, and C

2. What are the effective doses associated with radiographic examinations?
   A. Very low (<0.1 mSv)
   B. Low (0.1 to 1 mSv)
   C. Moderate (1 to 10 mSv)
   D. A and B

3. What is the most likely range of patient effective doses (mSv) encountered in fluoroscopy-guided GI/genitourinary (GU) examinations?
   A. <0.1
   B. 0.1 to 1
   C. 1 to 10
   D. >10

4. How many chest x-ray examinations would most likely deliver the same amount of radiation to a patient as a TIPS procedure with an effective dose of 50 mSv?
   A. 10
   B. 100
   C. 1000
   D. 10,000

CASE 13.5

Effective Dose in Radiography and Fluoroscopy

Fig. 13.9 Scout radiograph of the abdomen (A) and digital subtraction venogram (B) obtained prior to removal of an inferior vena cava filter.

1. **D.** To convert the radiation that is incident on any given patient (KAP) into the corresponding patient effective dose would require explicit consideration (by the medical physicist) of the beam quality, body region, and projection used.

2. **D.** Virtually all radiographic examinations are either very low dose, with effective doses less than 0.1 mSv, or low-dose examinations, with effective doses between 0.1 and 1 mSv. A two-view chest x-ray (0.05 mSv) is very low dose, and an AP abdomen (0.3 mSv) is low dose.

3. **C.** Virtually all fluoroscopy-guided GI and GU examination have effective doses between 1 and 10 mSv and would be considered to be moderate-dose examinations.

4. **C.** A two-view chest x-ray examination is a very-low-dose examination, with a typical effective dose of 0.05 mSv. A TIPS procedure with an effective dose of 50 mSv imparts approximately 1000 times more radiation than would be received from one two-view chest x-ray examination.

Comments

Effective doses are computed from organ doses obtained from the KAP that is *incident* on a patient. Nominal conversion factors (i.e., E/KAP) that may be used for normal-sized adults are depicted in Table 13.2. For a body examination, for example, doubling the patient size from a nominal 75 to 150 kg would likely halve the magnitude of the E/KAP conversion factor. This occurs because the energy deposited for a given incident KAP is essentially unchanged (two-thirds absorbed), but the absorbing mass has been doubled, which will halve the resultant doses.

Effective doses in radiographic imaging are always either very low (i.e., <0.1 mSv) or low (i.e., 0.1 to 1 mSv). Radiographic examinations with effective doses less than 0.1 mSv include the chest x-rays and extremities. Examples of radiographic examinations with effective doses between 0.1 and 1 mSv include most of the principal body regions (e.g., skull, cervical spine, abdomen, and pelvis). Thoracic and lumbar spine examinations have the highest effective doses, which can approach 1 mSv.

The average patient effective dose in fluoroscopy-guided examinations that use an incident KAP of 20 Gy-cm² is generally 4 mSv. Small bowel follow-through (SBFT) is an example of an average dose procedure, whereas esophagrams, upper GI studies, and most urologic examinations have lower doses. However, barium enemas have higher doses. GI and GU examinations therefore have a range of effective doses between 1 and 10 mSv (Fig. 13.10).

Median effective dose for IR patients treated at AMCs is approximately 30 mSv. Approximately half of these patients would have effective doses between 10 and 50 mSv. Examples of low-dose interventional radiology procedures include nephrostomy (stone access) (6 mSv) and pulmonary angiography (IVC filter) (15 mSv). Renal angioplasty (or stent) would likely have an effective dose of 30 mSv. Examples of high-dose procedures include TIPS (50 mSv) and spine embolization (90 mSv).

References

Huda W. Radiography. In: *Review of Radiological Physics*. 4th ed. Philadelphia: Wolters Kluwer; 2016:100–101.

Huda W. Kerma-area product in diagnostic radiology. *AJR. Am J Roentgenol.* 2014;203:W565–W569.

Radiological Society of North America. Radiation dose in x-ray and CT exams. http://www.radiologyinfo.org/en/info.cfm?pg=safety-xray. Accessed February 27, 2018.

TABLE 13.2 NOMINAL EFFECTIVE DOSE PER UNIT KERMA AREA PRODUCT (mSv/Gy-cm²) CONVERSION FACTORS FOR NORMAL-SIZED ADULTS UNDERGOING RADIOGRAPHIC OR FLUOROSCOPY IMAGING EXAMINATIONS

Body Region	Projection	Effective Dose per Unit Kerma Area Product (mSv/Gy-cm²)
Head	All	0.03
Body	AP	0.25
	PA	0.15
	Lateral	0.1
Extremities	All	<0.01

AP, Anteroposterior; *PA,* posteroanterior.

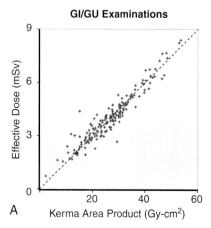
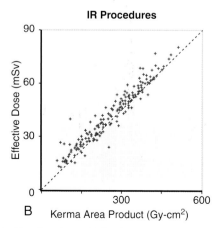

Fig. 13.10 Effective dose versus kerma area product data for gastrointestinal/genitourinary *(GI/GU)* examinations (A) and interventional radiology procedures (B).

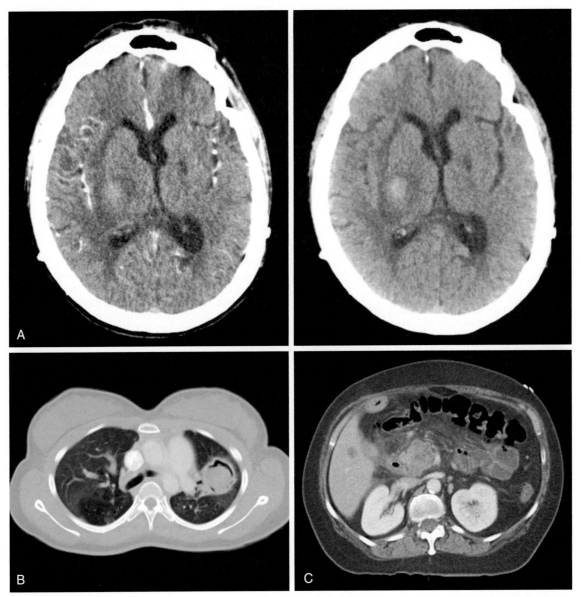

Fig. 13.11

1. What do effective dose per unit DLP conversion factors (*k*-factors), expressed in mSv/mGy-cm, take into account?
   A. Body region
   B. Patient age
   C. Patient size
   D. A, B, and C

2. Adult undergoing routine head CT examinations use _____ radiation and have _____ effective doses than body CT?
   A. More and higher
   B. More and lower
   C. Less and higher
   D. Less and lower

3. Which type of body CT scan likely results in the highest patient effective dose?
   A. Chest
   B. Abdomen
   C. Pelvis
   D. A, B, and C are similar

4. What is the likely dose category of a four-phase abdomen/pelvic CT scan?
   A. Very low dose (<0.1 mSv)
   B. Low dose (0.1 to 1 mSv)
   C. Moderate dose (1 to 10 mSv)
   D. High dose (>10 mSv)

CASE 13.6

Effective Dose in CT

Fig. 13.11 Three studies, each with effective dose of 5 mSv for the complete examination. CT head with and without contrast in a patient with right basal ganglia hemorrhage (A) (dose length product [DLP] = 2000 mGy-cm). CT chest (B) showing aspergilloma (DLP = 250 mGy-cm). CT abdomen (C) showing pancreatic adenocarcinoma (DLP = 340 mGy-cm).

1. **D.** CT effective dose per unit DLP conversion factors account for the body region, as well as patient age and size. For technical reasons, these k-conversion factors just happen to be independent of x-ray beam quality (kV).

2. **B.** Head CT scans use a $CTDI_{vol}$ of 60 mGy, twice as much as in body CT, even though the head transmits much more than an abdomen. Effective doses in head CT scans are approximately 1 to 2 mSv and much lower than body scans (5 mSv) because the head is much less radiosensitive than the body.

3. **D.** Effective dose for chest, abdomen, and pelvic examinations are all generally close to 5 mSv, with little difference between them. However, all these body scans are markedly higher than a typical head CT, which has an effective dose of typically 1 to 2 mSv.

4. **D.** The effective dose for a single body CT scan is most likely 5 mSv so that a four-phase examination that covered the same anatomy would likely be 20 mSv. As such, most multiphase CT scans likely fall into the high-dose category (i.e., >10 mSv).

Comment

CT E/DLP conversion coefficients, generally known as k-factors, enable the total amount of radiation used to perform any given CT scan into the corresponding patient effective dose. All E/DLP conversion coefficients will depend on the scanned anatomy and be different for head and body scans. The k-factors also depend on patient age and the corresponding patient size but have been empirically found to be independent of the x-ray tube voltage (i.e., kV). Using a large phantom rather than a small phantom will halve measured values of $CTDI_{vol}$ and DLP, so E/DLP conversion factors need to be doubled, which makes the patient dose independent of phantom size (Fig. 13.12).

E/DLP conversion factors for normal-sized adult patients are 0.0025 mSv/mGy-cm (small phantom) for head scans. E/DLP conversion factors are 0.020 mSv/mGy-cm for a chest CT and 0.015 mSv/mGy-cm for abdominal and pelvic CT examinations (large phantom). Table 13.3 shows effective doses when normal-sized patients are scanned using the American College of Radiology (ACR) CT Accreditation Program Reference Dose values. These values would be considered an "upper limit." For an adult head CT, a facility using a $CTDI_{vol}$ between 75 and 80 mGy for a routine head would be notified of the high value, and any submission with a $CTDI_{vol}$ more than 80 mGy would automatically "fail." For a routine abdomen on a normal-sized adult, a submission with a $CTDI_{vol}$ in excess of 30 mGy would automatically "fail."

Representative effective doses to a normal-sized adult patient are 2 mSv for head CT scans and 5 mSv for scans of a single body region such as the chest or abdomen. It is of interest to note that head CT scans use more radiation than body scans and transmit more radiation than do body scans. This is to reduce the amount of mottle and permit visualization of subtle features (gray and white matter). Head CT effective doses are low because of the absence of radiosensitive organs. However, when multiphase scans in the body are performed, patient effective doses may exceed 10 mSv and thus be classified as being high, not moderate.

When the CT output is kept constant, effective doses change with patient size in the same manner as organ doses, so that larger patients have lower doses and vice versa. Nowadays, most CT scanners operate an AEC system whereby larger patient receive more radiation, and vice versa. It is generally difficult to predict how patient doses are affected when patient size increases, because the E/DLP conversion factors in larger patients are generally reduced. Modest increases in radiation output for larger patients could result in lower patient doses compared with normal-sized adults, whereas more aggressive AEC systems could result in higher patient doses.

References

Deak PD, Smal Y, Kalender WA. Multisection CT protocols: sex- and age-specific conversion factors used to determine effective dose from dose-length product. *Radiology.* 2010;257(1):158–166.

Huda W. CT II. In: *Review of Radiological Physics.* 4th ed. Philadelphia: Wolters Kluwer; 2016:168–171.

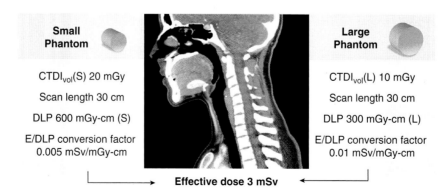

Fig. 13.12 The radiation output for a cervical spine examination can be expressed using a small phantom *(left)* or a large phantom *(right)*. For small phantoms, $CTDI_{vol}$ values are typically double of those for a large phantom, but the effective dose *(E)*/dose length product *(DLP)* conversion factors are half of those for large phantoms so that the resultant effective doses are always the same irrespective of which phantom size is used.

TABLE 13.3 AMERICAN COLLEGE OF RADIOLOGY CT ACCREDITATION PROGRAM REFERENCE DOSE VALUES FOR ADULT CT SCANS AND CORRESPONDING EFFECTIVE DOSES[a]

Parameter	Head	Chest	Abdomen
Reference dose ($CTDI_{vol}$) (mGy)	75	12.5[b]	25
Scan length (cm)	16	40	26
Dose length product (DLP) (mGy-cm)	1200	500	650
Effective dose (E)/DLP (mSv/mGy-cm)	0.0025	0.02	0.015
Effective dose (mSv)	3	10	10

[a]For patients who are scanned at these (upper) dose limit values. Note that the CTDI and DLP for head CT are measured in small phantoms, whereas those for body CT are measured in a large phantom.
[b]Taken as 50% of the abdomen value.

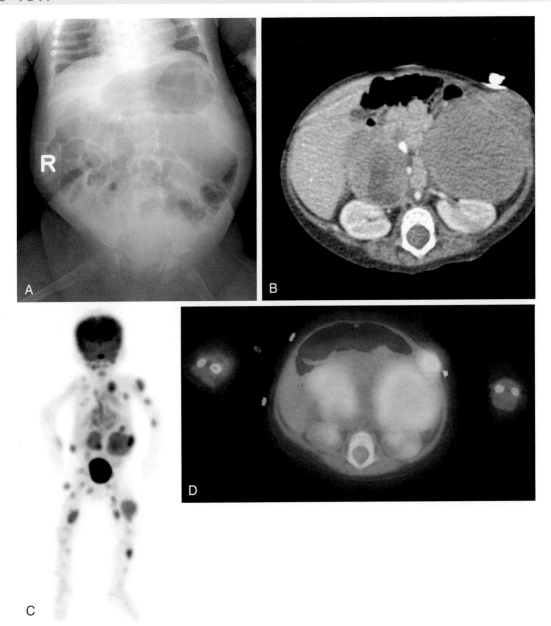

Fig. 13.13

1. How do pediatric patient dose per unit radiation used conversion factors compare with those of adults?
   A. Smaller than
   B. Similar to
   C. Bigger than

2. How do pediatric effective doses in radiography and fluoroscopy most likely compare with those of adults undergoing similar types of examination?
   A. Less than
   B. Similar to
   C. More than

3. What is the ACR CT Accreditation Program Dose limit for a head CT scan in a 1-year-old child (mGy)?
   A. 10
   B. 20
   C. 40
   D. 80

4. Increasing which factor would most likely reduce the dose in a pediatric patient undergoing a diagnostic nuclear medicine scan?
   A. Administered activity
   B. Radionuclide half-life
   C. Biological clearance
   D. Imaging time

CASE 13.7

Pediatric Effective Doses

Fig. 13.13 Five-month-old female with metastatic sarcoma. Multiple imaging modalities using ionizing radiation were used in assessing this patient, including x-ray (A), CT (B), and PET/CT (C and D).

1. **C.** Patient dose (organ and effective) per unit administered activity conversion factors generally increases as the patient gets smaller. The largest dose conversion factors are in infants and the smallest ones in oversized adults.

2. **A.** For similar examinations, effective doses in pediatric patients are lower than those in adults. Both the amount of radiation used (K_{air}) and the x-ray beam areas are much lower, so the incident radiation falls much more than the corresponding increases in the E/KAP conversion factors.

3. **C.** The ACR dose limit for a 1-year-old child having a head CT study is currently 40 mGy in a small phantom. This is half the value of 80 mGy applied to an adult.

4. **C.** Patient doses in nuclear medicine imaging will be reduced when biologic clearance increases because the activity stays in the patient for a shorter period. Increasing the administered activity and the radionuclide half-life both increase patient dose. Imaging time has no effect on dose.

Comment

When pediatric patients undergo any radiologic examination, the effective dose is the product of the amount of radiation used and the corresponding effective dose per unit radiation conversion factor. The amount of radiation used is reduced in pediatric imaging, but the effective dose per unit radiation conversion factors generally increases. Pediatric doses always depend on whether the increase in conversion factor is bigger (or smaller) than the reduction in the amount of radiation being used.

Conversion factors (i.e., E/KAP) in children always increase as patient age and size are reduced (Table 13.4). However, the radiation intensity required to achieve adequate penetration through smaller patients is markedly reduced, as is the corresponding x-ray beam area. The reduction in the amount of radiation used is generally much greater than the increase in the E/KAP conversion factors. Therefore effective doses to pediatric patients undergoing both radiographic and fluoroscopic examinations are markedly lower than those of adults undergoing similar examinations.

Pediatric E/DLP conversion factors exhibit large increases with decreasing age, as shown by the data in Table 13.4. However, the amount of radiation required to generate satisfactory CT images is also reduced, and scan lengths in children are shorter than in adults. Pediatric effective doses in CT are similar to those of adults undergoing similar examinations when reductions in radiation used (DLP) are offset by increases in E/DLP conversion factor (Fig. 13.14). When pediatric protocols adopt more aggressive reductions in x-ray technique factors, modest dose reductions for pediatric CT scans can be achieved.

Organ and effective dose for nuclear medicine studies in infants and young children increase as patient age is reduced per unit administered activity (see Table 13.5 in Case 13.8). However, the activity required to generate satisfactory nuclear medicine studies is also reduced, with most protocols using an amount that is directly related to the patient mass. To make a

TABLE 13.4 PEDIATRIC EFFECTIVE DOSE PER UNIT DOSE-LENGTH

Patient Age	E/DLP$_{pediatric}$/E/DLP$_{adult}$
10	1.6
5	2.2
1	2.3
Newborn	5.0

Conversion factors (*k*-factors) increase as patient age is reduced.
DLP, Dose length product; *E*, effective dose.

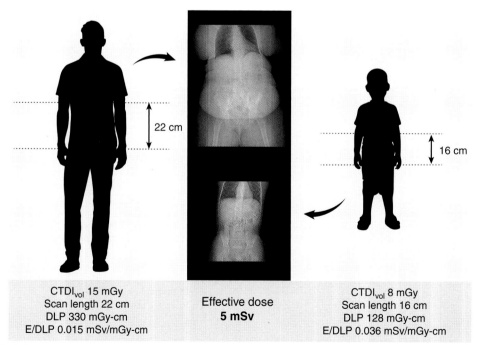

Fig. 13.14 An adult abdomen CT examination (*left*) with a CTDI$_{vol}$ of 15 mGy and a total dose length product (*DLP*) of 330 mGy-cm results in an effective dose of 5 mSv. The CTDI$_{vol}$ for a 5-year-old child is reduced to 8 mGy (*right*), and the total DLP is markedly lower (128 mGy-cm), but the resultant effective dose is still 5 mSv because of the corresponding increase in E/DLP *k*-factor (i.e., 0.036 vs. 0.015).

good approximation, the increase in E/MBq conversion factor is offset by the reduction in the administered activity (MBq). Pediatric effective doses in nuclear medicine are thus generally taken to be approximately similar to those of adults undergoing similar examinations.

References

Huda W. CT II. In: *Review of Radiological Physics*. 4th ed. Philadelphia: Wolters Kluwer; 2016:170-171.

Huda W, Ogden KM. Computing effective doses to pediatric patients undergoing body CT examinations. *Pediatr Radiol*. 2008;38(4):415-423.

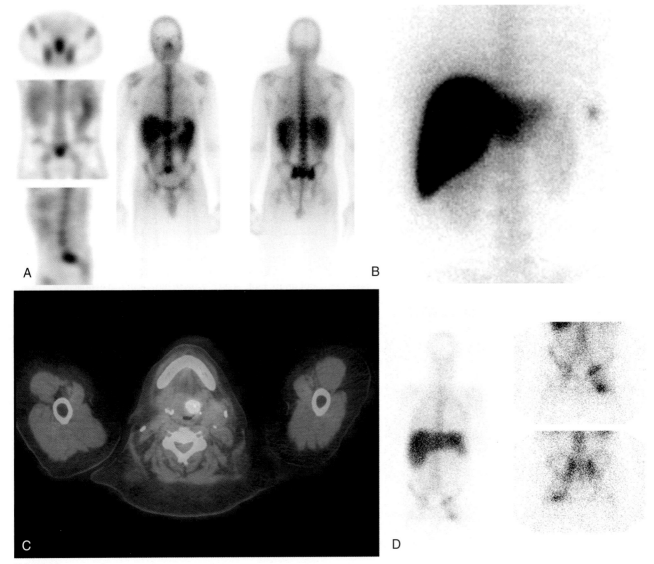

A

B

C

D

Fig. 13.15

1. Which radionuclide is most likely to result in the lowest patient dose in diagnostic nuclear medicine imaging?
   A. Gallium 67
   B. Technetium-99m
   C. Indium-111
   D. Thallium-201

2. What is the most likely patient effective dose from a technetium-99m–labeled pharmaceutical used in diagnostic nuclear medicine (mSv)?
   A. 1
   B. 4
   C. 16
   D. 64

3. Which radionuclide likely has the highest patient dose for the same amount of administered activity?
   A. Technetium-99m
   B. Iodine-123
   C. Fluorine-18
   D. A, B, and C

4. What is the best descriptor of the patient dose for a PET/CT procedure?
   A. Very low (<0.1 mSv)
   B. Low (0.1 to 1 mSv)
   C. Moderate (1 to 10 mSv)
   D. High (>10 mSv)

CASE 13.8

Nuclear Medicine

Fig. 13.15 Gallium-67 scan (A) in a patient with L5–S1 discitis, technetium-99m sulfur colloid scan (B) demonstrating residual splenic tissue following splenectomy, fluorodeoxyglucose PET/CT (C) showing recurrence of laryngeal cancer, indium-111tagged white blood cell scan (D) in a patient with left hip infection.

1. **B.** The amount of activity administered to patients undergoing diagnostic nuclear medicine examinations is approximately 500 MBq to generate the required number of counts in any acquired nuclear medicine image. Technetium-99m has a half-life that is approximately 10 times shorter than the other three nuclides and will decay more quickly, which will markedly reduce patient doses.

2. **B.** Nuclear medicine studies performed with pharmaceuticals labeled with technetium-99m generally have effective doses of approximately 4 mSv. This makes the patient doses fall into the moderate category, such as GI/GU fluoroscopy-based studies, as well as simple CT scans of one body region.

3. **C.** Doses to patients from positron emitters are typically twice as high as those from gamma emitters such as technetium-99m and iodine-123. Positron emitters have energetic positrons (MeV) that deposit their kinetic energy in the uptake organ before undergoing an annihilation event to release two 511 keV photons.

4. **D.** The effective dose from ^{18}F- fluorodeoxyglucose (FDG) is typically 8 mSv, and the nondiagnostic CT scan would likely have an effective dose in the range of 2 to 4 mSv. A PET/CT scan would therefore be at least 10 mSv (i.e., high dose).

Comment

For a given patient, doses from any radiopharmaceutical depend on the decay characteristics of the radionuclide being used. When high-energy beta particles and positrons are present, this energy will always be deposited locally at the uptake site. Short radionuclide physical half-life and rapid biologic clearance will both help to reduce patient radiation doses. The most important characteristic of any radiopharmaceutical is the biodistribution within the patient. When radiosensitive organs and tissues have a high uptake of the administered activity, patient organ and effective doses will increase, and vice versa.

The Society of Nuclear Medicine has a formal Medical Internal Radiation Dosimetry (MIRD) group that publishes organ and effective dose conversion factors for virtually any radiopharmaceutical and for a wide range of patient ranging from the newborn to an adult. Table 13.5 shows typical effective dose per unit administered activity for three radionuclides when administered to normal-size adults, as well as for pediatric patients ranging from a 1-year-old to a 15-year-old. Positron emitters have higher effective doses per unit administered activity, and radiopharmaceuticals that are taken up by radiosensitive organs such as the lung will be higher than those that are taken up by less radiosensitive organs such as bone.

The average effective dose for the technetium-99m radiopharmaceuticals is 4 mSv, which corresponds to a moderate dose procedure (Fig. 13.16). Relatively low effective doses occur in lung scans, which have effective doses of 1.5 mSv. By contrast, cardiac studies generally have much higher effective doses of approximately 9 mSv per procedure. Doses for iodine-123–labeled pharmaceuticals are similar to those of technetium-99m because these have short half-lives of 13 and 6 hours, respectively. A gallium-67 citrate scan (200 MBq) has an effective dose of 20 mSv, primarily because the radionuclide

TABLE 13.5 EFFECTIVE DOSE PER UNIT ADMINISTERED ACTIVITY CONVERSION FACTORS (mSv/100 MBq) FOR THREE RADIOPHARMACEUTICALS (100 MBq = 2.7 mCi)

Patient Age	^{99m}Tc-Labeled MDP[a]	^{99m}Tc-Labeled MAA[a]	^{18}F-Labeled FDG[a]
Adult	0.6	1.1	1.9
15	0.7	1.6	2.4
10	1.1	2.3	3.7
5	1.4	3.4	5.6
1	2.7	6.3	9.5

FDG, Fluorodeoxyglucose; *MAA,* macro aggregated albumin; *MDP,* methylene diphosphonate.
[a]mSv/100 MBq administered activity.

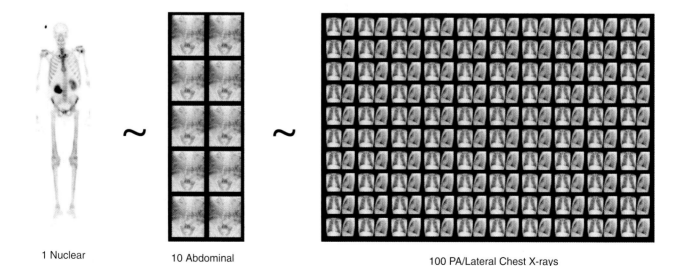

1 Nuclear Medicine Scan   10 Abdominal X-rays   100 PA/Lateral Chest X-rays

Fig. 13.16 The effective dose from a typical nuclear medicine scan performed with technetium-99m has an effective dose that is comparable to 10 abdominal x-rays or 100 chest x-ray examinations. In a given patient, the total detriment of the nuclear medicine scan is taken to be (approximately) comparable with 10 abdominal x-rays or 100 chest x-ray examinations. *PA,* Posteroanterior.

half-life is 10 times longer than that of technetium-99m. The use of thallium-201 has fallen dramatically because of the high dose of this long-lived radionuclide, and it is likely that the same fate will eventually befall gallium-67 and indium-111.

PET using 800 MBq ^{18}F-labeled FDG has an effective dose of approximately 8 mSv and will be directly proportional to the amount of activity administered to the patient. In PET/CT, the total dose depends on the type of CT scan used. Attenuation correction applications only use a relatively low-dose CT scan, which should be performed with a $CTDI_{vol}(L)$ as low as 2 mGy. Such CT scans, for attenuation correction and fusion purposes alone, would have relatively low effective doses of approximately 2 mSv. To generate diagnostic images requires higher-dose CT scans that use a $CTDI_{vol}(L)$ of 15 mGy and have effective doses of up to 15 mSv. When the CT component is added to the PET scan, the resultant combined patient effective dose is generally a little above 10 mSv and thus categorized as a "high-dose" procedure.

References

Huda W. Nuclear medicine II. In: *Review of Radiological Physics.* 4th ed. Philadelphia: Wolters Kluwer; 2016:199–201.

Mettler FA, Huda W, Yoshizumi TT, Mahesh M. Effective doses in radiology and diagnostic nuclear medicine: a catalog. *Radiology.* 2008;248(1):254–263.

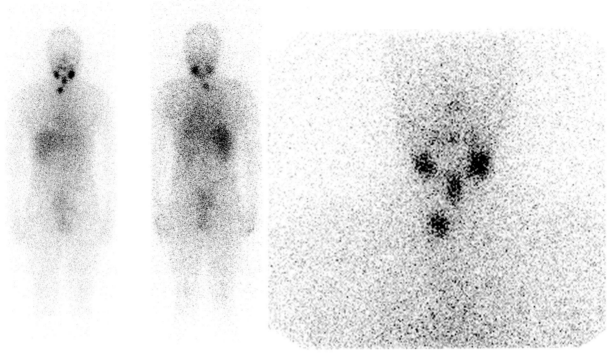

Fig. 13.17

1. What type of "unsealed" radionuclide would likely be ideal for therapeutic applications in a nuclear medicine department?
   A. α
   B. β−
   C. β+
   D. γ

2. Which is a better container for reducing operator exposures from phosphorus 32?
   A. 1-cm plastic
   B. 2-mm lead
   C. A and B are equivalent

3. What is the threshold amount of activity in a patient's thyroid for being discharged to go home from a medical hospital (MBq)?
   A. 10
   B. 100
   C. 1000
   D. 10,000

4. Can a patient who is undergoing iodine 131 treatment continue to breastfeed her infant?
   A. Yes
   B. No
   C. Perhaps

CASE 13.9

Therapeutic Agents

Fig. 13.17 5850 MBq (158 mCi) of iodine-131 was administered to this patient for ablation of the thyroid remnant. An involved lymph node can also be seen in the lower right neck.

1. **B.** A beta minus is the ideal radionuclide for therapy because the emitted electrons have a very short range and deposit virtually all their kinetic energy in the organ of uptake. Positron emitters and gamma ray emitters would make patients a "walking source of radiation," and unsealed alpha sources would be extremely difficult to handle because of concerns about "contamination."

2. **A.** The presence of 2 mm of lead or 1 cm of plastic would both stop the beta particles from escaping. However, energetic electrons will interact with high-Z lead nuclei to produce bremsstrahlung radiation, whereas there is virtually no such bremsstrahlung in a plastic container. Beta emitters should therefore be stored in plastic containers, not lead ones!

3. **C.** The Nuclear Regulatory Commission (NRC) requires that patients with more than 1 GBq (30 mCi) be kept in hospitals to minimize radiation doses to family members, as well as to members of the public.

4. **B.** No, patients undergoing treatment with iodine 131 must not breastfeed their current child. They may breastfeed children in the future with subsequent pregnancies. Radioactive iodine should never be administered to pregnant women.

Comment

The goal of therapeutic nuclear medicine is generally to destroy unwanted tissue (Fig. 13.18). The emitters need to deposit a large dose locally to kill tissue. Furthermore, isotope emissions should not be too penetrating, to limit dose to other organs and to nearby persons. For therapy applications, a beta minus emitter is perfect because virtually all of the beta particle energy is deposited in the organ taking up the radionuclide while sparing surrounding organs and other persons. Table 13.6 shows examples of four beta emitters that are commonly used for therapeutic applications of nuclear medicine.

Iodine-131 is used for treatment of hyperthyroidism and thyroid cancer. Thyroid doses are extremely high. A patient receiving 370 MBq of iodine-131 with a 50% thyroid uptake would likely receive a total thyroid dose of 200 Gy (200,000 mGy). Conversely, dose rates 1 m from the patient are between approximately 50 and 300 μSv/h for hyperthyroidism and cancer, respectively. Patients are generally retained at the hospital until their activity falls to less than 1.2 GBq, but these patients need to be warned that they may activate radioisotope security alarms and are given instruction on minimizing dose to family and the surrounding public. Breastfeeding is contraindicated for radioiodine therapy and should be discontinued several weeks before therapy to reduce the mother's breast dose.

Yttrium-90 is a pure beta emitter that is formed into microspheres and used in selective internal radiation therapy (SIRT). Yttrium-90 microspheres are selectively injected into the right or left hepatic artery to embolize and radiate primary or metastatic hepatic tumors. Preferential arterial flow causes the microspheres to accumulate in hypervascular malignant tissue.

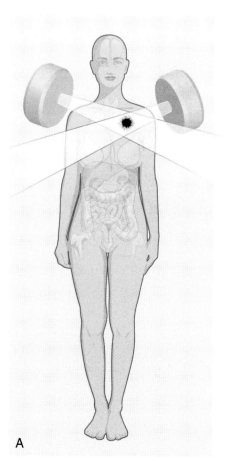

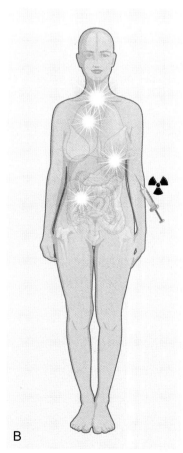

Fig. 13.18 External radiation therapy is ideal for a single lesion (A) which can be targeted using multiple external beams. A radioactively labeled biologic agent (e.g., monoclonal antibody) is ideal for diffuse disease (B).

TABLE 13.6 FOUR OF THE MOST COMMON RADIONUCLIDES USED FOR THERAPEUTIC APPLICATIONS OF NUCLEAR MEDICINE (UNSEALED SOURCES)

Radionuclide	Half-Life (Days)	Mean Beta Energy (MeV)
Phosphorus-32 ($Z = 15$)	14	0.70
Strontium-89 ($Z = 38$)	51	0.58
Yttrium-90 ($Z = 39$)	2.7	0.93
Iodine-131[a] ($Z = 53$)	8.0	0.19

[a]Also emits 365 keV gamma rays.

External dose is very low after injection because yttrium-90 is a pure beta emitter with beta particle penetration of only a few millimeters in soft tissue. Other historically notable therapeutic agents from nuclear medicine include strontium-89 and phosphorus-32. The treatment of painful osseous metastases from prostate and breast cancer can be performed using strontium-89, which is administered intravenously. Phosphorus-32 has been used to treat polycythemia vera.

A written directive is always mandatory for any therapy procedures in a nuclear medicine department. Furthermore, the NRC has to be notified if any unintended radiation exposures exceed 50 mSv effective dose or 500 mGy organ doses. In areas where written directive procedures are carried out, surveys are performed daily to monitor for spills. A major radiation spill involves the release of more than 40 MBq of iodine-131 and will require the presence of the radiation safety officer to direct clean-up. Volatile radionuclides such as liquid iodine-131 is generally stored in fume hoods. Nuclear medicine operators handling liquid iodine-131 are required to undergo mandatory bioassay to have their thyroids monitored for iodine uptakes.

References

American College of Radiology. Resolution 49 ACR practice parameter for the performance of therapy with unsealed radiopharmaceutical sources; 2015. https://www.acr.org/Clinical-Resources/Practice-Parameters-and-Technical-Standards/Practice-Parameters-by-Modality. Accessed February 27, 2018.

Huda W. Nuclear medicine II. In: *Review of Radiological Physics*. 4th ed. Philadelphia: Wolters Kluwer; 2016:200–201.

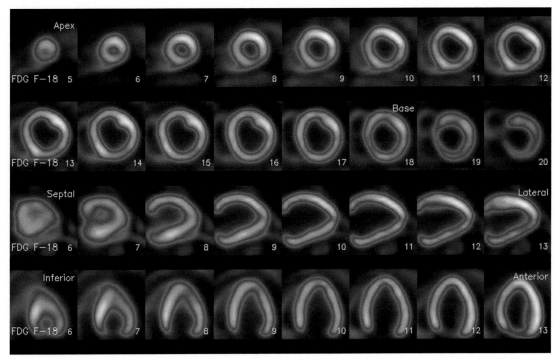

Fig. 13.19

1. Which category would the background radiation received by an American in 1 year fall?
   A. Very low (<0.1 mSv)
   B. Low (0.1 to 1 mSv)
   C. Moderate (1 to 10 mSv)
   D. High (>10 mSv)

2. Which background source contributes the highest doses to an average American?
   A. Cosmic
   B. Terrestrial
   C. Internal
   D. Radon

3. In 2006, what percentage of ionizing radiation received by the US population was most likely from medical exposures (%)?
   A. 10
   B. 25
   C. 50
   D. 75

4. Which modality contributes the highest dose of ionizing radiation to the population?
   A. Conventional radiography
   B. Interventional fluoroscopy
   C. Nuclear medicine
   D. Computed tomography

CASE 13.10

US Population Doses

Fig. 13.19 Cardiac single photon emission CT in a patient with congestive heart failure.

1. **C.** The average American is exposed to approximately 3 mSv every year from natural background. This effective dose would be deemed moderate because it falls in the range of 1 to 10 mSv and is like a low-end GI/GU fluoroscopy-guided procedure.

2. **D.** Radon contributes an average effective dose of 2 mSv/year, whereas the remaining three sources (cosmic, terrestrial, and internal) contribute a total of 1 mSv/year.

3. **C.** In 2006, the National Council on Radiation Protection and Measurements issued Report 160, which showed that the average US medical dose was approximately 3 mSv/year, which is virtually the same as that from natural background.

4. **D.** CT accounts for approximately 15% of all diagnostic imaging procedures but contributes approximately 50% of the total medical dose (see figure).

Comment

Ubiquitous natural background radiation consists of cosmic, internal, and terrestrial radioactivity. Cosmic radiation from outer space has an average E of 0.4 mSv/year. Internal radionuclides include potassium-40, with a half-life in excess of 1 billion years, and carbon-14, which is produced by cosmic radiation interacting with atmospheric nitrogen. Internal radionuclides have average E values of 0.4 mSv/year. Terrestrial radioactivity includes naturally occurring radionuclides such as radium-226, which is a decay product of uranium-238. Terrestrial radioactivity results in an average E of 0.3 mSv/year. The average ubiquitous background annual dose in the United States is thus approximately 1 mSv.

Ubiquitous background is received by everyone, but Americans are also exposed to radon. Radon-222 is a radioactive gas produced during the decay of radium that emits alpha particles. Radon daughters are also radioactive and can attach to aerosols that, when inhaled, will be deposited in the lungs. Exposure of the lung results in elevated levels of lung cancer, similar to the excess lung cancers that are observed in miners that inhale large quantities of alpha emitters (e.g., uranium miners). The US Environmental Protection Agency believes that 15% of lung cancers may be a result of domestic exposure to radon.

Air crews receive approximately 5 mSv because they fly approximately 1000 hours each year at heights greater than 30,000 ft, where cosmic radiation is much higher. Leadville, CO, at a 3000-m elevation, has an additional 0.9 mSv/year attributed to higher cosmic radiation and also has elevated levels of terrestrial radioactivity that give an additional 0.7 mSv/year to its inhabitants. One of the highest background radiation levels is in Ramsar, Iran, mainly because of the use of naturally radioactive limestone for building dwellings. It has been estimated that the most exposed residents will receive an effective dose from external sources of approximately 6 mSv each year. This exposure level is six times the regulatory public dose limit exposure in the United States.

The US population average dose from 400 million diagnostic medical examinations in 2006 was 3 mSv, much higher than in the 1980s, when it was only 0.6 mSv. Medical doses have thus increased sixfold in one generation, and medical doses (per capita) will soon exceed doses from all sources of natural background combined. CT scans contribute nearly one-fifth of all diagnostic x-ray exams but account for approximately half the population medical dose (Table 13.7). Nuclear medicine accounts for another 25% of the medical dose, so that three-quarters of this medical dose is attributable to nuclear medicine and CT (Fig. 13.20). It is estimated that 85% of the nuclear medicine dose is attributable to cardiac imaging.

References

Huda W. Patient dosimetry. In: *Review of Radiological Physics.* 4th ed. Philadelphia: Wolters Kluwer; 2016:54–57.

National Council on Radiation Protection and Measurements (NCRP). *NCRP Report No. 160: Ionizing Radiation. Exposure of the Population of the United States. National Council on Radiation Protection and Measurements.* Bethesda, MD: NCRP; 2009.

TABLE 13.7 ESTIMATED NUMBER OF PROCEDURES, AND THE CORRESPONDING DOSES, OF DIAGNOSTIC EXAMINATIONS CURRENTLY IN THE UNITED STATES

Procedure	Number of Procedures (Millions)	Average Effective Dose (mSv)	Collective Effective Dose (Person-Sieverts)
CT	80	5	400,000
Radiography/fluoroscopy	300	0.3	100,000
Interventional	20	7.5	150,000
Nuclear	20	10	250,000

The total collective effective dose is 900,000 person-Sieverts, the same to that from natural background.

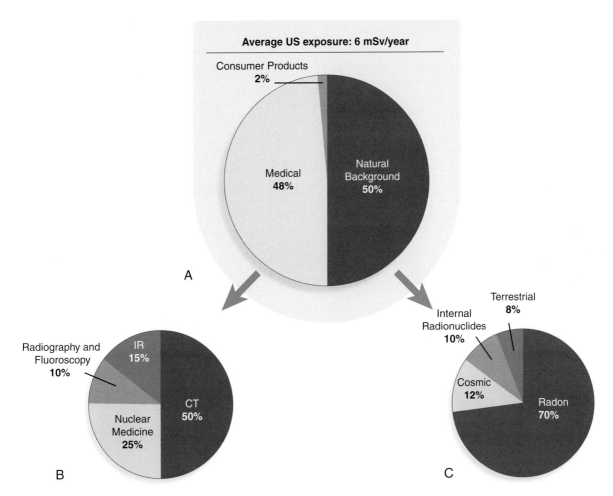

Fig. 13.20 In 2006 the average American was exposed to approximately 6 mSv, with half being from natural background and the man-made exposures dominated by medical exposures (A). CT and nuclear medicine exams account for three-quarters of the medical exposures (B), and radon dominates the natural background exposures (C).

CASE 14.1

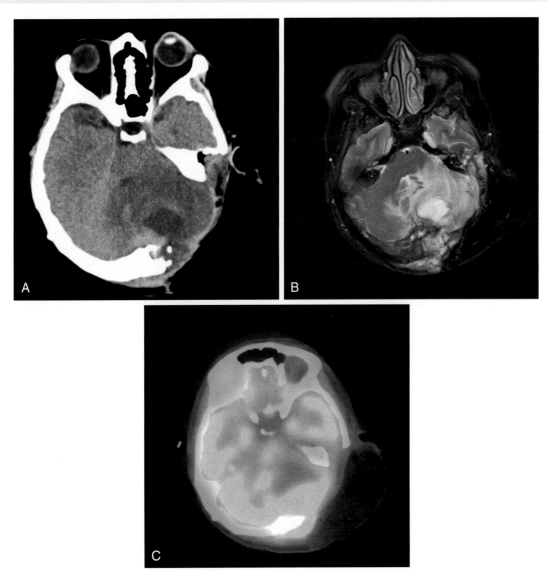

Fig. 14.1

1. Which is the foremost international agency that currently provides specific recommendations on radiation protection issues?
   A. International Commission on Radiological Units and Measurements (ICRU)
   B. International Commission on Radiological Protection (ICRP)
   C. United Nations Scientific Committee on Effects of Atomic Radiation (UNSCEAR)
   D. National Council on Radiation Protection and Measurements (NCRP)

2. Which organization provides estimates of radiation carcinogenic risks?
   A. ICRP
   B. UNSCEAR

   C. National Academy of Sciences (NAS) Committee on the Biological Effects of Ionizing Radiation (BEIR)
   D. A, B, and C

3. Which agency do US regulatory bodies turn to for advice on regulatory dose limits?
   A. NCRP
   B. ICRP
   C. UNSCEAR
   D. NAS Committee on the BEIR

4. Which of the following are examples of non-ionizing radiations?
   A. Ultrasound
   B. Radio waves
   C. Infrared
   D. A, B, and C

CASE 14.1

Advisory Bodies

Fig. 14.1 Imaging of a patient with recurrent epidermoid may include examinations with ionizing radiation using CT (A) and PET/CT (B) in addition to nonionizing studies such as MRI (C).

1. **B.** The ICRP is the foremost international radiation protection organization. ICRU addresses only scientific measurements, and UNSCEAR addresses only the issue of radiation risks. The NCRP is a specifically US organization whose recommendations are generally adopted by US regulatory agencies.

2. **D.** The ICRP, UNSCEAR, and BEIR are all valuable sources of scientific information pertaining to current estimates of radiation risks. In general, the risk estimates provided by each of these independent bodies are very similar (and authoritative).

3. **A.** US regulatory agencies generally look to the NCRP for specific recommendations on radiation protection issues and especially on regulatory dose limits. In general, the recommendations of the NCRP are similar to those of the ICRP.

4. **D.** Ultrasound, radio waves, and infrared are all examples of nonionizing radiations, and each of these is used in medical imaging. Safety of these types of nonionizing radiations is addressed by the international level by the International Commission of Non-Ionizing Radiation (ICNRP).

Comments

The ICRU makes general recommendations about scientific issues such as measurement units in radiology (e.g., Kerma). The ICRP was founded in 1928 at the second International Congress of Radiology in Stockholm, Sweden. The ICRP is an independent scientific agency that provides general recommendations pertaining to radiation workers, as well as members of the public. The ICRP also provides specific recommendations pertaining to patients and subjects in research studies. The ICRP published its most recent general recommendations in ICRP Publication 103 (2007).

The UNSCEAR assesses quantitatively issues that relate to radiation risk that pertain to international populations. The United States has the NAS, which set up a committee pertaining to the BEIR that generates estimated radiation risks applicable to an American population. Together with the ICRP, these are the scientific bodies that have published the most authoritative radiation risk estimates. In general, the radiation risk estimates provided by the ICRP, UNSCEAR, and BEIR are very similar and are valuable scientific sources regarding ionizing radiation (Fig. 14.2).

The major radiation protection body in the United States is the NCRP. Federal and state regulators are provided advice on radiation protection by the NCRP. NCRP Publication 160 is a definitive study into the exposure of the US population to both natural background radiation and man-made sources, which are primarily medical. The most important finding of the NCRP is that, from the 1980s to 2006, per capita US medical doses have increased by approximately 600%. The NCRP also makes recommendations on dose limits to regulatory agencies, and its recommendations are generally very similar to those that are currently made by the ICRP.

The American College of Radiology (ACR) publishes on MR safety what is generally considered the "gold standard" in the field. Research in the field of MR safety is conducted by members of the International Society for Magnetic Resonance in Medicine (ISMRM) and published in its journals. The American Institute of Ultrasound in Medicine (AIUM) is a multidisciplinary association dedicated to advancing the safe and effective use of ultrasound in medicine. In the United Kingdom, the British Medical Ultrasound Society provides advice on ultrasound, which includes explicit quantitative guidelines of limits in clinical ultrasound (e.g., thermal index [TI] in obstetrics). The ICNRP publishes advisory documents pertinent to MRI and is currently working on documents for ultrasound.

References

Huda W. Radiation protection. In: *Review of Radiological Physics*. 4th ed. Philadelphia: Wolters Kluwer; 2016:77–78.

Huda W. Radiation risks: what is to be done? *AJR Am J Roentgenol.* 2015;204:124–127.

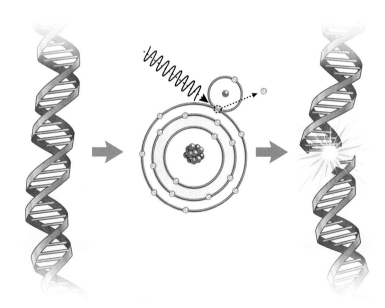

Fig. 14.2 Ionizing radiation may have photon energies that are greater than a few electron volts and can thereby eject outer shell electrons from atoms and molecules. Ionizing radiation thus can break apart biological important molecules such as DNA, which cannot be achieved by nonionizing radiation such as radio waves and visible light.

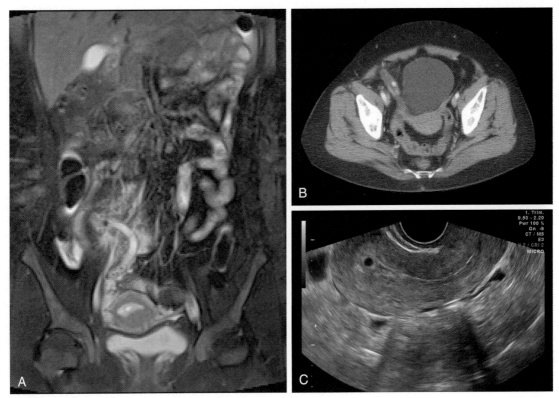

Fig. 14.3

1. Which are possible effects when a patient undergoes a PET/CT examination?
   A. Nonfatal cancers
   B. Fatal cancers
   C. Hereditary effects
   D. A, B, and C

2. What is the most appropriate radiation risk model for predicting radiation risks of solid tumors at doses currently encountered in CT imaging?
   A. Threshold
   B. Linear–no threshold
   C. Supralinear
   D. Hormesis

3. For a population exposed to 10 mSv of uniform whole-body x-ray radiation, what is the most likely cancer induction risk (%)?
   A. 0.01
   B. 0.1
   C. 1
   D. 10

4. What is the most likely frequency of embryo doses exceeding 100 mGy in a US academic medical center?
   A. Monthly
   B. Annually
   C. Decade
   D. Generation

CASE 14.2

Radiation Risks

Fig. 14.3 Acute appendicitis seen on MRI (A) in a patient with early pregnancy. A gestational sac can be seen on transvaginal ultrasound (C). An appendicolith and dilated appendix can be seen on CT in a different patient (B).

1. **D.** Following a PET/CT scan, with an effective dose more than 10 mSv, it is possible to estimate the risk of nonfatal cancers, fatal cancers, and the induction of genetic effects. The average detriment associated with genetic effects is currently believed to be very low (<10%).

2. **B.** Virtually all authoritative radiation protection bodies use the linear–no threshold for the risk of solid tumors, mainly on the grounds of being prudent (cautionary principle). There are advocates for a threshold, as well as those who believe current risks are underestimates. Hormesis is a theory that proposes that exposure to low doses is beneficial.

3. **B.** The average risk from an exposure of 10 mSv is currently believed to be 0.1%. Adopting this value requires practitioners to ensure there is a net patient benefit (i.e., benefit > risk) and to eliminate unnecessary radiation.

4. **D.** Embryo doses that exceed 100 mGy are very rare. A medical physicist working at a large US academic medical center might encounter one such event in their professional lifetime.

Comment

For most patients undergoing diagnostic radiological examinations with x-rays, organ doses will be well below the threshold dose for the induction of deterministic risks. Accordingly, effects such as radiation burns, cataracts, epilation, and sterility are not normally expected. Patient risks in radiology primarily pertain to the stochastic processes of carcinogenesis (fatal and nonfatal), as well as the induction of genetic effects that would be manifested in the offspring of exposed individuals. There is plentiful evidence of the carcinogenic effects of radiation at high doses, and most current risk estimates have been obtained from studies of atomic bomb survivors. There are no human data pertaining to genetic effects of radiation, and current genetic risk estimates are based on experiments performed in animals.

There are only limited data on the effects of radiation at the doses normally encountered in radiology, where organ doses are generally less than 100 mGy. In the absence of definitive data at low doses, the scientific community must make decision on how it should act in practice. Most scientific agencies make use of the linear–no threshold model to extrapolate established risks at higher doses to those encountered in medical imaging. This is a prudent approach because the alternative of assuming that there are no risks (which might be subsequently prove to be erroneous) would be socially unacceptable. Practitioners should always remember that current radiation risks estimates at "low doses" are generally believed to have uncertainties of factors of two to three in both directions and may even be zero.

A 25-year-old exposed to 10 mSv is believed to increase their cancer risk by approximately 1 in 1000 (i.e., 0.1%). For young children, this radiation risk is taken to be approximately three times higher, and in retirees the risk would be reduced by approximately a factor of three (Fig. 14.4). In the absence of any exposure to ionizing radiation, it is currently estimated that the average risk of cancer in the US population is approximately 40%. A 25-year-old patient undergoing a chest/abdomen/pelvic CT examination with an effective dose of 10 mSv might be increasing their cancer risk from 40% to 40.1%. Because of this nonzero risk, radiologists must make sure that patient benefits are greater than this assumed patient cancer risk. Furthermore, given that any exposure is taken to increase the cancer risk, there is a requirement to eliminate unnecessary radiation exposures (i.e., keep exposures As Low As Reasonably Achievable [ALARA]). Four hundred million radiological examinations are performed each year, resulting in a huge net benefit to the US population.

At less than radiation doses of approximately 100 mGy, it is currently believed that there are no deterministic effects such as embryonic death and congenital malformations. At less than 100 mGy, there is a small risk of an increase in childhood cancer. There are good scientific data that fetal doses as low as 10 mGy can increase the incidence of childhood cancer. The

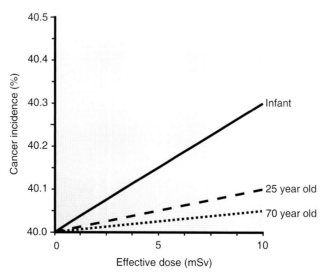

Fig. 14.4 For a 25-year-old, the cancer risk from a whole-body dose of 10 mSv is currently estimated at 0.1% *(dashed line)*. For the same exposure, the risk in an infant *(solid line)* is approximately three times higher, and the risk in a 70-year-old is estimated to be three times lower *(dotted line)*.

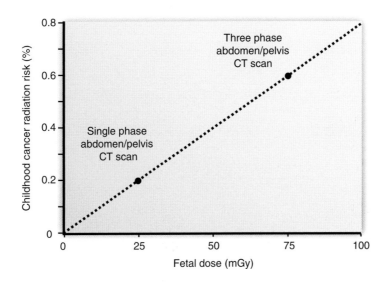

Fig. 14.5 A fetal dose of 25 mGy is believed to carry a risk of childhood cancer of 1 in 500, which is also the background incidence in the absence of radiation exposure. A single-phase abdomen/pelvis CT scan likely results in a conceptus dose of 25 mGy, which would be increased threefold in a full three phase study.

best current risk estimate of childhood cancer risks are 1 in 500 of childhood cancer (0.2%) following a dose of 25 mGy in late pregnancies. In the absence of any x-ray exposure, the background incidence of childhood cancer in the United States is 1 in 500 (0.2%). A CT scan of the abdomen/pelvis of pregnant woman would likely result in a fetal dose of approximately 25 mGy and which would therefore most likely double the (small) background incidence of childhood cancer (Fig. 14.5).

References

Huda W. Radiation risks. In: *Review of Radiological Physics*. 4th ed. Philadelphia: Wolters Kluwer; 2016:63–70.

Huda W. Radiation risks: what is to be done? *AJR Am J Roentgenol.* 2015;204(1):124–127.

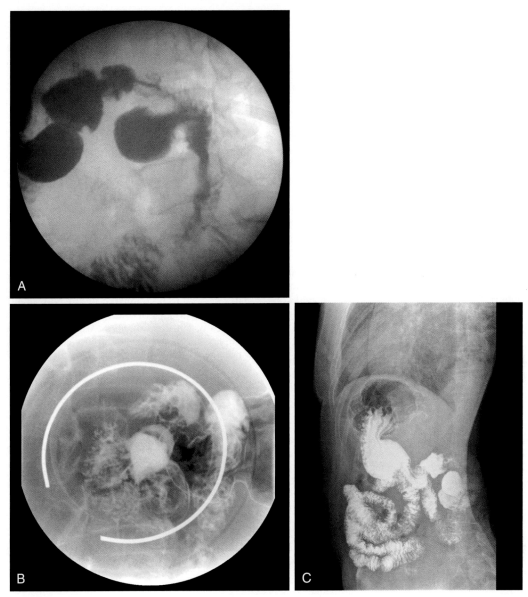

Fig. 14.6

1. Which is required by the US Food and Drug Administration (FDA) of all x-ray imaging equipment that is currently sold in the United States?
   A. Safety
   B. Effectiveness
   C. Accuracy
   D. A and B

2. Who regulates the clinical use of a CT scanner?
   A. FDA
   B. "State regulatory agency" (SRA)
   C. Nuclear Regulatory Commission (NRC)
   D. NCRP

3. Who regulates mammography in the United States?
   A. FDA
   B. SRA
   C. NRC
   D. Conference of Radiation Control Program Directors (CRCPD)

4. Who regulates the use of therapeutic agents (e.g., iodine 131) in nuclear medicine?
   A. FDA
   B. SRA
   C. NRC
   D. ICRP

CASE 14.3

US Regulatory Bodies

Fig. 14.6 Duodenal diverticulum seen on upper gastrointestinal examination with small bowel follow-through. Exposure index for the spot (C) and photospot (B) images is approximately 300, and that for the fluoroscopy image (A) is approximately 100 times lower.

1. **D.** The FDA is a US regulatory agency that requires all medical devices sold in the United States to be "safe" and "effective." However, the FDA does not regulate the use of imaging systems, and these are the responsibility of individual states.

2. **B.** Individual states are responsible for regulating the use of x-ray emitting devices such as fluoroscopy systems and CT scanners. Consistency in regulations is achieved through an annual meeting of the CRCPD.

3. **A.** Individual states generally regulate x-ray emitting devices such as conventional x-ray systems. The one exception to this general rule is mammography, which is regulated at the federal level through the Mammography Quality Standards Act (MQSA), which is enacted by the FDA.

4. **C.** The use of any radioactive material such as iodine 131 is the responsibility of the NRC. The NRC can "delegate" the implementation of these regulations to the states (agreement states) or regulate a state directly (nonagreement states).

Comment

One important federal agency is the FDA, which requires that all medical devices are deemed to be "safe and effective" and is also responsible for regulating the quality and purity of radio pharmaceuticals. Equipment sold in the United States must meet specified safety standards, where the data in Table 14.1 show examples that pertain to medical imaging. The MQSA was passed in 1992 and is the only example of direct regulation of x-ray emitting devices at the federal level. The 10,000 operational mammography systems in the United States need to be certified by the FDA and accredited by an appropriate agency such as the ACR. Implementation of the MQSA requirements is frequently subcontracted to appropriate state agencies.

Individual states regulate how x-ray emitting devices are actually used by practitioners. Regulations will specify, for example, a minimum x-ray beam quality, which is approximately 3 mm aluminum at 80 kV, which ensures that there is adequate patient penetration that minimizes unnecessary patient dose (Fig. 14.7). The maximum entrance air kerma rate in fluoroscopy will be measured by state inspectors by imaging a lead apron that drives the exposures to the maximum possible. Operator exposures and safety practices are also the responsibility of state agencies, which have power to impose significant financial penalties for noncompliance. In practice, the regulatory requirements in all 50 US states are very similar. The reason for this is that all 50 states coordinate their radiation protection activities via an annual meeting of the Conference of Radiation Control Program Directors, known as the CRCPD.

The federal NRC has the responsibility to regulate all materials that are radioactive. An agreement state will arrange with the NRC to self-regulate licensing and inspection requirements for nuclear materials, including medically related issues. Other states, which are thus categorized as nonagreement states, choose to be regulated by the NRC in a direct manner. The Nuclear Regulatory Agency defines the roles and responsibilities of authorized users (AUs) in nuclear medicine. The accuracy of the amount of activity administered to a patient, for example, has to be within ±20% of the amount specified in the protocol.

TABLE 14.1 EXAMPLES OF US FOOD AND DRUG ADMINISTRATION REQUIREMENTS FOR RADIOGRAPHY AND FLUOROSCOPY UNITS SOLD IN THE UNITED STATES

Imaging Modality	Requirement
Radiography	X-ray tube leakage <1 mGy/h
	Half value layer >2.5 mm aluminum (at 70 kV)
	Table thickness <2 mm aluminum equivalence (80 kV)
Fluoroscopy	Audible alarm after 5 min
	Fluoroscopy activation by dead man's switch
	Last image hold

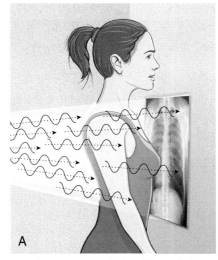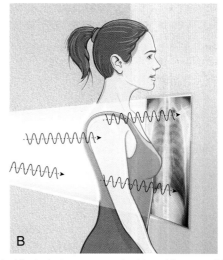

Fig. 14.7 Low-energy x-rays have poor penetration (A), so that many incident photons are required to achieve a satisfactory intensity at the image receptor (e.g., exposure index of 300). When the x-ray energy increases (B), less of this more penetrating radiation is required to achieve the same intensity at the image receptor. Requiring a minimum beam quality in radiography and fluoroscopy helps to ensure that patients are not subject to unnecessary exposures.

Current US regulations frequently require an imaging modality to be accredited by an appropriate agency. In effect, the accreditation process is subcontracted to bodies, such as the ACR, or to individual states. Mammography accreditation can be provided by bodies who meet minimum requirements that are specified by the FDA. Most facilities get accredited by the ACR. However, mammography accreditation is also available by several states, including Texas, Iowa, and Arkansas.

References

Huda W. Radiation protection. In: *Review of Radiological Physics*. 4th ed. Philadelphia: Wolters Kluwer; 2016:77-78.

US Food and Drug Administration. Radiation-emitting products; 2018. https://www.fda.gov/Radiation-EmittingProducts/default.htm. Accessed February 27, 2018.

US Nuclear Regulatory Commission. Medical uses of nuclear materials; 2017. https://www.nrc.gov/materials/miau/med-use.html. Accessed February 27, 2018.

CASE 14.4

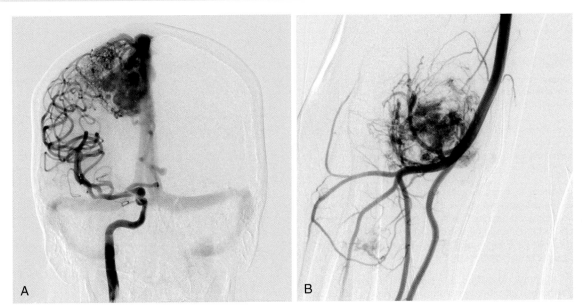

Fig. 14.8

1. What is the current regulatory effective dose limit to an interventional radiology (IR) fellow (mSv/year)?
   A. 0.5
   B. 5
   C. 50
   D. 500

2. How did the ICRP in 2011 recommend that eye lens dose limits be adjusted?
   A. Reduced
   B. Unchanged
   C. Increased

3. What is the current regulatory extremity dose limit for a radio pharmacist (mSv/year)?
   A. 0.5
   B. 5
   C. 50
   D. 500

4. What is the current regulatory dose limit to the fetus of a radiation worker (mSv/month)?
   A. <0.5
   B. 0.5
   C. >0.5

CASE 14.4

Regulatory Dose Limits

Fig. 14.8 Digital subtraction angiography images showing a right frontoparietal arteriovenous malformation (A) and an arteriovenous malformation adjacent to the elbow (B).

1. **C.** The current effective dose limit for occupational exposure is 50 mSv per year. An IR fellow would likely be exposed to 10% of this limit at most US academic medical centers.

2. **A.** The current US regulatory eye lens dose limit is 150 mSv/year and is designed to ensure that operators do not get a radiation induced cataract. In 2011, the ICRP recommended that this limit should be reduced to 20 mSv/year because the threshold dose for cataract induction is now lower (0.5 Gy) than previously believed.

3. **D.** The extremity dose limit is currently 500 mSv per year. A radio pharmacist who continuously handles radioactivity would likely wear an extremity dosimeter to ensure that this dose limit is not exceeded.

4. **B.** The current fetal dose limit for a radiation worker is 0.5 mSv per month. This implies that a fetus might be exposed to more than the 1 mSv annual dose limit that currently applies to "members of the public."

Comment

Regulatory (i.e., legal) effective dose limit in the US for radiation workers is currently 50 mSv/year. Actual occupational exposures in medicine, including IR, are less than or equal to 10% of the regulatory limits. These exposures are monitored by various dosimeters worn by radiation workers (Fig. 14.9). The current US public regulatory effective dose limit for members of the public is 1 mSv/year. X-ray facilities are built to ensure that exposure to members of the public cannot exceed 1 mSv/year. All regulatory dose limits explicitly exclude natural background radiation and all medical exposures. This latter requirement refers to all exposures the worker may receive as a patient. Medical x-rays are excluded because diagnostic information from radiological examinations confers a benefit to the exposed individual.

Eye lens dose limit for occupational exposed US workers is currently 150 mSv/year. This limit was designed to prevent the induction of eye cataracts for exposures over a working lifetime. The ICRP (2011) recommended reducing this regulatory eye lens dose equivalent limit (i.e., 150 mSv) to 20 mSv/year. In 2017 the NCRP recommended the eye lens dose limit be reduced to 50 mSv/year, but no US regulations have hitherto (2018) been introduced. Although the United States has not reduced lens dose limits, countries in Europe, such as the United Kingdom, have implemented the ICRP recommendations into their regulatory framework.

Dose limits to the extremities and skin are designed to ensure that deterministic effects cannot occur. Because skin has been shown to tolerate therapy fractionated doses of 20 Sv

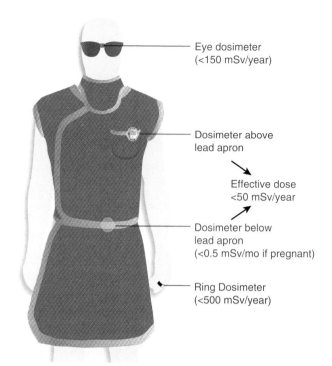

Fig. 14.9 Four dosimeters would be required to demonstrate full compliance will all US regulatory dose limits. Two dosimeters, worn above and below a lead apron, are required to obtain an estimate of the operator effective dose. A dosimeter should be worn near the eyes to estimate the eye lens dose, and a ring dosimeter should be worn on a finger to estimate the extremity dose. For pregnant workers, a fetal dose badge worn under the lead at waist level is used to estimate fetal dose.

of x-rays, a regulatory limit of 500 mSv/year guarantees lifetime occupational doses less than this value. Accordingly, the dose limit to the skin of a radiation worker is 500 mSv/year. The dose limit for hands of radiation workers is also 500 mSv/year, which guarantees that there will be no deterministic risks over a working lifetime.

Regulatory dose limits in the United States to a fetus of a radiation worker is 0.5 mSv/month, but this is only applicable to women who have "declared a pregnancy" in writing to their employer. Accordingly, the US maximum fetal dose is 5 mSv during pregnancy, which is higher than the public limit of 1 mSv/year. If the maximum fetal dose limit was set to 1 mSv, this would potentially deprive women of reproductive capacity the chance of employment as radiation workers.

References

Huda W. Radiation protection. In: *Review of Radiological Physics.* 4th ed. Philadelphia: Wolters Kluwer; 2016:77–78.

Sensakovic WF, Flores M, Hough M. Occupational dose and dose limits: experience in a large multi-site hospital system. *J Am Coll Radiol.* 2016;13(6):649-655.

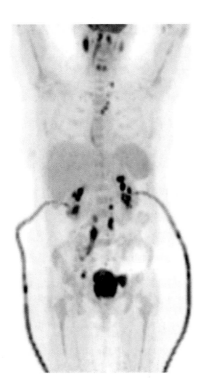

Fig. 14.10

1. What are the goals of a radiation protection program for radiologists and technologists?
   A. Eliminate deterministic risks
   B. Minimize stochastic risks
   C. Reduce operating expenses
   D. A and B

2. What is the radiation intensity at 1 m from the patient, relative to the radiation intensity that is incident on the patient (%)?
   A. 0.01
   B. 0.1
   C. 1
   D. 10

3. Increasing which of the following will most likely reduce operator doses?
   A. Distance from source
   B. Shielding thickness
   C. Exposure time
   D. A and B

4. Which protection strategy is most effective at reducing operator doses?
   A. Distance
   B. Shielding
   C. Time
   D. A, B, and C are similar

CASE 14.5

Protecting Workers

Fig. 14.10 PET/CT in a patient with cervical cancer and bilateral nephrostomy tubes. Note the presence of excreted FDG in the drainage catheters.

1. **D.** The goal of any radiation protection program should always be to prevent the induction of deterministic effects and to minimize stochastic effects. To achieve this will involve the expense of significant monies for equipment and personnel.

2. **B.** The scatter intensity is typically 0.1% of the radiation incident on the patient. In a lateral skull x-ray, where 1 mGy is incident on the patient, the scatter intensity at 1 meter is 1 μGy.

3. **D.** Increasing the amount of shielding and the distance from a source of radiation will both reduce operator doses. However, increasing the exposure time will increase operator doses, and this increase is directly proportional to the exposure time.

4. **B.** Increased shielding is by far the most effective way of reducing operator doses. Increasing distance, which obeys the inverse square law, is also more effective than reducing the exposure time, which is directly proportional to operator doses.

Comment

Radiation protection of workers should strive to prevent deterministic effects as well as minimize all stochastic risks. All reductions of stochastic risks should be "reasonable" by considering all the benefits gained by the radiation workers, including their incomes. This approach is essentially the same as the ALARA principle applied to patient exposures where unnecessary radiation is to be minimized. Occupationally exposed individuals are usually monitored using personnel dosimeter such as thermoluminescent dosimeters (TLDs) or optically stimulated luminescent (OSL) dosimeters. Radiation intensities can be measured by a medical physicist using an ionization chamber. The NCRP explicitly stresses that the ALARA principle must always be applied and that dose limits are upper limits and not "allowable worker doses."

X-ray personnel can be exposed if they are close to the patient undergoing the x-ray. Operator doses are always directly proportional to patient doses. To estimate an operator exposure, a good rule of thumb is that the scatter dose level from patients at a distance of 1 m is approximately 0.1% of the entrance dose incident on the patient. Radiation workers do not restrain a patient for a study, because doses from many examinations could be substantial. Infants should be held by a parent or relative and be given a lead apron to wear and may be monitored with a pocket dosimeter to reassure them that exposures will be small.

The amount of time that the x-ray beam is on should be minimized (Fig. 14.11). In practice this means minimizing the amount of fluoroscopy time and the number of photospot images that are acquired. The distance between the source of radiation and the operator should be as large as possible. In practice, where the x-ray beam enters the patient is the largest source of operator exposure. Shielding should also be used to reduce operator doses. In fluoroscopy and IR procedures, it is essential that operators wear appropriate protective clothing including lead aprons, thyroid shields, and leaded glasses. Minimizing operator dose using time, distance, and shielding always needs to be achieved without compromising the acquired diagnostic information.

Decreasing exposure time will reduce operator doses in a linear manner. Doubling the distance from the radiation source will reduce the exposure by a quarter, according to the inverse square law. Using appropriate lead shielding is by far the best way of reducing the operator doses. A 0.5-mm lead apron, for example, will reduce the operator effective dose by at least 90% and must always be worn by operators in rooms where x-rays are being used. One additional consideration is that any steps to minimize operator exposures must never be at the expense of diagnostic performance. It would be a serious error to reduce the exposure intensity for a scan of a pregnant patient if this could result in a "missed diagnosis!"

References

Huda W. Radiation protection. In: *Review of Radiological Physics.* 4th ed. Philadelphia: Wolters Kluwer; 2016:81–84.

Huda W. Interventional radiology. In: *Review of Radiological Physics.* 4th ed. Philadelphia: Wolters Kluwer; 2016:141–142.

Radiological Society of North America. RSNA/AAPM physics modules. Radiation protection: fundamentals of radiation protection. https://www.rsna.org/RSNA/AAPM_Online_Physics_Modules_.aspx. Accessed February 27, 2018.

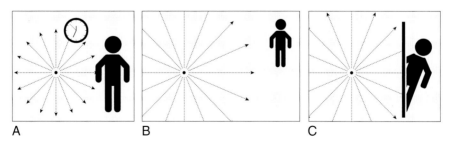

Fig. 14.11 There are three ways of reducing operator doses, which involve the reduction of exposure time (A), increase of the distance from the radiation source (B), and the use of shielding (C). Shielding is the best way of reducing operator doses.

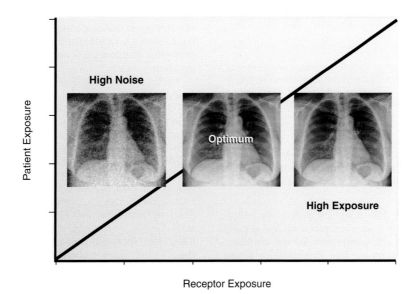

Fig. 14.12

1. What is the best definition of a worthwhile examination?
   A. High sensitivity
   B. High accuracy
   C. Low cost
   D. Benefit > risks

2. What is the regulatory entrance air kerma limit for diagnostic spot/photospot image in a gastrointestinal/genitourinary examination (mGy)?
   A. 1
   B. 10
   C. 100
   D. No limit

3. What is the most likely problem when a chest x-ray is generated with an exposure index of 1000?
   A. Excessive blur
   B. Reduced contrast
   C. Increased mottle
   D. Overexposed patient

4. What is of paramount importance when imaging a pregnant patient?
   A. Maintain diagnostic performance
   B. Reduce patient dose
   C. Increase lesion contrast
   D. Reduce image mottle

CASE 14.6

Protecting Patients

Fig. 14.12 Using too little radiation *(left)* will increase image mottle and may result in a drop of diagnostic performance when important lesions are "missed," whereas high exposures do not affect image quality *(right)* but result in unnecessary patient radiation exposure. The goal of radiation protection is to ensure that patients receive optimal exposures in between these two extremes *(middle)*.

1. **D.** A worthwhile examination is one in which there is a net patient benefit. What this means in practice is that the perceived benefit to the patient will exceed (any) corresponding examination risks.

2. **D.** There are *no* patient dose limits for *diagnostic imaging*. MQSA has a limit for a standard phantom but not for patients. There is also a regulatory limit for the entrance air kerma rate in fluoroscopy, but this imaging mode is not deemed to be "diagnostic," because fluoroscopy frames have too much mottle.

3. **D.** The exposure index for chest x-ray examinations is generally approximately 300, so this examination would have exposed the patient to more than three times the radiation needed to achieve a satisfactory diagnosis. All the key image quality metrics (i.e., resolution, mottle, and contrast) would most likely be satisfactory.

4. **A.** In any examination, it is essential to maintain diagnostic performance. A focus on dose reduction might result in an increase in image mottle and thereby result in a missed diagnosis, which is clearly counterproductive. Radiation intensity never affects contrast, and to reduce mottle would most likely require more radiation!

Comment

All exposures are assumed to be associated with possible radiation risk. This means that radiological examinations will always need to be justified. As a result, patients should never be exposed to x-ray radiation unless they are expected to obtain a net benefit from the radiological examination (Fig. 14.13). One way of reducing patient doses is to eliminate unnecessary examinations. The radiology community should encourage referring physicians to practice "Don't Order Tests that Don't Affect Management" (i.e., .DAM) philosophy. Radiologists need to understand the magnitude of radiation risks to be able to identify indicated examinations where the patient benefits exceed any possible risk. Value judgments are essential when attempting to balance risks and benefits in medical imaging. For example, a radiologist might consider whether they would perform a diagnostic test on a family member to solve a clinical problem.

There are no dose limits for patients undergoing x-ray-based examinations, and it is the responsibility of the radiologist to ensure that the correct amount is being used. Using too little radiation will raise the level of mottle in x-ray images and would therefore run the risk of reducing diagnostic performance by missing a lesion. Using too much radiation does not usually result in image quality degradation but unnecessarily irradiates the patient. Tailoring techniques to the diagnostic task is what the ALARA principle requires. Guidance on how examinations can be optimally performed can be obtained from websites that include Image Wisely, Image Gently, and Step Gently. Additional useful resources include the ACR Choosing Wisely guidance, ACR Appropriateness Criteria, and the CT dose index registry.

There are many ways to implement the ALARA principle in clinical practice. Radiographical technique (kV and mAs) must be modified to take the size of the patient into account, so that techniques would be reduced in smaller patients and increased in large patients. Sensitive organs, including the eye lens and breasts, may be shielded for specific examinations, provided that the shielding will not hide potentially useful diagnostic information. The use of breast shields during scoliosis examination is normal practice because these will not affect the diagnostic information which relates to the spine curvature and does not require the breasts to be irradiated. Shielding may be used when the gonads are directly irradiated by the x-ray beam but only when this does not interfere with obtaining a satisfactory diagnosis.

Before a pregnant patient is exposed to x-rays, fetal or embryo doses and risks should be established by the radiologists. Embryo doses, together with an assessment of any corresponding radiation risks, must always be determined by a qualified medical physicist, whenever this is requested by a radiologist. If an x-ray beam does not *directly* irradiate the embryo, the embryo dose can always be assumed to be low, and of no clinical importance. The diagnostic information from a radiological examination must exceed any possible risks to the patient and conceptus. Patients may be undergoing a radiological examination when they are not aware that they are pregnant. In such circumstances, no medical action is needed at unintended doses up to 100 mGy. At less than this radiation dose, any risks would only be stochastic (not deterministic) and deemed low in comparison with normal risks of pregnancy.

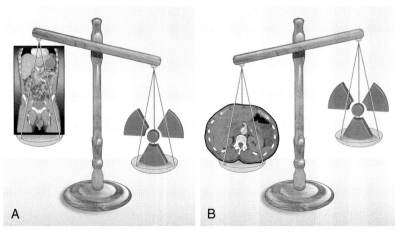

A   B

Fig. 14.13 A CT scan performed on an asymptomatic adult taken in a shopping mall is unlikely to be worthwhile (A). An abdominal CT exam with a proper indication recommended by an American Board of Radiology–certified radiologist who is knowledgeable about both the benefits and risks of medical imaging is virtually guaranteed to be worthwhile (B).

References

Huda W. Radiation protection. In: *Review of Radiological Physics*. 4th ed. Philadelphia: Wolters Kluwer; 2016:78-81.

National Council on Radiation Protection and Measurements (NCRP). *NCRP Report No. 168. Radiation Dose Management for Fluoroscopically-Guided Interventional Medical Procedures*. Bethesda, MD: NCRP; 2010.

National Council on Radiation Protection and Measurements (NCRP). *NCRP Report No. 107. Implementation of the Principle of as Low as Reasonably Achievable (ALARA) for Medical and Dental Personnel*. Bethesda, MD: NCRP; 1990.

Sensakovic WF, Warden DR, Bancroft LW. The link between radiation optimization and quality. *J Am Coll Radiol*. 2017;14(6):850-851.

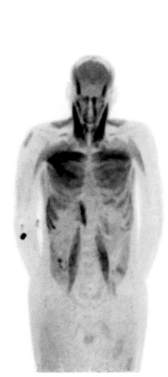

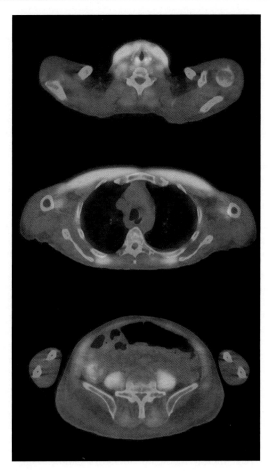

Fig. 14.14

1. What is the most likely annual dosimeter badge reading for a nuclear medicine technologist (mSv)?
   A. 1
   B. 3
   C. 10
   D. 30

2. What type of radiation detector would most likely be used to assess the extremity dose of a radio pharmacist?
   A. Geiger counter
   B. Ring dosimeter
   C. Pocket dosimeter
   D. Ionization chamber

3. What is the threshold technetium-99m (^{99m}Tc) activity that constitutes a major spill (MBq)?
   A. 40
   B. 400
   C. 4000
   D. 40,000

4. Can radioactive excreta be disposed of via a public sewer system?
   A. Yes
   B. Probably
   C. Possibly
   D. No

CASE 14.7

Protection in Nuclear Medicine

Fig. 14.14 Nondiagnostic PET/CT due to prominent muscular uptake.

1. **B.** A nuclear medicine technologist will most likely receive (badge reading) approximately 3 mSv/year.

2. **B.** Extremity doses are generally measured with a ring dosimeter that is small and conveniently carried on a finger (like a ring). Geiger counters, pocket dosimeters, and ionization chambers are all totally impractical devices for this radiation measurement task.

3. **C.** A spill of technetium-99m that is greater than 4 GBq (4000 MBq) is considered a major spill and would generally require the *presence* of a radiation safety officer (RSO).

4. **A.** Excreta from people who have received radioactive materials can be disposed via any sewer system. Radioactive human excrement is simply not regulated by the NRC and poses essentially no risk to the public if disposed of via conventional sewers.

Comment

The major sources of operator exposure in nuclear medicine are *patients* with several hundred MBq of administered ^{99m}Tc. Overall, technologist's effective doses in nuclear medicine in the United States range between 2 and 4 mSv/year. After injection of most diagnostic radiopharmaceuticals, radiation exposure levels from patients at 1 m are approximately 10 μSv/h. For fluorine 18 patients, the dose rate at 1 m is 10 time higher (i.e., approximately 100 μSv/h). Nuclear medicine operators handling volatile liquid radionuclides such as iodine 131 undergo mandatory bioassay to have their thyroids monitored for iodine uptakes. Lead aprons are less effective in nuclear medicine than x-ray imaging because of the much higher gamma ray photon energies (140 keV vs. 40 keV), and therefore nuclear medicine technologists do not routinely wear lead aprons.

To minimize contamination, protective clothing and handling precautions are used. When handling radionuclides, operators wear gloves and dispose of them in radioactive waste receptors. Volatile radionuclides such as iodine-131 and xenon-133 are generally stored in fume hoods. Leaded syringe shields are used when handling radionuclides to help minimize extremity doses and reduce extremity doses threefold. Extremity dose rates when using radiopharmaceutical syringes are 0.2 mGy/h per MBq and are monitored using ring dosimeters worn on a finger. When leaving a radiation area, it is essential that workers check for radioactive contamination using appropriate instrumentation such as Geiger-Müller counters. A contamination of approximately 37 kBq (1 μCi) will likely result in a detected count rate of 1000 counts (clicks) per second.

Radioactive packages need to be monitored for surface contamination upon receipt, and wipe tests are performed wherever and whenever radionuclides are used. A wipe test involves small dampened filter papers that are wiped over the area being checked and counted for selected radionuclides using a well counter. In areas where written directive procedures are carried out, surveys are performed daily. A minor spill requires containment and notification of the RSO, as well as subsequent decontamination. A major radiation spill involves the release of more than 4 GBq of technetium-99m, or more than 40 MBq of ^{131}I, and will require the presence of the RSO. Virtually all iodine-131 spills are likely major, whereas only a spill of "bulk" ^{99m}Tc would be considered major.

Radionuclide generators must be correctly packaged and labeled when these are returned to the supplier. Radioactive waste is stored for 10 half-lives so that the activity has decayed to approximately 0.1%. After such storage, this waste must be surveyed and can then be disposed of as regular waste if its contamination reading is no different than background. Although dumping of radioactive material by the hospital is regulated, excreta from people who have received radioactive materials (diagnosis or therapy) are not regulated by the NRC and can be disposed via any sewer system without worrying about dumping limits. It is expected that an AU knows the regulatory requirements surrounding radiopharmaceutical use, safety, and disposal. Radiology residents must pass the American Board of Radiology Radioisotope Safety Examination (RISE) to be AU eligible (Fig. 14.15).

References

Bushberg JT, Seibert JA, Leidholdt Jr EM, Boone JM. Radiation protection. In: *The Essential Physics of Medical Imaging.* 3rd ed. Philadelphia: Wolters Kluwer; 2012:880–892.

Huda W. Nuclear medicine II. In: *Review of Radiological Physics.* 4th ed. Philadelphia: Wolters Kluwer; 2016:200–201.

Mettler Jr FA, Guiberteau MJ. Authorized user and radioisotope safety issues. In: *Essentials of Nuclear Medicine Imaging.* 6th ed. Philadelphia: Elsevier; 2012:421–442.

Fig. 14.15 Radiology residents who obtain a passing score on the American Board of Radiology (ABR) Radioisotope Safety Examination (RISE) are considered authorized user eligible by the Nuclear Regulatory Commission (NRC). This eligibility is reflected on the ABR certificate.

Fig. 14.16

1. Increasing which factor will increase cell killing?
   A. Absorbed dose
   B. Dose rate
   C. Cell oxygenation
   D. A, B, and C

2. Which would likely indicate that an individual has been exposed to a high level of ionizing radiation exposure?
   A. Diarrhea
   B. Hypotension
   C. Fever
   D. A, B, and C

3. What *uniform whole-body* dose would likely kill half of the exposed population (i.e., LD_{50}) assuming medical intervention (Gy)?
   A. 0.5
   B. 5
   C. 50
   D. 500

4. Approximately how long would it take an individual exposed to 100 Gy of uniform whole-body radiation to die?
   A. Hour
   B. Day
   C. Week
   D. Month

CASE 14.8

Radiation Accidents and Terrorism

Fig. 14.16 High activity sources found in radiation oncology. (A) Leksell gamma knife perfexion houses 200 cobalt 60 sources with a combined activity of 220,000 GBq (6000 Ci). (B) High dose rate brachytherapy afterloader houses radioactive seeds with combined activities exceeding 560 GBq (15 Ci). ([A] Courtesy Elekta, Stockholm, Sweden.)

1. **D.** Studies in radiobiology have clearly shown that increases in the absorbed dose, the rate at which radiation is delivered (i.e., dose rate), and cell oxygenation would all likely increase the amount of cell killing.

2. **D.** Medical staff involved in the triage of potentially exposed individuals to high levels of radiation would look for the telltale signs of diarrhea, hypotension, and/or fever. Absence of these signs would indicate an individual had not been "highly exposed."

3. **B.** The best current estimate of the LD_{50} dose is 5 Gy when medical care is available. There would be no deaths at a whole-body dose of 0.5 Gy, and at 50 Gy the exposed population would be dead within a week or so from the gastrointestinal syndrome.

4. **B.** A whole-body dose of 100 Gy or greater kills everyone in 1 day or so (central nervous system syndrome). Neurological symptoms include dizziness and headache and are expected to occur within a few hours of exposure.

Comment

The ability of radiation to kill cells is dependent on absorbed dose, dose rate, and inherent cell sensitivities. Cell killing increases as the amount of energy deposited in the cell increases, and thus high doses are more effective at killing cells. Similarly, as the dose rate decreases, so will the probability of killing a cell. This reduction is because damage repair occurs before enough damage accumulates to kill the cell. This is illustrated using fractionated exposures in radiotherapy, which has been shown to reduce cell killing and helps to protect normal tissues. The most radiosensitive cells are undifferentiated and have high mitotic rates. Thus, stem cells and progenitor cells are more easily killed by radiation (i.e., less dose is necessary to kill them). By contrast, highly differentiated and nonproliferating cells such as neurons are least sensitive and require very high doses to achieve cell killing. An important exception to this rule is lymphocytes that are sensitive to radiation even though they are differentiated and do not divide.

High amounts of radiation can cause serious or even lethal biological effects. Acute radiation syndrome (ARS) is the result of exposure to high doses of radiation. The disease development and impact are predictable and depend on the amount of radiation encountered. A 2- to 10-Gy whole-body dose will sterilize red marrow stem cells and reduce circulating blood elements within 2 weeks. This is the called hematopoietic syndrome (a form of ARS). The LD_{50} is the uniform whole-body dose that kills half the population, estimated at 3 to 4 Gy for young adults. Medical intervention can increase the LD_{50}, which may be increased using medications such as antibiotics to stave off opportunistic infection after immunosuppression from exposure. However, at whole body doses of 8 Gy and higher, survival is unlikely. The LD_{50} is generally reduced for children or the very old. ARS is based on whole-body doses.

A 10- to 50-Gy whole-body dose will kill crypt cells in the gastrointestinal tract. This leads to denudation of villi, which leads to electrolyte loss, endotoxemia, bacteremia, and then death. This is called gastrointestinal syndrome (a form of ARS). All patients exposed to whole-body doses higher than 10 Gy die within 5 to 10 days. The final form of ARS is cerebrovascular or central nervous system syndrome. This occurs at 50 Gy or higher. Cerebral edema and microvasculitis and neuron necrosis eventually kill everyone in 1 or 2 days. Patients with gastrointestinal syndrome also have hematopoietic syndrome and those with cerebrovascular syndrome also have both gastrointestinal and hematopoietic syndromes.

Radiologists will not encounter high whole-body doses in their clinical practice but need to have a basic understanding of high dose effects in the event of a major nuclear accident or a terrorist action (Fig. 14.17). In accidents that involve high-energy (penetrating) radiation, the whole body is exposed to radiation and not just a limited region (as in diagnostic imaging). In addition, the level of exposure may be 10 to 100 times higher than encountered in diagnostic imaging. Peripheral lymphocyte analysis from blood samples of exposure victims can be used for patient triage. Indications of a high whole-body exposure to radiation include immediate diarrhea, hypotension, and fever. The first action to take after separating from the source of the radiation is to stabilize the patient, even before decontamination begins.

References

Gusev I, Guskova AK, Mettler Jr FA, eds. *Medical Management of Radiation Accidents.* 2nd ed. Boca Raton, FL: CRC Press; 2001.

Huda W. Radiation risks. In: *Review of Radiological Physics.* 4th ed. Philadelphia: Wolters Kluwer; 2016:61–70.

Wolbarst AB, Wiley Jr AL, Nemhauser JB, et al. Medical response to a major radiologic emergency: a primer for medical and public health practitioners. *Radiology.* 2010;254(3):660–677.

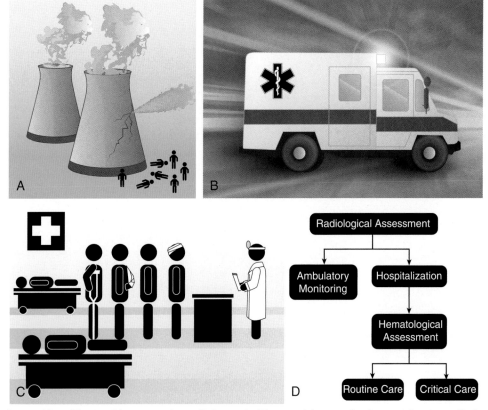

Fig. 14.17 In a nuclear accident (A), casualties exposed to radiation and with potential contamination are taken to medical centers (B). Radiologists (C), who are one of the few medical personnel trained to understand radiation issues, are likely to be involved with triaging patients (D).

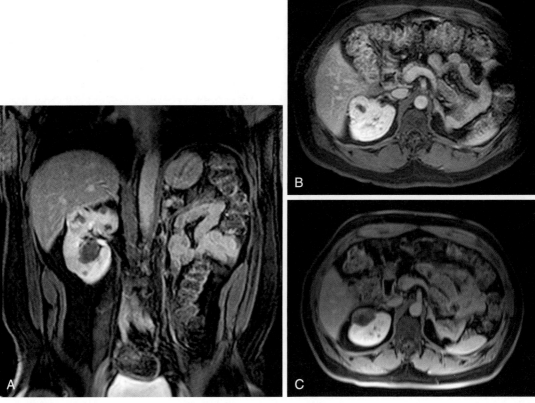

Fig. 14.18

1. Besides contrast reactions, what is the most common adverse effect from MR scans as reported to the FDA?
   A. Cardiac arrhythmias
   B. Thermal burns
   C. Acute hearing loss
   D. Nephrogenic systemic fibrosis (NSF)

2. What should be documented before a pregnant woman undergoes an MR scan?
   A. Cannot use ultrasound
   B. Cannot delay scan
   C. Scan medically necessary
   D. A, B, and C

3. What is the maximum temperature rise that is permitted in normal mode MR scanning (°C)?
   A. 0.1
   B. 0.2
   C. 0.5
   D. 1.0

4. What is the maximum specific absorption rate (SAR) allowed by the FDA for the whole body (W/kg)?
   A. 0.4
   B. 4
   C. 40
   D. 400

CASE 14.9

MR Safety

Fig. 14.18 Gadolinium-enhanced MRI (A and B) in a patient with von Hippel-Lindau disease and history of left nephrectomy demonstrates an additional renal cell carcinoma. Percutaneous radiofrequency ablation was performed, with no gadolinium enhancement of the mass seen on follow-up MRI (C).

1. **B.** The FDA reports that approximately 70% of adverse MRI events relate to radiofrequency burns.

2. **D.** Prior to performing an MRI scan on a pregnant patient, it is prudent to document that the patient will not benefit from an ultrasound scan, that the scan cannot be delayed until after the birth of the child, and the information from the MRI scan is medically necessary.

3. **C.** The maximum allowed temperature rise in normal mode MR imaging is 0.5°C, but rises up to 1°C are permitted in level 1 mode, where there is medical supervision of the procedure. To exceed 1°C would require institutional review board (IRB) approval.

4. **B.** The maximum average body SAR is currently 4 W/kg and slightly lower at 3 W/kg for the head.

Comment

The FDA has previously recommended a limit of 20 T/s, which is expected to prevent peripheral nerve stimulation (PNS), but current regulations are not numerical and are simply required not to induce painful PNS. Conductive structures, bone screws, (some) tattoos, and EKG leads can result in RF heating. The number of adverse events reported to the FDA pertaining to RF burns is currently estimated to be 70% of all adverse events. MR harm evidence for conceptus is very limited, but nevertheless, pregnancy is considered a contraindication for scanning the abdomen unless no safer alternatives exist. When scanning pregnant women, attempts should be made to document that diagnostic information is medically necessary. It is also useful to address issues of whether this information can be achieved by ultrasound and cannot be delayed until after birth.

Absorption of RF power will increase tissue temperatures, which is currently quantified by the SAR. The SAR estimates the power absorbed per unit of mass of tissue in watts per kilogram.

The average SAR must not exceed 3 W/kg head MRI and 4 W/kg in body using the MRI body (volume) coil. The maximum temperature rise in normal mode limit that is deemed suitable for all patients is currently 0.5°C. Some protocols that deposit more energy in the patient are classified as first level controlled mode, which require medical supervision. In this first level control protocols, the maximum temperature rise is 1°C and requires technologist acknowledgment before scanning begins. There is also a second level controlled mode, applicable where temperature rises could exceeds 1°C, but these are typically used in research and will always requires IRB approval in US medical centers.

Four zones around MR systems require demarcation and should be marked accordingly: zone 1, zone II, zone III, and zone IV (Fig. 14.18). Materials can be classified as MR safe (e.g., plastics) or unsafe (e.g., iron). Materials classified as MR conditional are those where there is no known hazard in a specified environment but only when operated under well-defined conditions. For example, a new cardiac pacemaker was introduced in 2011 that can be scanned on 1.5 T when switched to an MR safe mode of operation by a cardiac technician. The FDA has set the safe limit for static field exposure to the public at 0.5 mT (i.e., 5 gauss), which is contained within zones III and IV. There must be a physical barrier preventing access by members of the public wherever the fringe MR field exceeds 0.5 mT, and considerable effort is expended to limit the extent of fringe fields by the use of magnetic shielding.

Hearing loss is a major concern because noise levels can range from 65 to 120 dB in MR systems. Hearing protection in the form of earplugs and/or headphones is generally mandatory. Administration of gadolinium contrast agents should be avoided in patients with reduced kidney function or kidney failure. NSF has occurred in a small number of patients, generally those who have preexisting severe kidney function abnormalities. Gadolinium is currently under scrutiny because it has been shown that it may be retained in the brain and accumulates with each contrast-enhanced scan, although no harm has yet been attributed to this finding.

References

Expert Panel on MR Safety, Kanal E, Barkovich AJ, et al. ACR guidance document on MR safe practices: 2013. *J Magn Reson Imaging*. 2013;37(3):501-530.

Huda W. Magnetic resonance II. In: *Review of Radiological Physics*. 4th ed. Philadelphia: Wolters Kluwer; 2016:232-233.

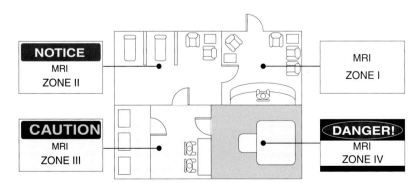

Fig. 14.19 MR suites contain four zones, including of zone I where any members of the public can be present and zone II where unscreened patients await screening before entering zone III. The MR scanner room is zone IV.

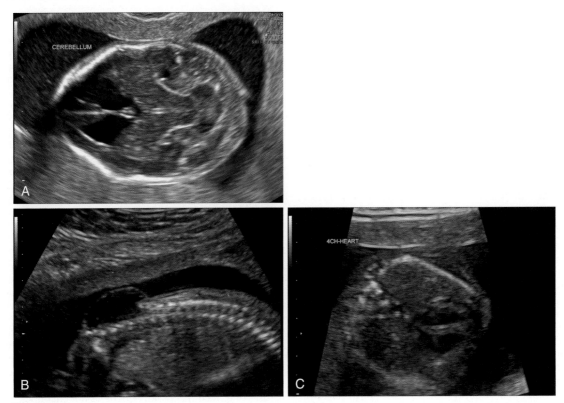

Fig. 14.20

1. What is the most likely maximum temperature rise for an ultrasound scan where the thermal index (TI) is given to be 1 (°C)?
   A. <1
   B. 1
   C. >1

2. The ultrasound "thermal index" is specified in which type of tissue?
   A. Soft tissue
   B. Bone
   C. Cranial bone
   D. A, B, and C

3. In which type of exam would a mechanical index (MI) greater than 1 likely be of greatest concern?
   A. Obstetrics
   B. Pediatric
   C. Neurological
   D. Contrast

4. Which should always be followed in obstetrical examinations?
   A. TI < 1
   B. MI < 1
   C. TI and MI < 1
   D. ALARA

CASE 14.10

Ultrasound Safety

Fig. 14.20 Obstetrical ultrasound at 19 weeks in a patient with Chiari II malformation would use thermal index values for cranial bone (TIC) in A, bone (TIB) in B, and soft tissue (TIS) in C.

1. **B.** A TI of 1 in a specified material means that the maximum temperature rise in this material will be 1°C.

2. **D.** TI values can be obtained in soft tissue (TIS), bone (TIB), or cranial bone (TIC).

3. **D.** High MI values would run the risk of cavitation of small air bubbles and would therefore be of primary concern in contrast agent ultrasound.

4. **D.** Values of TI and MI can exceed 1 if this is indicated. All imaging, including ultrasound, should always strive to eliminate "unnecessary exposures." This approach is commonly referred to as the ALARA principle.

Because ultrasound is a nonionizing radiation, it does not have the same risks as ionizing radiations such as x-rays and gamma rays. Ultrasound imaging has been used for more than a generation with a very good safety record. Ultrasound imaging is considered safe when used by appropriately trained healthcare providers. Many states regulate diagnostic ultrasound by recommendations and/or requirements for personnel qualifications, quality control programs, and facility accreditation. For all ultrasound examinations the FDA recommends that patients talk to their healthcare provider to learn why an examination has been requested and how the medical information that will be obtained will be used to affect patient management.

Ultrasound deposits energy into the body, and diagnostic levels of ultrasound can produce physical effects in tissue, such as pressure oscillations with subsequent mechanical effects and rise in temperature. Ultrasound waves can produce small pockets of gas in body fluids or tissues, including cavitation, which is the collapse of microscopic bubbles. The MI can be computed from the transducer intensities, which estimates the chance of inducing cavitation effects. Gas-containing structures, such as the lungs and the gastrointestinal tract, are susceptible to the effects of acoustic cavitation. Ultrasound waves can heat the tissues slightly. The TI is a computed parameter that predicts the rise in tissue temperature in degrees Celsius (°C). A TI of 1 is estimated to increase tissue temperatures by up to 1°C. TI values are generally specified for soft tissue (TIS), bone (TIB), and cranial bone (TIC). Because of the aforementioned effects, guidelines have been developed for general ultrasound scanning and obstetrical ultrasound (Fig. 14.21).

Healthcare providers should consider examinations such as ultrasound or MRI, if medically appropriate, because these use no ionizing radiation. The FDA recommends that operators always reduce unnecessary risks from medical imaging, including ultrasound, while maintaining diagnostic quality. As with all other imaging modalities, the principles of ALARA should be practiced by healthcare providers.

Because ultrasound does not involve ionizing radiation, it is useful for women of childbearing age when x-ray–based imaging might result in exposure to radiation. Long-term consequences of ultrasound are unknown, and therefore the American Institute of Ultrasound in Medicine (AIUM) advocates prudent use of ultrasound imaging in pregnancy. More than half of the pregnant women in the United States undergo ultrasound during pregnancy, with no scientifically supported evidence of detrimental effects. The AIUM considers ultrasound as being safe with minimal, if any, risk. Use of ultrasound for nonmedical uses (e.g., fetal "keepsake" videos) is not encouraged. Keepsake images or videos should only be obtained if produced during a medically indicated exam and when no additional exposure is required.

References

Huda W. Ultrasound II. In: *Review of Radiological Physics*. 4th ed. Philadelphia: Wolters Kluwer; 2016:259–260.

Radiological Society of North America. RSNA/AAPM physics modules. Ultrasound: image quality, artifacts, and safety. https://www.rsna.org/RSNA/AAPM_Online _Physics_Modules_.aspx. Accessed February 27, 2018.

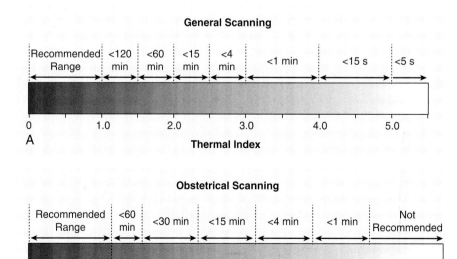

Fig. 14.21 Guidance on thermal index limits provided by the British Medical Ultrasound Society for general ultrasound scanning (A) and for obstetrical scanning (B).

Note: Page number followed by *t* or *f* indicates table or figure, respectively.